Applied Statistical Inference
with MINITAB®

STATISTICS: Textbooks and Monographs

Recent Titles

Applied Statistical Inference
with MINITAB®

Sally A. Lesik

Central Connecticut State University

New Britain, Connecticut, U.S.A.

CRC Press
Taylor & Francis Group
Boca Raton London New York

CRC Press is an imprint of the
Taylor & Francis Group, an **informa** business

A CHAPMAN & HALL BOOK

Chapman & Hall/CRC
Taylor & Francis Group
6000 Broken Sound Parkway NW, Suite 300
Boca Raton, FL 33487-2742

© 2010 by Taylor and Francis Group, LLC
Chapman & Hall/CRC is an imprint of Taylor & Francis Group, an Informa business

No claim to original U.S. Government works

Printed in the United States of America on acid-free paper
10 9 8 7 6 5 4 3 2 1

International Standard Book Number: 978-1-4200-6583-1 (Hardback)

Library of Congress Cataloging-in-Publication Data

Lesik, Sally A.
    Applied statistical inference with MINITAB / Sally A. Lesik.
        p. cm. -- (Statistics : textbooks and monographs)
    Includes bibliographical references and index.
    ISBN 978-1-4200-6583-1 (hardcover : alk. paper)
    1. Mathematical statistics--Data processing. 2. Minitab. I. Title. II. Series.

    QA276.3.L47 2010
    519.5'40285--dc22   2009022577

Visit the Taylor & Francis Web site at
http://www.taylorandfrancis.com

and the CRC Press Web site at
http://www.crcpress.com

Dedication

To DHK ... for all the laughter and the joy.
To NNK ... for believing.

Contents

Preface

There are numerous statistics books that are available for readers who are learning statistics for the first time. What distinguishes this book from many is that it presents statistics with an emphasis toward applications for readers who are not experts in statistics, but who want to learn about basic inferential techniques and be able to implement such techniques in practice using a statistical software package. One characteristic of this text is that it is written in such a way that the material is presented in a seamless manner, thus making it easier to read and follow. By using a seamless step-by-step approach, readers are introduced to a topic, presented with the calculations in detail, provided with how to interpret the findings, and given an illustration of how to perform the same analysis with a statistical software program. Although this approach may be somewhat different than is presented in other texts, readers may find it easier to learn statistics by being exposed to all the calculations and software details.

This text is written to be beginner-friendly and is oriented toward the practical use of statistics. The presentation leans less toward the theoretical side of statistics and is focused more on addressing the expectations of students and practitioners who are not experts in statistics but who are interested in getting an appreciation for applying statistical techniques using a statistical software program. One of the key features of this text is that the mathematical calculations are presented in step-by-step detail. Presenting such detail on how the calculations are actually done by hand and the kinds of inferences that can be made comes from teaching a course on applied inference to undergraduate and graduate students who understood only the most basic statistical concepts, but who plan to use statistics in their senior or master's theses. Many beginning readers of statistics tend to struggle when they are not presented with step-by-step details and are left to fill in the gaps of how a particular statistic is calculated. Although the prerequisite level of mathematics for this text is intermediate algebra, many novices still like to see the nuts-and-bolts of the calculations so they can get a better understanding of the concepts and to connect with what the software program is actually doing.

Another key feature of this text is that instructions on how to use the statistical software package MINITAB® are incorporated in detail immediately following a topic. By presenting the software explanations immediately following a given topic, this allows the reader to learn about the topic and then see how to use a statistical package to arrive at the same conclusions as found when doing the calculations by hand. This style of presentation comes from watching many beginners become increasingly frustrated when trying to read through statistics text books where it was necessary to flip through the pages to try to relate the presentation of the topic with the appropriate software package and commands. A part of creating a seamless presentation required using only a single statistical software package, specifically MINITAB. Using only a single software package provides the reader with the opportunity to focus on the details of a specific package without having to filter through the commands or output from other programs.

The audience for this text can come from diverse disciplines. I chose to write this text to be useful for just about any field of study and especially for those students who attend colleges and universities that may not offer discipline-specific statistics courses. By not targeting any specific discipline, this allowed me to present examples and discussions based on data and scenarios that are common to many students in their everyday lives. Perhaps the biggest challenge in writing this text is that many disciplines have their own spin on what topics should be presented as well as how they should be presented. Though such differences in emphasis and presentation across the various fields can often be seen as a source of tension, it pushed me to think more deeply about the subject and how to best express the concepts of applied statistical inference in a general and understandable way. In this text I tried to present the material as simply as possible without sacrificing the necessary technical details. Also, given differences in notation and terminology across fields, on many occasions I incorporated the notation and terminology that is used by MINITAB.

The choice to use MINITAB as the single statistical software program for this text was an easy one. MINITAB is a very simple and easy to use statistical package. But yet, MINITAB is also very sophisticated and many fields such as business and engineering actively use it. The clear menus and informative dialog boxes make it a natural choice for a text such as this which is written for a novice with little or no experience using a statistical software package. The printouts are clear and easy to follow, while still presenting enough relevant information, and the graphics are excellent.

Like many things in life, good data is hard to find. For the data sets used in this text, some were intentionally contrived and modified in order to be manageable enough to illustrate the step-by-step calculations and the inferences that can be made. By using smaller data sets, this allowed for the mathematical calculations to be done out in their entirety so that readers can follow through the step-by-step calculations if they wish. Other sets of data presented are either entire sets of actual data that are available from the public domain or subsets of data available from the public domain. Each chapter has a set of homework problems that were created to give the reader some practice in using the techniques and methods described in the text with real data.

This text was written to establish the foundation for students to build on should they decide to study more advanced inferential statistics. Virtually every type of statistical inference in practice, from beginning to advanced, relies on confidence intervals, hypothesis testing, validating model assumptions, and power analysis. Since this book was written with these considerations emphasized throughout, it is my hope that readers will be able to generalize the basic framework of applied inference at just about any level. The topics covered and the order with which the topics are presented in this text may not follow most traditional texts. However, I decided to write a text that was oriented toward the practical use of statistics for those who may be contemplating using statistics in their own work.

Chapter 1 presents a basic introduction to some common terminology that one is likely to encounter when learning statistics. Although conventions and definitions may differ across disciplines, I tried to use more common definitions and terminology throughout the text while also trying to stay consistent with notation and terminology that is used in MINITAB.

Chapter 2 presents some basic graphs as well as how to create such graphs using MINITAB. The presentation of graphs such as the stem-and-leaf plot and the box plot are aligned with the conventions used in MINITAB. For instance, the stem-and-leaf plot generated in MINITAB will have an extra column that includes the cumulative frequencies below and above the median, and the quartiles for the box plot are calculated by using interpolation.

Chapter 3 presents basic descriptive statistics using both traditional hand-calculations along with MINITAB. The calculations are done out in detail to give students the chance to feel more comfortable with the notation and symbols that are introduced. Although somewhat untraditional, I also introduce random variables and sampling distributions in this chapter as I saw it as a natural extension of a way to describe variables.

Chapter 4 presents basic statistical inference. I begin by deriving confidence intervals using the t-distribution, and I also emphasize the interpretation of confidence intervals as students seem to get confused with what a confidence interval is really estimating. I also begin the discussion about hypothesis tests by testing a single population mean. I repeatedly elaborate on how inferences are made with confidence intervals and hypothesis testing by referring back to the sampling distribution of the sample mean. This chapter also covers basic inferences for proportions. Chapter 4 also provides a conceptual introduction to power analysis as well as how to use MINITAB to conduct a power analysis.

Chapter 5 describes simple linear regression. To understand simple linear regression, one must have a good intuitive feel for what the line of best fit is. I elaborate on this topic by first presenting how a line for two random points can be used to express the relationship between two variables. I then show how the line of best fit is "better" in the sense that it is the single line that best fits the data. This chapter also provides an introduction to the model assumptions for simple linear regression and how to make inferences with the line of best fit.

Chapter 6 provides more detail for simple linear regression by describing the coefficient of determination, the sample correlation coefficient, and how to assess model assumptions. One key feature of this chapter is that it introduces the Ryan–Joiner test as a formal test of the normality assumption. There is also a discussion on how to assess outliers by using leverage values, studentized residuals, and Cook's distances.

Chapter 7 provides an introduction to multiple regression analysis. The ANOVA table and the issue of multicollinearity are introduced. Chapter 8 provides more detail for multiple regression by introducing how to include categorical predictor variables, how to pick the best model, and how to assess outliers.

Chapter 9 provides a conceptual introduction to basic experimental design and the basics of a one-way ANOVA. This chapter introduces Bartlett's and Levene's tests as a formal way to establish the assumption of constant variance. Multiple comparison techniques are also introduced as well as power analysis for a one-way ANOVA.

Chapter 10 provides a discussion of a two-way ANOVA in addition to some basic non-parametric analyses and basic time series analysis. The calculations for the test statistics for the Wilcoxon signed-rank test and the Kruskall–Wallis test are worked out in great detail to help the reader gain a greater understanding of the complexities of these tests.

Acknowledgments

There are so many people who contributed to this project over the past few years. I am particularly grateful to my friend and colleague Frank Bensics, who graciously agreed to edit many versions of this manuscript. His many suggestions, comments, and corrections brought a new perspective to the work. I am also grateful to my friend and colleague Zbigniew Prusak, who not only provided valuable comments about the content and presentation of the text, but who also sat in on my class on numerous occasions and provided valuable feedback about how students struggle when learning statistics.

David Grubbs, Susan Horwitz, and the staff at Taylor & Francis were most helpful. Not only was I encouraged to write the text as I saw fit, I also received continuous support, patience, and guidance along the way. I would also like to thank the reviewers for the numerous comments and corrections that help guide my writing.

Finally, I would like to extend my gratitude and appreciation to my fiends, family, and colleagues who directly and indirectly contributed to this project.

Correspondence

Although a great amount of effort has gone into making this text clear and accurate, if you have any suggestions or comments regarding errors, content, or feel that some clarification is needed, please contact me at lesiks@ccsu.edu. I am interested in hearing your feedback and comments.

1

Introduction

1.1 What This Book Is About

Statistical inference involves collecting and analyzing data in order to answer a meaningful question of interest. For instance, a researcher in education may want to know if using computers in an algebra classroom is effective in helping students build their mathematical skills. A researcher in psychology may want to know whether children who play violent video games tend to have more disturbing thoughts than children who do not play violent video games. In other fields, such as environmental science, researchers may want to know what factors contribute to global warming by asking questions such as which makes and models of automobiles emit larger amounts of greenhouse gas.

Once a researcher has described a problem he or she wishes to investigate, he or she will set out to collect or identify a set of data that consists of information about a variable or variables of interest. There are two basic types of data that can be collected, quantitative data and qualitative data.

Quantitative data is numeric in form. The main purpose of collecting quantitative data is to describe some phenomenon using numbers. For example, quantitative data could be collected to assess the effect of advertising on gross product sales.

On the other hand, *qualitative data* is categorical in nature and describes some phenomenon using words. For instance, qualitative data can be used to describe what the learning environment is like in a given mathematics classroom by using words to describe the types of interactions between the students and the teacher and how students appear to be engaged in learning.

Determining whether to collect quantitative or qualitative data is typically driven by the characteristic or relationship that is being assessed and the type of data that is available. The purpose of this book is to introduce some of the different statistical methods and techniques that can be used to analyze data in a meaningful way. The methods and techniques that we will be considering in this book are broadly categorized as follows:

Graphical displays of data

Descriptive representations of data

Basic statistical inference

Regression analysis

Analysis of variance

1.1.1 Graphical Displays of Data

Graphical displays of data visually describe some of the characteristics of a set of data by using different types of charts and graphs. The advantage to using charts and graphs to display data is that a large amount of information can be displayed in a concise manner. For example, suppose you are interested in comparing the fuel efficiency for the following four different makes and models of vehicles: the Toyota Corolla®, the Honda Civic®, the Ford Focus®, and the Chevrolet Aveo®. You could obtain the average miles per gallon for both city and highway driving for each of the models you are interested in and then graph the comparison between the different brands of cars by using a bar chart, as illustrated in Figure 1.1.

Notice that Figure 1.1 graphically displays the city and highway miles per gallon for each of the different makes and models of cars, and thus allows you to make comparisons between the different cars.

1.1.2 Descriptive Representations of Data

Descriptive representations of data consist of methods and techniques that can be used to describe and summarize data. For instance, if you have ever shopped for a new car you may have noticed that the sticker on the window

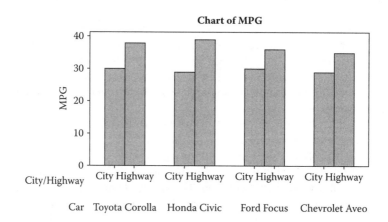

FIGURE 1.1

Bar chart comparing city and highway mileage per gallon based on the different brands of cars.

of the car provides the average miles per gallon of gasoline for both city and highway driving. This value describes, on average, the gas mileage that you can expect from the vehicle. For example, the sticker on the window of the 2007 Ford Focus® suggests that, on average, the vehicle will get 27 miles per gallon of gasoline driving in the city and 37 miles per gallon driving on the highway (http://www.fueleconomy.gov/feg/byclass.htm).

1.1.3   Basic Statistical Inference

Basic statistical inference relies on estimating or predicting an unknown characteristic of interest by using available data. For example, suppose that a criminologist wants to know whether the average yearly crime rate in the United States has increased over the last year. Because it can be very difficult and time-consuming to obtain the yearly crime rate for every single community in the United States, the criminologist may decide to collect a representative sample of the communities in the United States along with their respective crime rates for the past 2 years. Then by using the available information obtained from this representative sample of communities, the criminologist could then try to make an inference or prediction about whether the crime rate has increased over the past year for the entire United States.

1.1.4   Regression Analysis

Regression analysis is a statistical technique that consists of developing and validating models that can be used to describe how one variable is related to another variable or collection of different variables. For instance, the price of a house is determined by many factors, such as square footage, lot size, number of bedrooms, number of bathrooms, and age. Regression analysis would allow you to develop a model that describes how the price of a house is related to these factors.

1.1.5   Analysis of Variance

Analysis of variance is a statistical technique that can be used to estimate whether there are differences in averages between more than two groups based on some characteristic. For example, suppose you are interested in determining whether there is a difference in the number of pages you can print with four different brands of printer toner. One way to assess this could be to set up an experiment where you have four identical brands of printers and a total of sixteen printer cartridges (four printer cartridges of each brand). Then you could put the printer cartridges in each of the printers and count up the total number of pages that each printer printed with the given cartridge. An analysis of variance could then be used to see if there is a difference in the average (or mean) number of pages printed across the different brands of printer cartridges.

Each of these methods and techniques can be used in their own unique way to describe some characteristic of a variable or to assess the relationships among a set of variables. When to use which method or technique will depend on the type of study that is being done and the type of data that is collected.

1.2 Types of Studies

Studies that provide quantitative data can be described as true experimental studies, observational studies, or retrospective studies. In *true experimental studies,* subjects are assigned to participate in a treatment program or intervention based only on chance, and this can be done by using a random assignment process. For example, one could conduct a true experimental study to determine if students who use computers in their statistics class perform better on a learning task than students who do not use computers in their statistics class. In order to conduct a true experimental study, students would have to be assigned to either the computer statistics class (the treatment group) or the noncomputer statistics class (the control group) based only on chance. To create treatment and control groups based on chance, we could flip a coin for each prospective participant in the study, and those who receive heads could be assigned to the treatment group, and those who receive tails could be assigned to the control group (or vice versa). Then a comparison of how these two groups perform on some learning task could be used as an indication of whether students who use computers perform better than students who do not use computers.

The benefit of conducting a true experimental study is that a random assignment process creates treatment and control groups that are not systematically different with respect to all other relevant factors that could impact performance on the given learning task. By determining group assignment based only on chance, this makes it less likely that any factors other than the group assignment are having an impact on how students perform on the learning task. However, one major problem with conducting a true experimental study is that it can be difficult to assign participants to treatment and control groups by using only a random assignment process.

In *observational studies,* data on subjects are collected by observation and subjects are not assigned by a random assignment process or in any other prescribed manner. For example, we could conduct an observational study to see if students who use computers in their statistics class perform better on some learning task than students who do not use computers in their statistics class by collecting data for two existing groups of students: one group who elected to enroll in the computer-based statistics course and another group who elected to enroll in the non-computer-based statistics course.

Although it is much easier to collect data for an observational study versus an experimental study, observational studies are often plagued with *selection bias.* In other words, because students self-selected into either of these two types of courses, any findings regarding the effectiveness of the use of computers in a statistics class on student performance on a learning task may be biased because of the possible differences between those students who elected to participate in the computer-based statistics course versus those students who elected not to participate in the computer-based statistics course.

Retrospective studies look at data that already exist or have been previously collected. Data for retrospective studies can be experimental or observational in nature. For instance, a retrospective study could entail analyzing observational data from the 2000 national census survey (http://www.census.gov/main/www/cen2000.html).

Now that we have discussed some of the different ways that data can be collected, we will describe some of the terminology that can be used to describe the three broad areas of statistics that are used to study and analyze data.

1.3   What Is Statistics?

Statistics essentially involves analyzing and studying data to answer a question of interest. Generally, a set of data represents a collection of variables that represent different characteristics. Data sets consist of numerous observations, and each observation has measures on the variable or variables of interest. For instance, Table 1.1 gives an example of a data set that describes the number of credits attempted, number of hours worked each week, major, and gender for a random sample of five freshmen at a university.

There are three broad areas of statistics that we will use to study data: *graphical methods*, which involve ways to present and illustrate data; *descriptive statistics*, which involve methods and techniques that can be used to numerically summarize data; and *inferential statistics*, which involve methods and techniques that can be used to make inferences that are generalizations or conclusions about a larger group of interest based upon a sample of observations.

TABLE 1.1

Example Data Set Showing Different Characteristics of Interest for Five Entering Freshmen

Observation Number	Number of Credits	Hours Worked Each Week	Major	Gender
1	15	22	Business	Male
2	12	20	Engineering	Male
3	9	0	Education	Female
4	18	18	Business	Female
5	15	9	Science	Male

Throughout this text we will be examining many different statistical methods and techniques that can be used to study data. However, before we begin, we first need to present some basic terminology regarding how data are described and the different types of data that we are likely to encounter.

A *population* is a collection of data that consists of every possible observation of interest. For instance, a population could be described as all of the students at a particular college or university. Any value that represents a characteristic or describes an attribute of a population is called a *parameter*. For example, the average age of all the students at a particular college or university is considered a population parameter because it is a value that represents a characteristic or attribute of a given population.

A *sample* is a collection of data that consists of a subset, or portion, of a population of interest. For instance, a sample from the population of students at a particular college or university could be all of the seniors at the college or university. A sample could also be a subset that is selected at random from a given population. Any variable that describes some characteristic or attribute of a sample is called a *statistic*. For example, the average age of all of the seniors at a particular college or university would be a statistic because it represents a characteristic or attribute of a given sample. A statistic could also be the average age for a sample of students who were selected at random from the population of students at a particular college or university.

We will be describing many different methods and techniques that can be used to make an inference, or prediction, about an unknown population parameter of interest based on information contained in a sample. These different methods and techniques rely on using statistics collected from a representative sample to make generalizations about an unknown population parameter of interest. One way to collect a representative sample is to collect *a random sample*. A random sample is taken from a population such that each sample of a given size from the population has the exact same chance of being selected as does any other sample of the same size.

1.4   Types of Variables

Before we can start to analyze data in a meaningful way, it is important to understand that there are many different types of variables that we are likely to encounter. One way to distinguish between the different types of variables is to consider whether additional observations can exist between any two values of a variable. We will begin by first describing two general types of variables: discrete and continuous.

We say that a variable is *discrete* if between any two successive values of the variable, other values *cannot* exist within the context they are being used. Discrete variables are often used to represent different categories.

Example 1.1

Suppose we code the variable of gender as male = 0 and female = 1. This variable is a discrete variable because between the values of 0 and 1 there cannot exist any other values that have meaning within the context they are being used.

Example 1.2

Consider three different categories of political party affiliation coded as follows:

Republican = 0   Democrat = 1   Independent = 2

This variable would also be a discrete variable even though between the categories of 0 and 2 another possible observation could exist (Democrat = 1). However, in order to be a discrete variable, there cannot be any observations that exist between any pair of categories, such as between the categories coded as 1 and 2. On the other hand, a variable is *continuous* if between any two different observations of the variable, other observations can exist.

Example 1.3

Consider two observations of the heights for two different individuals. If one individual is 5′11″ and the other individual is 5′6″, then we could possibly observe yet another individual whose height lies somewhere between these two values.

1.5   Classification of Variables

There are four different classifications of variables that can be used to describe the nature of the information that the variable represents. These classifications can also be used to describe the mathematical properties of a variable. The different classifications of variables are described as nominal, ordinal, interval, or ratio.

The weakest classification represents variables that have no mathematical properties at all. *Nominal variables* are the weakest classification because the values of a nominal variable represent different categories and, therefore, they cannot be manipulated using basic arithmetic.

Example 1.4

Suppose we consider three different categories of political party affiliation and code them as follows: Republican = 0, Democrat = 1, and Independent = 2. This variable is nominal because it describes the different categories of political affiliation and there are no mathematical properties that can be associated with them. For instance, if we take any two observations, such as

0 = Republican and 1 = Democrat, when we add them together, 0 + 1 = 1, the sum does not make sense within the context that these variables are being used. Also, neither multiplication nor division makes any sense within the context that these variables are being used. Furthermore, there is not a natural ordering of the categories of the variable because it is generally not accepted that Republican = 0 is in some way "less than" Democrat = 1, which is "less than" Independent = 2.

The second classification of a variable is an *ordinal variable*. An ordinal variable can be grouped into separate categories, but ordinal variables are different from nominal variables in that there is a natural ordering of the categories. Other than the ordering of the categories, ordinal variables have no numerical properties.

Example 1.5

Suppose we are looking at the four different ways in which economic status can be classified and described:

0 = poverty   1 = lower class   2 = middle class   3 = upper class

Notice that there is a natural ordering among the different categories, since 1 = lower class can be seen as "less than" 2 = middle class. Thus, the inequality 1 < 2 is sensible within the context that this variable is being used. However, notice that there are no mathematical properties that are associated with these categories. If we add two values that represent any two of these categories, their sum or difference does not have any meaning within the context they are being used, nor does multiplication or division.

With ordinal variables it can be very difficult to measure the distance between two different categories. For instance, if you are considering the ordinal variable that describes military rank (private, corporal, sergeant, captain, major, and general), it may be difficult to measure or quantify the distance between any two values of military status, such as the distance between captain and major.

The third classification of a variable is an *interval variable*. With interval variables, equal distances between observations represent equal-sized intervals.

Example 1.6

Consider the variable of weight in pounds. This variable would be an interval variable because the measure of distance between two individuals who weigh 165 and 170 pounds would represent the same-sized interval for two individuals who weigh 181 and 186 pounds. Another example would be the number of students studying in various classrooms. We can add and subtract interval data, and we can measure and quantify such sums or differences within the context that the data are being used.

Finally, the fourth classification of a variable is where the value of 0 has meaning by indicating the absence of the quantity of interest. This type of

TABLE 1.2

Mathematical Properties of Nominal, Ordinal, Interval, and Ratio Variables

Classification of Variable	Mathematical Properties
Nominal	None
Ordinal	Inequalities
Interval	Addition and subtraction
Ratio	Addition, subtraction, multiplication, and division

variable is called a *ratio* or *ratio–scale variable*. In other words, for ratio data, quotients or ratios can be formed that have meaning within the context that they are being used, and the value of 0 represents the absence of the quantity of interest.

Example 1.7

Consider the amount of money that two people have in their pockets. If one person has $100 and the other person has $200, then the ratio of the two amounts of money has meaning because $200 is twice as much as $100. Also, the value of 0 is meaningful because it indicates having no money.

All ratio variables also have the properties of interval variables, but not all interval variables have the properties of ratio variables. For example, consider temperature in degrees Fahrenheit. If it is 84° Fahrenheit on one day and 42° Fahrenheit on another day, it does not make sense to say that 84° is twice as hot as 42° even though we can form the quotient 84/42 = 2. Furthermore, the value of 0 on the Fahrenheit scale does not represent the absence of temperature.

Table 1.2 illustrates the four different classifications of variables based on the mathematical properties that were described.

1.6 Entering Data into MINITAB

MINITAB is a statistical software program that we will be using throughout this text that can simplify analyzing a set of data. One characteristic of MINTIAB is that the data can easily be entered in the form of a spreadsheet. MINITAB is very easy to use and has some very nice features that will be described throughout this book.

Figure 1.2 provides an illustration of what the basic MINITAB session and worksheet windows look like. The worksheet section at the bottom of Figure 1.2 is used to display the data that are entered, and the session window at the top of Figure 1.2 displays the results of any calculations or analyses. The top bar of the MINTAB screen provides all the different types of pull-down menus that are available.

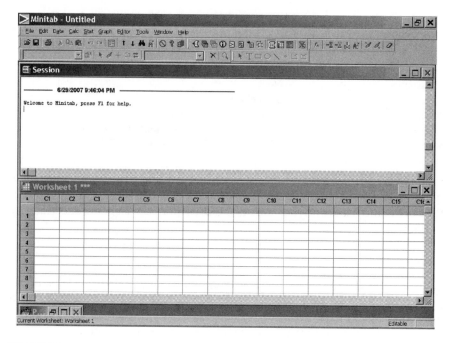

FIGURE 1.2
MINITAB session and worksheet window.

Data are typically entered into a MINTAB worksheet in row and column form. Each column represents a variable, and a collection of variables can be represented by different columns, namely, C1, C2, ..., etc. The rows of a MINITAB worksheet consist of the individual observations for each of the variables, and these are labeled with the numbers 1, 2, 3, ..., etc.

Data can be entered into MINITAB by typing them in by hand, which is usually only appropriate if you have a small amount of data. If you have a large amount of data, you can also cut and paste data from another program or import it from another file.

Data that are entered into MINITAB can be numeric, text, or in the form of a date. For example, in Figure 1.3, a collection of sales data is entered that represents the date, name, and amount of sales in dollars for a sample of four salespeople. Notice in Figure 1.3 that the date column is C1-D (column one, date), the column that contains the names of the salespeople is a text column and is represented as C2-T (column 2, text), and the column that consists of numeric data is C3 (column 3).

Once your data are entered into a MINITAB worksheet you can save your project data from the session window and the data entered in the worksheet window. This is done by selecting **File** on the top menu bar, then **Save Project As**, as is illustrated in Figure 1.4.

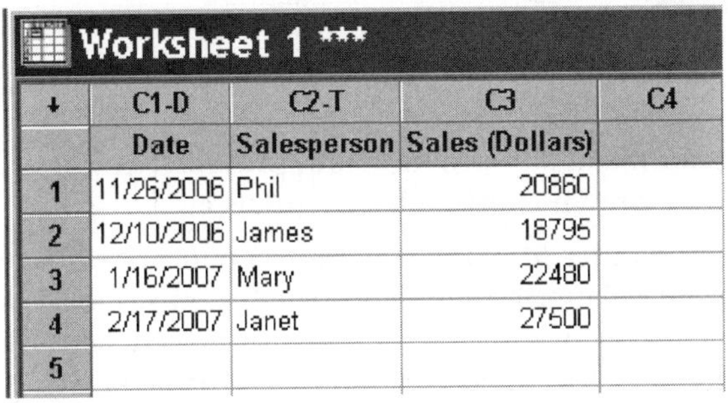

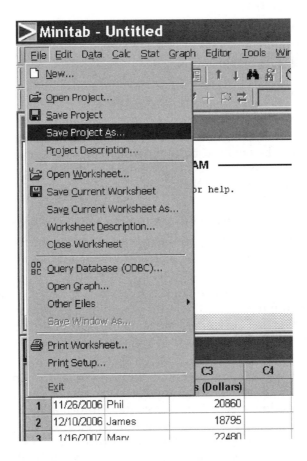

FIGURE 1.3
MINITAB worksheet illustrating date, text, and numeric forms of data.

FIGURE 1.4
How to save a project or worksheet in MINITAB.

Exercises

1. Give an example of a population and provide two different samples that could be drawn from such a population.

2. A survey asked fifty consumers to rate their level of satisfaction with a certain brand of laundry detergent. The responses to the survey were either very satisfied with the detergent, satisfied with the detergent, not satisfied with the detergent, or have not used the detergent. Would you classify this variable as nominal or ordinal? Justify your answer.

3. Classify each of the following variables as nominal, ordinal, interval, or ratio.
   a. Social security numbers
   b. Year of birth
   c. The amount of money in a bank account
   d. The different models of cars on the road
   e. The length of a fiber optic cable
   f. College classes (freshman, sophomore, junior, and senior)

4. A *stratified random sample* is obtained by first partitioning the population into disjoint groups, or strata, and then taking random samples from each of the strata. For instance, a stratified random sample can be obtained by first partitioning all the undergraduate students at a local university by their class (freshman, sophomore, junior, and senior), and then taking a random sample from each of the partitions.
   a. Describe a population and a stratified sample of the population.
   b. Can you think of a reason why a stratified random sample may improve on obtaining a representative sample, as compared to drawing a simple random sample? Explain.

5. The data set "Automobile Data" (http://www.fueleconomy.gov/feg/byclass.htm) represents the miles per gallon and amount of greenhouse gases that are emitted based on the type, make, and model of the vehicle. Classify each of the variables in this data set as nominal, ordinal, interval, or ratio.

6. A *code sheet* is often prepared with a set of data in order to describe more details about the given variable or variables in the data set. Code sheets can also contain information that relates a shortened variable name with a more detailed description of the variable. For instance, with the "Crime Data by County" data set (http://www.ojp.usdoj.gov/bjs/dtdata.htm#State), the variable "1989 median family income" could be also coded as "MFI 1989"

for brevity, and an entry on a code sheet for that variable could look something like:

Variable Name	Description
MFI 1989	1989 median family income

Before we begin graphing and analyzing data, we may be interested in having shortened versions of names of variables because it is much clearer to analyze large amounts of data that have abbreviated variable names.

a. For the "Crime Data by County" data set, develop a set of shortened variable names consisting of eight characters or less for each variable and develop a code sheet that relates the variable name to each of the original variables.

b. Also include a column on the code sheet that describes whether the variable is nominal, ordinal, interval, or ratio.

2

Graphing Variables

2.1 Introduction

Often we are interested in creating a graphical display of a given variable or set of variables. One advantage to graphically displaying data is that large amounts of information can be presented in pictorial form, and thus can be used to present a visual summary of the underlying characteristics of the variable of interest. There are many different types of graphs and charts that can be used to display data. In this chapter, we will be discussing six basic types of graphs:

Histograms

Stem-and-leaf plots

Bar charts

Box plots

Scatter plots

Marginal plots

This chapter describes these different types of graphs and provides some examples of how these graphs can be used in practice. In this chapter we will also describe how to use MINITAB to generate these different types of graphs.

2.2 Histograms

The first type of graph that we will be considering is called a *histogram*. A histogram shows the distribution of a single continuous variable, in which the *x*-axis consists of the data grouped into equal-sized intervals or bins that do not overlap each other, and the *y*-axis represents the number or frequency of the observations that fall into each of these bins. A histogram can be drawn from a *frequency distribution*, which is a table that illustrates the number or frequency of the observations within the data set that fall within a given range of values. The purpose of drawing a histogram is to see the pattern of how the data are distributed over a given range of values.

TABLE 2.1

Round-trip Commuting Distance (in miles) for
a Random Sample of Twenty-Five Executives

26	18	56	102	110
74	44	68	10	110
50	66	144	50	36
32	88	58	154	38
48	42	62	72	70

Example 2.1

The data set "Commuting Distance" provided in Table 2.1 describes the round-trip commuting distance (in miles) for a random sample of twenty-five executives.

To construct a histogram for the data in Table 2.1, we first need to create a frequency distribution. This is done by grouping the data into equal-sized intervals or bins that do not overlap each other, and then counting the number of observations that fall into each interval or bin. Table 2.2 illustrates what a frequency distribution with eight equal-sized bins would look like for the executive round-trip commuting data.

The histogram is then created by graphing these intervals on the x-axis and the number of observations that fall into each interval on the y-axis, as is presented in Figure 2.1.

So, just how many bins are needed to see how data are distributed in a histogram? There are no generally accepted rules for the number of bins that a histogram must have, but one general rule of thumb is that seven or eight bins are usually an appropriate amount that will allow you to see the shape of the distribution of the data, but fewer or more bins can be

TABLE 2.2

Frequency Distribution of the Round-Trip Commuting Distance
(in miles) Using Eight Equal-Sized Bins

Commuting Distance (in miles)	Frequency
10–29	3
30–49	6
50–69	7
70–89	4
90–109	1
110–129	2
130–149	1
150–169	1
Total	**25**

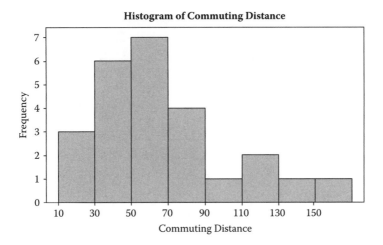

FIGURE 2.1
Histogram of the frequency distribution for the commuting distance data from Table 2.1.

used if needed. However, keep in mind that a histogram with too many bins may only show a random pattern, and a histogram with too few bins will not reveal the shape of the distribution of the data (see Exercise 1). Another rule of thumb for determining the number of bins for a histogram is to use the square root of the sample size.

Although graphing a histogram can easily be done by hand, it is much easier and more efficient to use a statistical software program such as MINITAB to create histograms.

2.3 Using MINITAB to Create Histograms

The histogram for the commuting distance data can easily be created using MINITAB by first entering the data in a single column in a MINITAB worksheet, as illustrated in Figure 2.2.

To draw the histogram, first select **Graph** from the top menu bar, and then **Histogram**, as is illustrated in Figure 2.3.

We can then select a simple histogram, a histogram with a normal fit superimposed, a histogram with outline and groups, or a group of histograms with a normal curve superimposed, as illustrated in Figure 2.4.

After selecting a simple histogram, we then need to specify the variable that we wish to draw the histogram for. This requires highlighting the variable of interest and hitting **Select**, as is illustrated in Figure 2.5.

Then selecting **OK**, the histogram for the commuting distance data is presented in Figure 2.6.

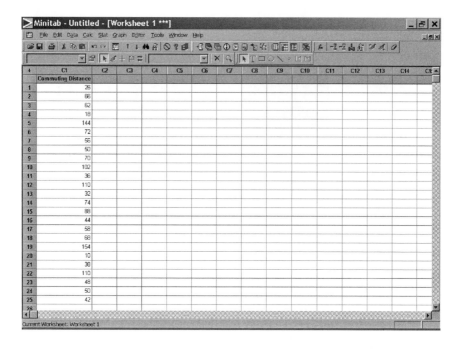

FIGURE 2.2
MINITAB worksheet of the round-trip commuting distance data from Table 2.1.

We can change the shape of a histogram by selecting different widths for the intervals or bins, or by specifying more or fewer bins. This can be done by right-clicking on the bars of the histogram to obtain the **Edit Bars** dialog box that is presented in Figure 2.7.

To change the number of bins for a histogram, select the **Binning** tab in the **Edit Bars** dialog box and specify the number of intervals (say 10), as is illustrated in Figure 2.8. You can also select to draw a histogram based on the midpoint or cut-point positions of the histogram.

This gives a histogram of the round-trip commuting distance data from Table 2.1 with ten bins, as is illustrated in Figure 2.9.

Notice in Figure 2.9 that the histogram has only eight bars, but there are ten bins with respect to the x-axis. This is because there are no observations in the bin that have a center point of −10 and there are no observations in the bins with a center point of 130 or 170.

2.4   Stem-and-Leaf Plots

Stem-and-leaf plots are graphs that can be used to simultaneously show both the rank order and the shape of the distribution of a continuous variable.

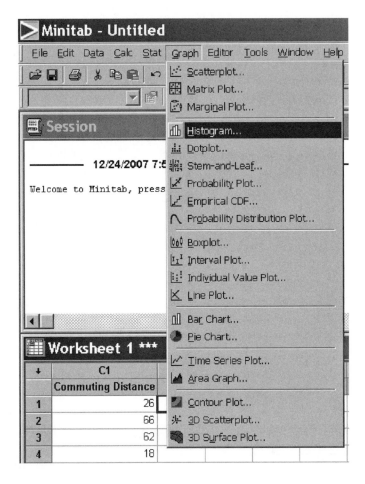

FIGURE 2.3
MINITAB commands for drawing a histogram.

Example 2.2

Suppose we wish to create a stem-and leaf plot for the commuting distance data given in Table 2.1. In order to do this we need to create both the stem portion and the leaf portion of the graph. To create the stem portion, we list all of the possible leading digits for the values in the data set to the left of a vertical line. For example, the stem for the commuting distance data given in Table 2.1 is illustrated in Figure 2.10.

Notice in Figure 2.10 that the leading digits go from 1 to 15. This is because the range of the data is from 10 to 154.

To create the leaf portion of the stem-and-leaf plot, begin by listing each observation to the right of the vertical line for its respective leading digit. For example, the observation 26 would have the value 2 appear in the

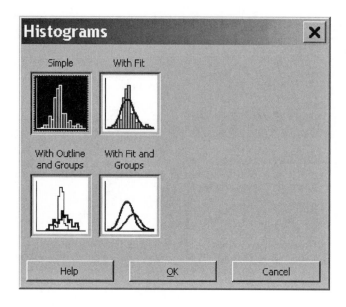

FIGURE 2.4
MINITAB dialog box to select the type of histogram.

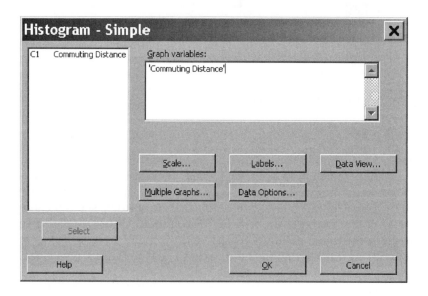

FIGURE 2.5
MINITAB dialog box to select the variable for the histogram.

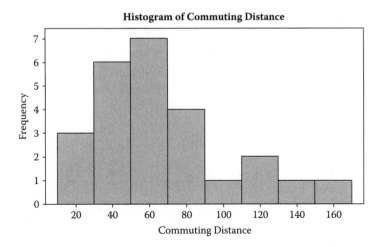

FIGURE 2.6
Histogram of the commuting data from Table 2.1 using MINITAB.

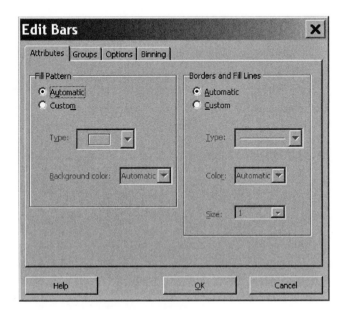

FIGURE 2.7
MINITAB dialog box to edit the bars of the histogram.

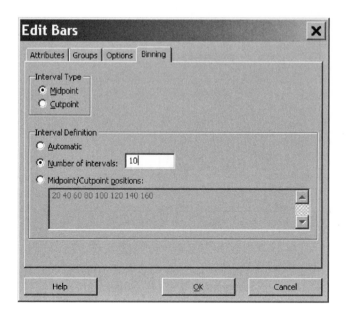

FIGURE 2.8
MINITAB dialog box to specify the number of intervals.

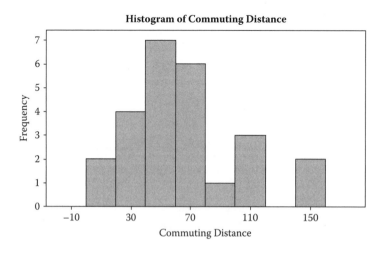

FIGURE 2.9
MINITAB histogram of the round-trip commuting distance using ten bins.

stem portion and 6 in the leaf portion of the stem-and-leaf plot, as illustrated in Figure 2.11.

Creating the rest of the leaves for the commuting distance data gives the plot that is illustrated in Figure 2.12.

The final step in creating the stem-and-leaf plot is to put all of the leaves in numerical order, as can be seen in Figure 2.13.

Notice that in Figure 2.13, the stem-and-leaf plot shows the distribution of the data set along with the rank ordering of the data.

2.5 Using MINITAB to Create a Stem-and-Leaf Plot

To create a stem-and-leaf plot using MINITAB, select **Stem-and-Leaf** under the **Graphs** tab, as can be seen in Figure 2.14.

This brings up the stem-and-leaf plot dialog box that is presented in Figure 2.15. Notice that we put the value 10 in the increment box. The increment represents the difference between the smallest values on adjacent rows. For this example, the smallest value on adjacent rows would be 10 because we want the stem-and-leaf plot to have adjacent rows separated by 10.

By selecting the commuting distance data, this gives the MINITAB output shown in Figure 2.16.

Notice in Figure 2.16 that the MINITAB printout adds an additional column to the left of the stem-and-leaf plot. This column represents the cumulative count of the number of observations that fall below or above the median. The median is the value such that 50% of the observations fall below this value and 50% of the observations fall above this value. For the commuting distance data, the median is 58. If you look in the fifth row in Figure 2.16, you will see that the count value is 4, since there are four observations that fall above or below the median in that given row. Also notice that the value 4 is given in parentheses in the fifth row. This means that the median value for the sample is included in that particular row. The count for a row below the median represents the total number of observations for that row and for the rows below the median. Similarly, the count for a row above the median represents the total number of observations for that row and for the rows above the median. For instance, if you

FIGURE 2.10

Stem of the commuting distance data, which consist of all possible leading digits to the left of a vertical line.

look at the third row of Figure 2.16, you will see that the cumulative count for that row is 6, which means that there are a total of six observations at or below the third row that fall below the median. Similarly, if you look at the ninth row of Figure 2.16, you will see that the cumulative count for that row is 5, which means that five observations at or above the ninth row fall above the median.

2.6   Bar Charts

Continuous variables can also be treated as categorical variables by partitioning the continuous data into nonoverlapping categories. For example, the data for the round-trip commuting distance for the sample of twenty-five executives from Table 2.1 can be categorized as "near" if the round-trip commuting distance is less than 50 miles, or "far" if the round-trip commuting distance is greater than or equal to 50 miles. Because discrete data are categorical in nature, we would use a bar chart instead of a histogram to illustrate how the data are distributed within these two distinct and nonoverlapping categories.

2.7   Using MINITAB to Create a Bar Chart

In order to create a bar chart to categorize the round-trip commuting distance as "near" or "far" using MINITAB, we first need to create a categorical variable that will be labeled "Distance" to represent whether the commuting distance is categorized as "near" or "far." This can be done using MINITAB by selecting **Code** and then **Numeric to Text** under the **Data** tab, as shown in Figure 2.17. This brings up the dialog box that is illustrated in Figure 2.18.

Notice in Figure 2.18 that we need to specify the data we wish to code, the column of the worksheet where we want to store the coded data, and the range of values that will be assigned to a given category. For our example, since we want to classify those round-trip commuting distances less than 50 miles as "near," we would specify that the values from 0 through 49 are to be coded as "near." Similarly, since round-trip commuting distances 50 miles or more are to be coded as "far," this is done by assigning the values of 50 through 154 as "far."

This gives the newly created categorical variable that we labeled as "Distance," which will be stored in column 2 of Figure 2.19.

FIGURE 2.11
Stem-and-leaf plot for the observation 26.

Notice that the second column heading in Figure 2.19 is C2-T, which means that this is a text column because the data under consideration represent the two distinct categories of "near" and "far."

We can then draw a bar chart using MINITAB by selecting **Graph** from the top menu bar and then **Bar Chart,** as can be seen in Figure 2.20.

We then need to select the type of bar chart that we want to draw. For instance, if we want to draw a simple bar chart, select **Simple**, as shown in Figure 2.21.

After hitting **OK**, we can then choose to graph the categorical variable, Distance, which describes if the round-trip commuting distance is "near" or "far," as is shown in Figure 2.22.

Selecting **OK** gives the bar chart for the round-trip commuting distance that is presented in Figure 2.23.

2.8   Box Plots

In addition to histograms, *box plots* (also referred to as box-whisker plots) can be created to illustrate the distribution of a single continuous variable. A box plot is constructed by first breaking the data into quarters, or *quartiles*, and then calculating upper and lower whiskers and identifying any outliers.

The general form of a box plot is illustrated in Figure 2.24, where the box portion consists of the first, second, and third quartiles, and the whiskers extend from the outer edges of the box to the largest value within the upper limit and the smallest value within the lower limit. Any outliers are identified with an asterisk (*), and they represent those observations that lie outside of the upper and lower limits.

Example 2.3

The data presented in Table 2.3 describe the average number of hours that a random sample of twelve full-time college students spend studying for their classes each week.

To create a box plot for this set of data we first need to find the quartiles. This is done by first making sure that the data are listed in numerical order. The first quartile, or Q_1, is the value that partitions the data set so that 25% of the

```
 1 | 8 0
 2 | 6
 3 | 6 2 8
 4 | 4 8 2
 5 | 6 0 0 8
 6 | 8 6 2
 7 | 4 2 0
 8 | 8
 9 |
10 | 2
11 | 0 0
12 |
13 |
14 | 4
15 | 4
```

FIGURE 2.12
Leaves for all of the observations in the commuting distance data set from Table 2.1.

TABLE 2.3

Average Number of Hours a Random Sample of 12 Full-Time
College Students Spend Studying for Their Classes per Week

5	6	7	8	8	9	10	12	15	18	19	40

observations lie at or below Q_1 and 75% of the observations lie at or above Q_1. The second quartile, or Q_2, is the value that partitions the data such that 50% of the observations lie at or below Q_2 and 50% of the observations lie at or above Q_2. Similarly, the third quartile, or Q_3, is the value that partitions the data such that 75% of the observations lie at or below Q_3 and 25% of the observations lie at or above Q_3.

The value of Q_1 is the value that is found in the $(n + 1)/4$ position. If this value is not an integer, then interpolation is used. Similarly, the value of Q_3 is the value that is found in the $3(n + 1)/4$th position, and if this value is not an integer, then interpolation is used.

For the data in Table 2.3, in order to find Q_1, we would locate the value in the $(12 + 1)/4 = 3.25$th position, but since this position is not an integer value, we would interpolate as follows:

$$Q_1 = x_3 + 0.25(x_4 - x_3) = 7 + 0.25 \cdot (8 - 7) = 7.25$$

where x_3 is the value of the observation in the 3rd position, x_4 is the value of the observation in the 4th position, and 0.25 is the decimal portion of the 3.25th position.

Similarly, for Q_3, we would locate the value in the $3(12 + 1)/4 = 9.75$th position, but since this position is not an integer we would interpolate as follows:

$$Q_3 = x_9 + 0.75(x_{10} - x_9) = 15 + 0.75 \cdot (18 - 15) = 17.25$$

where x_9 is the value of the observation in the 9th position, x_{10} is the value of the observation in the 10th position, and 0.75 is the decimal portion of the 9.75th position.

Therefore, $Q_1 = 7.25$ and $Q_3 = 17.25$.

The median position, or Q_2, is found by taking the value that partitions the ordered set of data in half. For a sample of size n, if there are an odd number of observations, the median is the observation in position $(n + 1)/2$, and if there are an even number of observations the median is the average of the observations that are in positions $n/2$ and $(n + 2)/2$.

```
 1 | 0 8
 2 | 6
 3 | 2 6 8
 4 | 2 4 8
 5 | 0 0 6 8
 6 | 2 6 8
 7 | 0 2 4
 8 | 8
 9 |
10 | 2
11 | 0 0
12 |
13 |
14 | 4
15 | 4
```

FIGURE 2.13
Stem-and-leaf plot for the round-trip commuting distance data given in Table 2.1.

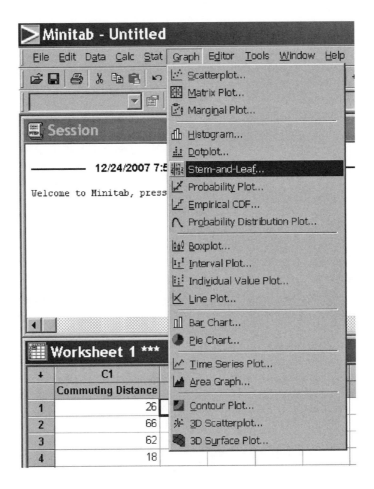

FIGURE 2.14
MINITAB commands to create a stem-and-leaf plot.

For the data in Table 2.3, because there is an even number of observations in this data set, the median would be the average of the observations that fall in the $n/2 = 12/2 = $ 6th and $(n + 2)/2 = (12 + 2)/2 = $ 7th position, as follows:

$$Q_2 = \frac{9 + 10}{2} = 9.50$$

Once we have found the quartiles, the upper and lower limits of the box plot can then be calculated by using the following formulas:

$$\text{Upper Limit} = Q_3 + 1.5\,(Q_3 - Q_1)$$

$$\text{Lower Limit} = Q_1 + 1.5\,(Q_3 - Q_1)$$

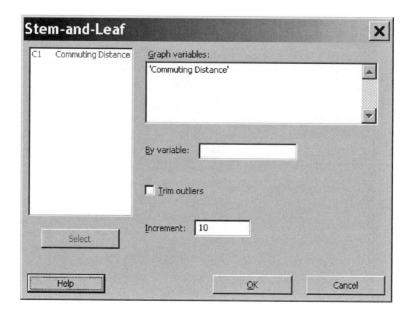

FIGURE 2.15
MINITAB dialog box for a stem-and-leaf plot.

Stem-and-Leaf Display: Commuting Distance

```
Stem-and-leaf of Commuting Distance N = 25
Leaf Unit = 1.0
  2  1 08
  3  2 6
  6  3 268
  9  4 248
 (4) 5 0068
 12  6 268
  9  7 024
  6  8 8
  5  9
  5 10 2
  4 11 00
  2 12
  2 13
  2 14 4
  1 15 4
```

FIGURE 2.16
MINITAB stem-and-leaf plot for the round-trip commuting data given in Table 2.1.

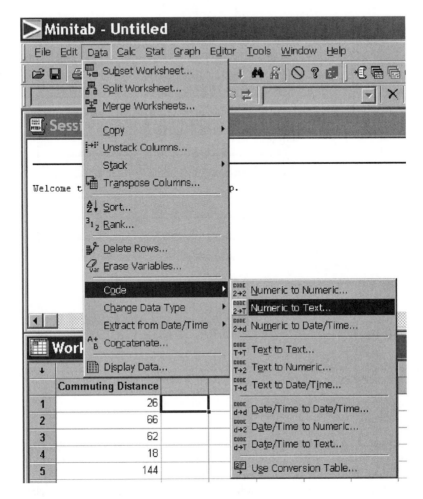

FIGURE 2.17
MINITAB commands to change data from numeric to text.

The whiskers of the box plot extend out to the largest and smallest observations that lie within *these upper and lower limits.* Outliers are then identified with an asterisk as those observations that lie outside of the whiskers.

Thus, for our example, the upper and lower limits are found as follows:

$$\text{Upper Limit} = Q_3 + 1.5 \, (Q_3 - Q_1) = 17.25 + 1.5 \cdot (17.25 - 7.25) = 32.25$$

$$\text{Lower Limit} = Q_1 - 1.5 \, (Q_3 - Q_1) = 7.25 - 1.5 \cdot (17.25 - 7.25) = -7.75$$

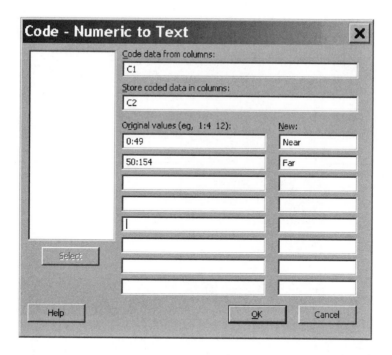

FIGURE 2.18
MINITAB dialog box to code the commuting distance data as a text variable.

The largest data value that lies within the upper limit is 19, and the smallest data value that lies within the lower limit is 5, so the whiskers extend out to these values that lie within the upper and lower limits.

Any observations that lie outside of either the upper or the lower whiskers are outliers, and they are identified with an asterisk. For this example, the only value that lies outside of the upper or lower whiskers is the observation 40. Putting this all together gives the box plot that is presented in Figure 2.25.

2.9 Using MINITAB to Create Box Plots

To create a box plot using MINITAB, select **Boxplot** under the **Graphs** bar to get the dialog box that is presented in Figure 2.26.

Notice that you can select to plot a single simple box plot or you can select to plot multiple box plots with groups. The dialog box for creating a simple box plot is presented in Figure 2.27.

After selecting the appropriate variable, MINITAB will create the box plot that is presented in Figure 2.28.

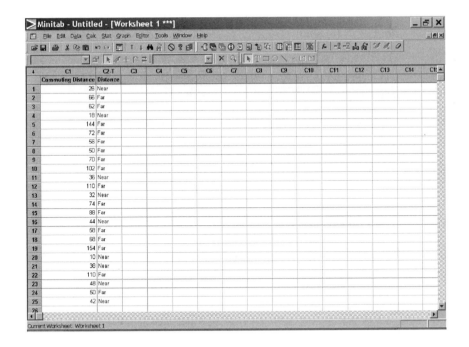

FIGURE 2.19
MINITAB worksheet containing a text column that indicates the category of whether the commuting distance is near or far.

2.10   Scatter Plots

Although histograms, bar charts, and box plots can be used to graphically summarize the properties of a *single* variable, often we may be interested in the relationship between *two different variables*. Graphically we can illustrate the relationship between two variables x and y by creating what is called a *scatter plot*. A scatter plot is simply the graph of the ordered pairs (x, y) of the observations for the two variables plotted on the Cartesian plane.

Typically we describe the relationship between two variables by how one variable influences the other. We call the *response* or dependent variable (the y-variable) that which is influenced by another variable, which is called the *predictor* or independent variable (the x-variable).

Example 2.4

Suppose we are interested in whether the score obtained on the mathematics portion of the SAT examination taken in high school is related to the first-year grade point average in college. For this particular situation, the score received on the mathematics portion of the SAT examination would be the

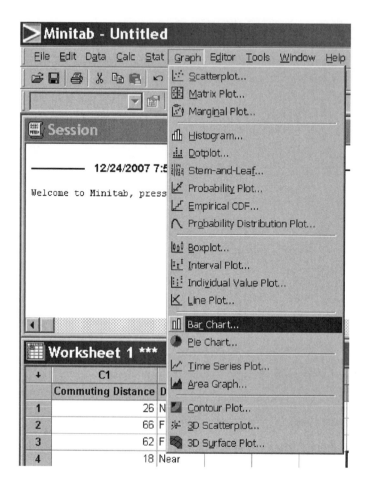

FIGURE 2.20
MINITAB commands to draw a bar chart.

predictor or independent variable (the *x*-variable), and the first-year grade point average on a scale of 0.00 to 4.00 would be the response or dependent variable (the *y*-variable).

Table 2.4 represents the data set "SAT–GPA," which consists of the score received on the mathematics portion of the SAT examination (SATM) and the associated first-year grade point average (GPA) for a random sample of ten first-year college students.

To draw a scatter plot by hand, we would simply plot each of the ordered pairs (*x*, *y*) on the Cartesian plane, or we could also use MINITAB to create the scatter plot for us.

TABLE 2.4

Table of SAT Mathematics Scores and First-Year Grade Point Average on a Scale of 0.00 to 4.00 for a Sample of 10 College Students

Observation Number	SAT Math Score (SATM) x	Grade Point Average (GPA) y
1	750	3.67
2	460	1.28
3	580	2.65
4	600	3.25
5	500	3.14
6	430	2.82
7	590	2.75
8	480	2.00
9	380	1.87
10	620	3.46

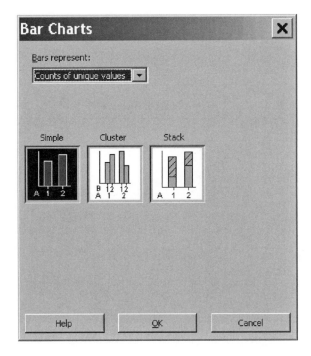

FIGURE 2.21

MINITAB dialog box to select the type of bar chart.

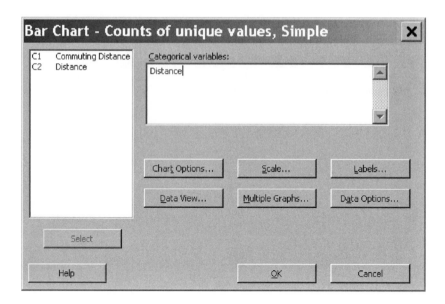

FIGURE 2.22
Bar chart dialog box.

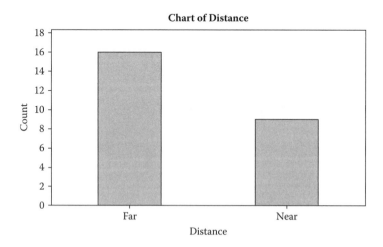

FIGURE 2.23
Bar chart that illustrates the number of executives that are categorized as commuting near (less than 50 miles) or far (50 or more miles).

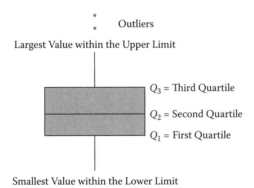

FIGURE 2.24
General form of a box plot.

2.11 Using MINITAB to Create Scatter Plots

Using MINITAB to create scatter plots is relatively easy, and requires that the data are entered in two separate columns in a MINITAB worksheet, one column that consists of the independent variable (or x-variable) and one column that consists of the dependent variable (or y-variable).

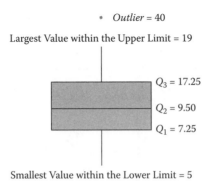

FIGURE 2.25
Box plot for the average number of hours a sample of twelve college students study per week.

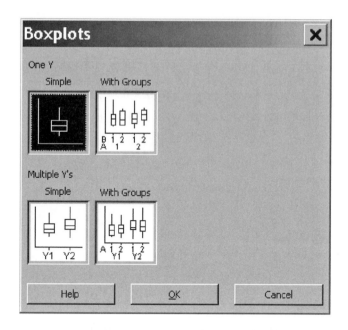

FIGURE 2.26
MINITAB dialog box for creating a box plot.

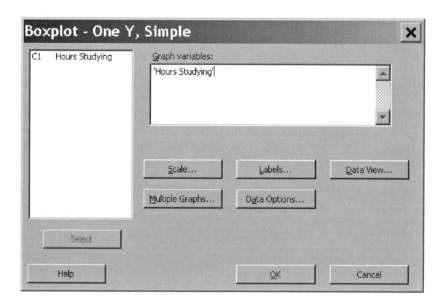

FIGURE 2.27
MINITAB dialog box for a simple box plot.

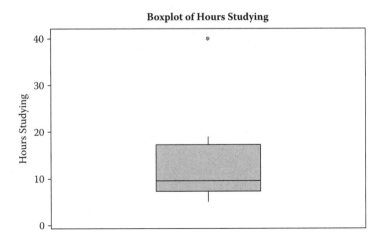

FIGURE 2.28
MINITAB box plot for the average number of hours a sample of twelve college students spend studying per week.

To draw a scatter plot, select **Scatterplots** from the **Graphs** menu bar to get the dialog box that is presented in Figure 2.29.

By selecting a simple scatter plot, this gives the dialog box in Figure 2.30, in which the dependent variable and independent variable need to be specified in the appropriate order to generate the graph of the scatter plot as in Figure 2.31.

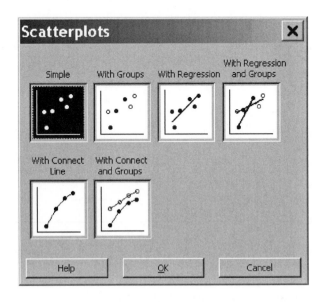

FIGURE 2.29
MINITAB dialog box for selecting the type of scatter plot.

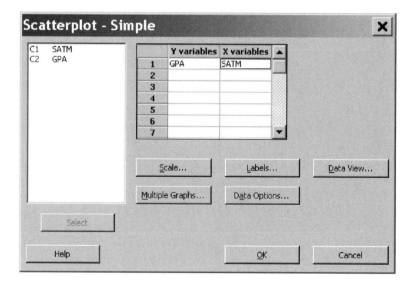

FIGURE 2.30
MINITAB dialog box to select the independent and dependent variables.

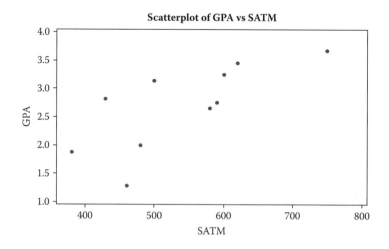

FIGURE 2.31
Scatter plot of first-year grade point average (GPA) versus SAT mathematics score (SATM).

2.12 Marginal Plots

Another interesting plot that can be used to assess the relationship between two variables is called a *marginal plot*. A marginal plot consists of a scatter plot to assess the relationship between two variables along with a graph that shows the distribution for each of the two variables. A marginal plot consists of either a histogram for each variable along with a scatter plot, a box plot for each variable along with a scatter plot, or a dot plot for each variable along with a scatter plot.

2.13 Using MINITAB to Create Marginal Plots

Suppose that we want to create a marginal plot for the data given in Table 2.4. The MINITAB commands require selecting **Marginal Plot** under the **Graph** tab, as is illustrated in Figure 2.32. This would give the marginal plots dialog box that is provided in Figure 2.33.

Notice that in Figure 2.33 we have the choice of creating a marginal plot with a histogram, box plot, or dot plot for each of the two variables along with the scatter plot showing the relationship between both of the variables.

Selecting the marginal plot with histograms gives the dialog box in Figure 2.34, and the respective graph is presented in Figure 2.35. The marginal plot in Figure 2.35 gives the histograms for SATM and GPA, respectively, along with the scatter plot of the relationship between first-year grade point average and SAT mathematics score. Figure 2.36 gives the marginal plot with box plots for each of the variables for the data given in Table 2.4.

Exercises

1. The data set "Company Profits," in Table 2.5, consists of the profits or losses (in millions of dollars) for a random sample of twenty companies.

   TABLE 2.5

   Data Set That Gives the Company Profits or Losses (in Millions) for a Sample of 20 Companies

24.1	105.8	−8.5	10.5	12.0
37.2	−27.8	26.4	226.7	−5.2
55.7	−67.3	19.5	−17.2	18.6
−37.1	104.5	1.3	38.7	−5.7

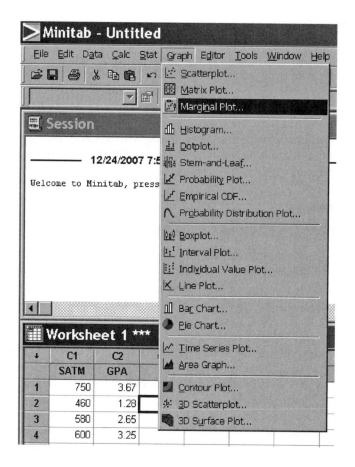

FIGURE 2.32
MINITAB commands to create a marginal plot.

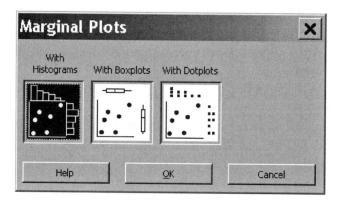

FIGURE 2.33
MINITAB dialog box to create a marginal plot.

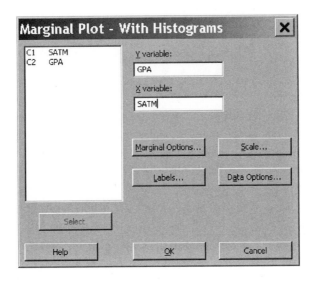

FIGURE 2.34
Marginal plot with histograms dialog box.

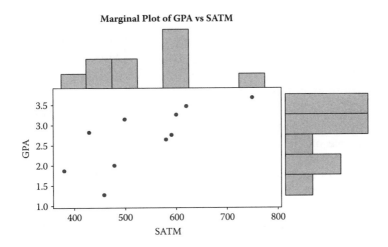

FIGURE 2.35
Marginal plot with histograms of first-year grade point average (GPA) versus SAT mathematics score (SATM).

a. Draw a histogram of the company profits for this sample of companies using the default settings in MINITAB.

b. Using MINITAB, draw a histogram with 20 bins of the company profits for this sample of companies using MINITAB.

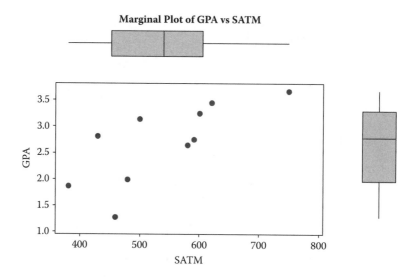

FIGURE 2.36
Marginal plot of GPA versus SATM with box plots.

    c. Using MINITAB, draw a histogram with ten bins of the company profits for this sample of companies using MINITAB.

    d. Which of these three histograms is more revealing as to the distribution of the company profits?

2. The data set "School Census Data" (http://www.census.gov/govs/ www/school04.html, All Data Items, Table Elsec04.xls) consists of a sample of the census data for all of the school districts in the United States. The variable TOTALEXP represents the total expenditures for each school district (descriptions of the other variables can be found at http://www.census.gov/govs/www/school04doc. html#layouts).

    a. Draw a box plot of the variable TOTALEXP using MINITAB. Why does this graph look strange?

    b. Draw a histogram of the variable TOTALEXP using MINITAB.

    c. What is the problem when using the default settings in MINITAB to draw the histogram and the box plot of the variable TOTALEXP?

    d. What happens when you try to adjust the number of bins of the histogram using MINITAB? Explain what you think the problem is.

3. Using MINITAB and the data set "Commuting Distance," draw a box plot. Calculate the values for Q_1, Q_2, and Q_3, the upper and lower limits, the upper and lower whiskers, and identify any outliers.

4. You may have heard that most car accidents happen within 25 miles of home. The data set in Table 2.6, "Car Accidents," gives a random sample of the mileage away from home for thirty-six car accidents in a given state over a 2-week period.

   a. Using MINITAB, draw a stem-and-leaf plot.

   b. Using MINITAB, draw a box plot.

   c. Using the graphs from a and b, comment on whether you would believe that most accidents happen within 25 miles of home.

5. The data set "Crime Data by County" (http://www.ojp.usdoj. gov/bjs/dtdata.htm#State) gives the crime rate and other related variables for a sample of sixty large counties in the United States. Using MINITAB, draw a box-whisker plot of the crime rate for the years 1991 and 1996 on the same graph. This can be done by selecting a simple box plot with multiple y's, as illustrated in Figure 2.37. Does this box plot suggest that there is a difference in the crime rates for these two years? Explain.

6. a. Using the data set "Expensive Homes" and MINITAB, draw a scatter plot of the asking price (y-axis) versus the square footage of the home (x-axis). Comment on the relationship between the square footage and asking price of a home.

   You can also use MINITAB to identify any outliers in a graph by clicking on a point on the graph, as illustrated in Figure 2.38.

   b. Using this technique, identify which points you would consider outliers or extreme observations.

TABLE 2.6

Mileage Away from Home for a Random Sample of Thirty-Six Car Accidents in a Given State

2	18	27	47	29	24
18	46	38	36	12	8
26	38	15	57	26	21
37	18	5	9	12	28
28	17	4	7	26	29
46	20	15	37	29	19

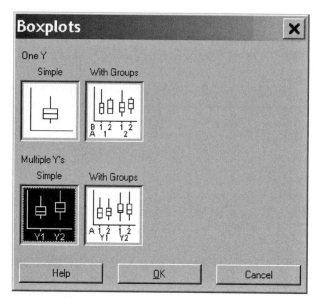

FIGURE 2.37
MINITAB dialog box to graph multiple box plots on the same graph.

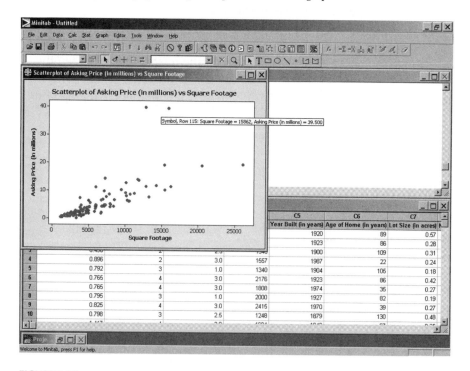

FIGURE 2.38
MINITAB commands to identify points in a graph by clicking on the point.

c. Using MINITAB, create a new variable that describes the size of a house as "small" if the square footage is less than 3,000 square feet and "large" if the square footage is at least 3,000 square feet.

d. Draw a bar graph that represents the number of homes that are described as "small" or "large."

7. Does the percent of unemployed civilians impact the crime rate?

a. Using the data set "Crime Data by County," draw a scatter plot to illustrate the relationship between the percent unemployed in 1991 (x-axis) and the crime rate for 1991 (y-axis).

b. Comment on the relationship between these two variables.

c. Draw a marginal plot with box plots to illustrate the relationship between the total number that are unemployed in 1991 and the crime rate for 1991.

8. Do cars that get better gas mileage emit less greenhouse gases?

a. Describe which variable would be the independent variable and which would be the dependent variable.

b. Using the "Automobile Data" data set and MINITAB, create a scatter plot to show the relationship between the city miles per gallon and the amount of greenhouse gas emitted.

c. How would you describe the relationship between city miles per gallon and the amount of greenhouse gas emitted?

d. Using MINITAB, draw a marginal plot with histograms to illustrate the relationship between the city miles per gallon and the amount of greenhouse gas emitted.

9. The data set "Freshmen" consists of a random sample of 350 first-year grade point averages for an entering class of freshmen at a local university.

a. Using MINITAB, draw a box-whisker plot to show the distribution of this sample of data.

b. Comment on what you think the box plot is illustrating.

c. Also draw a histogram with a normal curve superimposed to show the distribution of the first-year grade point averages for the given sample of freshmen, compared with a normal, or bell-shaped, distribution. This can be done by drawing a histogram **with fit**, as illustrated in Figure 2.39.

10. The data set "Executive Salary Data" provides the yearly salaries (in thousands) for a sample of fifteen executives by education, experience, gender, and political party affiliation.

a. Using MINITAB, draw a bar chart that shows the frequency of the sample based on political affiliation.

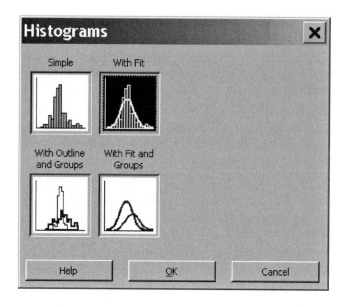

FIGURE 2.39
MINITAB commands to draw a histogram with a normal, or bell-shaped, curve superimposed.

b. Also draw a bar chart that shows the frequency of the sample based on gender.

c. Using the **Cluster** option of a bar plot as illustrated in Figure 2.40, draw a box plot that shows the frequency of the sample based on political affiliation and gender. This can be done by entering the categorical variables of gender and political party in the categorical variables box.

11. Suppose you want to compare the distributions for two different variables on the same graph. The data set in Table 2.7, "Coffee Prices," gives the prices (in dollars) for a specific brand of coffee from ten different stores in New York City and Kansas City, Missouri.

Using MINITAB, draw a normal, or bell-shaped, curve for the coffee prices in each of these different cities on the same graph by using the **With Fit and Groups** option in the histogram dialog box, as illustrated in Figure 2.41.

In order to draw these two histograms together on the same graph, you first need to put the data in two columns, one column for the coffee prices and the other column for the city.

TABLE 2.7

Coffee Prices from a Random Sample of Ten Different Stores in New York City and Kansas City

Price of Coffee	
New York City	**Kansas City**
6.59	6.80
8.47	6.75
7.24	6.99
9.81	7.25
7.87	6.90
9.25	5.99
8.50	7.81
7.90	6.55
7.79	6.20
10.25	6.59

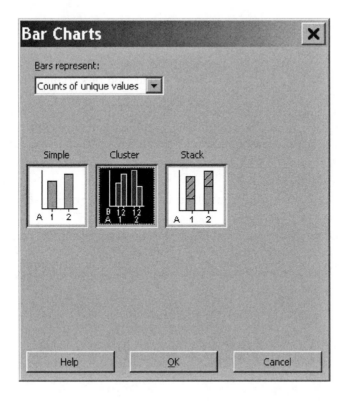

FIGURE 2.40

MINITAB command to draw a cluster of bar charts on a single graph.

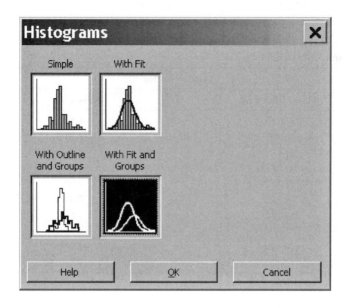

FIGURE 2.41
MINITAB commands to draw a normal curve for more than one sample of data.

12. The data set "Ferry" consists of a selection of variables for a ferry company that describe the revenue, time of day the ferry left the dock, number of passengers on the ferry, number of large objects on the ferry, the weather conditions, and the number of crew for a random sample of twenty-seven ferry runs over the course of a month.

   a. For the number of passengers, draw a stem-and-leaf plot using MINITAB.

   b. Draw a bar graph that illustrates the number of ferry trips based on the time of day.

   c. Draw a bar graph that illustrates the number of ferry runs based on the weather conditions.

   d. Draw a box plot for the number of large objects carried on the ferry.

3

Descriptive Representations of Data and Random Variables

3.1 Introduction

In this chapter we will begin by describing descriptive statistics that provide us with numerical measures calculated from a sample. We will present some typical measures of center, such as the mean, median, and mode. We will also describe how to measure the spread or variability of a variable by calculating the range, interquartile range, variance, and standard deviation. We will also explore how to use MINITAB to calculate measures of center and variability, and we will work through numerous examples.

In order to use sample statistics to make inferences about population parameters, we need to describe random variables and their corresponding distributions. In addition, the notion of sampling distributions will be described along with standard and nonstandard normal distributions, and a discussion of how to calculate probabilities for random variables that have a normal distribution.

3.2 Descriptive Statistics

A *descriptive statistic* is a numerical measure that is calculated from a sample. The purpose of calculating descriptive statistics is to summarize as succinctly as possible useful information by using a relevant numerical representation of a characteristic or attribute of interest for a given variable. For example, if we are interested in the first-year grade point averages for a sample of 100 students at a specific college, we may not want to see the entire list of the first-year grade point averages for all 100 students in order to get a good sense of some of the properties of this variable. Instead of looking at every single observation, we could simply calculate the mean, or average, first-year grade point average. Ideally, we like to be able to summarize different characteristics or attributes of a sample variable so we can succinctly convey useful information by using a relevant numerical representation.

There are two basic types of descriptive statistics that we will discuss:

Measures of center
Measures of spread

3.3   Measures of Center

Measures of center are descriptive statistics that represent the middle or center point of some characteristic or attribute of a sample variable. One common measure of center for numeric sample data is the *sample average* or *sample mean*. The sample mean is denoted by $\bar{x}$ and can be found by using the following formula:

$$\bar{x} = \frac{\sum\limits_{i=1}^{n} x_i}{n}$$

where n denotes the number of observations in the sample and $x_1, x_2, \ldots, x_n$ denote the individual observations with an assigned index number.

The summation notation, $\sum_{i=1}^{n} x_i$, is simply shorthand for summing the values x_1 through x_n. In other words, the summation notation represents the sum of the values x_1 through x_n, which would be $\sum_{i=1}^{n} x_i = x_1 + x_2 + \ldots + x_n$.

Example 3.1

Consider Table 3.1, which gives the first-year grade point averages for a random sample of five students.

To find the sample mean, we first begin by labeling the individual observations, and this is done by assigning an arbitrary index number to each observation as follows:

$$x_1 = 2.56$$
$$x_2 = 3.21$$
$$x_3 = 3.56$$
$$x_4 = 2.10$$
$$x_5 = 1.87$$

TABLE 3.1

First-Year Grade Point Averages for a
Random Sample of Five Students

| 2.56 | 3.21 | 3.56 | 2.10 | 1.87 |

Note that it does not matter the order in which the index numbers are assigned to the individual observations.

Then summing all these values we have

$$\sum_{i=1}^{5} x_i = x_1 + x_2 + x_3 + x_4 + x_5 = 2.56 + 3.21 + 3.56 + 2.10 + 1.87 = 13.30$$

The sample mean is then found by dividing this sum by the number of observations in the sample, which for this example is $n = 5$.

$$\bar{x} = \frac{\sum_{i=1}^{5} x_i}{5} = \frac{13.30}{5} = 2.66$$

The sample mean describes a characteristic or attribute of a given sample variable, namely, the average first-year GPA for the sample of five students. The mean is a single numerical representation of the center of a given variable.

Example 3.2

Table 3.2 gives the speeds (in miles per hour) for a random sample of seven cars driving on a particular stretch of a dangerous highway.

The mean for this sample would be calculated as follows:

$$\bar{x} = \frac{\sum_{i=1}^{7} x_i}{7} = \frac{68 + 72 + 73 + 84 + 67 + 62 + 74}{7} = 71.43$$

In addition to using descriptive measures to describe a sample variable, we can also use descriptive measures to describe characteristics or attributes of a population variable. For instance, the *population mean* is the average value of a numeric population variable, which is represented by the symbol μ. The population mean is calculated as follows:

$$\mu = \frac{\sum_{i=1}^{N} x_i}{N}$$

where N denotes the total number of observations in the population and $x_1, x_2, \ldots,$ x_N denote the N individual population values with an assigned index number.

TABLE 3.2

Sample of Speeds of Cars Driving on a Particular Stretch of Highway

68	72	73	84	67	62	74

TABLE 3.3

Population of Data That Consists of the Ten Even Numbers
from 2 through 20

2	4	6	8	10	12	14	16	18	20

Example 3.3

Table 3.3 represents the population of data that consists of the ten even numbers from 2 through 20.

The population mean would be found by adding up these values and dividing by the total number of data values in the population ($N = 10$) as follows:

$$\mu = \frac{2+4+6+8+10+12+14+16+18+20}{10} = 11$$

The *median* of a numeric variable is another measure of center and is defined as the numeric value that partitions the data set into two parts. The median is a number such that at least 50% of the observations lie at or below this number and at least 50% of the observations lie at or above this number. You may recall from Chapter 2 that the median is also described as the second quartile, or Q_2.

To find the median we first need to find the *median position*. The median position is found by first putting the data in numerical order and then finding the observation that partitions the ordered data set in half. For a sample of size n, if there are an odd number of observations, the median is the observation in position $(n + 1)/2$, and if there are an even number of observations, the median is the average of the observations that are in positions and $n/2$ and $(n + 2)/2$.

Example 3.4

Table 3.4 consists of the number of hours per week a random sample of five adults over the age of 50 spend exercising.

The median of the sample in Table 3.4 can be found by first arranging the data in numerical order as follows:

$$2 \quad 3 \quad 4 \quad 7 \quad 8$$

Since we have an odd number of observations ($n = 5$), the median of this sample will correspond to the observation that is in the $(n + 1)/2 = (5 + 1)/2 = 3$rd position. Therefore, the median for this sample is the number 4.

TABLE 3.4

Number of Hours Spent Exercising per Week for a
Random Sample of Five Adults over the Age of 50

2	4	7	3	8

TABLE 3.5

Number of Hours That a Sample of Ten Homeowners Spend
Working on Their Lawn Each Week

2	4	5	7	9	13	15	17	28	30

Example 3.5

Consider Table 3.5, which consists of the number of hours that a sample of ten homeowners spend maintaining their lawn per week (already in numerical order).

Since there is an even number of observations ($n = 10$), the median of this sample will be the average of the values in the $n/2 = 10/2 = 5$th and $(n + 2)/2 = (10 + 2)/2 = 6$th positions. The value in the fifth position is 9, and the value in the sixth position is 13. Therefore, the median corresponds to the average of these two values, or $9 + 13/2 = 11$. Notice that although the number 11 is not an observation in the actual data set, it still represents a numerical measure that cuts the data set in half because 50% (or five) of the observations lie at or below this value, and 50% (or five) of the observations lie at or above this value.

In some cases, the median may be the preferred measure of center as compared to the mean. This is because the median is less sensitive to extreme values, as will be illustrated in the next example.

Example 3.6

The mean and median for the sample of eleven observations that are presented in Table 3.6 are 58.55 and 22, respectively.

Notice that for the data in Table 3.6 the median may be the preferred measure over the mean because the median is less affected by extreme values or outliers. In this case, the median also more accurately represents the center of the underlying data set.

Another measure of center is called the *mode*. The mode can be found for a continuous or discrete variable, and it is defined as the value or values that occur most often.

Example 3.7

The data set given in Table 3.7 describes the number of hours that a sample of twelve college students spend studying per week.

The mode of the data in Table 3.7 is 8 since this is the value that occurs most often.

TABLE 3.6

Sample of Eleven Observations

0	12	17	18	20	22	24	25	27	28	451

TABLE 3.7

Sample of the Number of Hours a Random Sample of Twelve
College Students Spend Studying per Week

5	6	7	8	8	9	10	12	15	18	19	40

The mode can also be found for a discrete set of data, as is illustrated in the
following example.

Example 3.8

The data given in Table 3.8 consist of the political party affiliations for a sample
of five shoppers surveyed at random at a local shopping mall.

For this example, there would be two modes: Democrat and Republican.
This is because both of these values occur most often (namely, twice).

Although the mode can be used for both continuous and discrete variables, one limitation can arise when using the mode as a measure center for
continuous variables because many continuous variables do not have any
values that occur on more than one occasion.

TABLE 3.8

Political Affiliations for a Random Sample of Five Shoppers

Republican	Democrat	Democrat	Independent	Republican

3.4   Measures of Spread

Descriptive statistics that measure the spread of a variable are numeric
summaries that can be used to describe how the sample values differ from
each other. For instance, if I purchase ten 1-pound bags of my favorite brand
of coffee from ten different stores around town, I would expect that each
of the 1-pound bags would weigh approximately the same. In other words,
I would not expect much spread or variability between the weights of the
different bags of the same brand of coffee purchased at various locations.
However, if I asked ten of my friends to tell me how much money they
spent to put gasoline in their cars this week, I would expect there to be a
significant amount of variation because of differences in types of cars and
distances driven each week.

One simple measure of the spread of a variable is called the *range*. The
range is found by subtracting the smallest value from the largest value.

Example 3.9

Consider the data given in Table 3.1, which consists of the first-year grade point averages for a random sample of five students.

The range for this set of data can be found by taking the smallest value and subtracting it from the largest value as follows:

$$\text{range} = 3.56 - 1.87 = 1.69$$

Although the range is a very simple statistic to calculate, one concern with using the range to assess variability is that the range is based on only the two most extreme observations of the variable, and thus it does not reflect the variability between any of the other observations that are not at the extremes.

Another measure of variability that does not rely on the two most extreme observations is the *interquartile range*, or the *IQR*. The *IQR* is found by taking the difference between the third quartile and the first quartile, where the quartiles of a variable are values that cut the data into quarters:

$$IQR = Q_3 - Q_1$$

Recall from Chapter 2 that to find the quartiles of a data set, the data is first put in numeric order and the values for the quartiles are found such that a given percentage of the observations lie above or below the respective quartile. For instance, the first quartile, or Q_1, represents the value such that at least 25% of the data fall at or below this value and at least 75% of the data fall at or above this value. Similarly, the third quartile, or Q_3, represents the value such that at least 75% of the data fall at or below this value and at least 25% of the data fall at or above this value. The second quartile, or Q_2, is the median.

Once the data are put in numeric order, the value of Q_1 is the observation in the $(n + 1)/4$th position. Similarly, the value of Q_3 is the observation in the $3(n + 1)/4$th position. If these positions are not integer values, then interpolation is used.

Example 3.10

Suppose we want to find Q_1 and Q_3 for the data set in Table 3.5, which represents the number of hours a sample of ten homeowners spend maintaining their lawn each week. Since the sample is of size $n = 10$, Q_1 will be the value in the $(10 + 1)/4 = 2.75$th position. Because this value is not an integer, we would then interpolate as follows to find the value of Q_1:

$$Q_1 = x_2 + 0.75(x_3 - x_2) = 4 + 0.75(5 - 4) = 4.75$$

Similarly, to find Q_3, it will be the value in the $3(10 + 1)/4 = 8.25$th position, and since this is not an integer, we would interpolate as follows:

$$Q_3 = x_8 + 0.25(x_9 - x_8) = 17 + 0.25(28 - 17) = 19.75$$

Therefore, the interquartile range would be

$$IQR = Q_3 - Q_1 = 19.75 - 4.75 = 15.00$$

Similar to using the range to assess variability, the *IQR* tends not to be the preferred measure of variability because it also uses only a limited amount of the data values.

Two additional measures of spread for a sample variable are the *sample variance* and the *sample standard deviation*. These provide information that is generally more useful than the range or the *IQR*, because both the sample variance and the sample standard deviation use all of the individual observations in a given sample. Thus, these measures can be used to summarize how the individual sample observations vary about the mean. However, the sample mean and standard deviation can be sensitive to extreme values, especially for small sample sizes.

The sample variance, denoted as s^2, is calculated by using the following formula:

$$s^2 = \frac{1}{n-1} \sum_{i=1}^{n} (x_i - \bar{x})^2$$

where n is the sample size, $\bar{x}$ is the sample mean, and x_i represent the individual indexed observations.

The sample standard deviation, denoted as s, is the square root of the sample variance, and it is calculated as follows:

$$s = \sqrt{s^2} = \sqrt{\frac{1}{n-1} \sum_{i=1}^{n} (x_i - \bar{x})^2}$$

The sample variance measures "on average" the squared difference between each observation and the sample mean. Notice that this average is found by dividing by $n - 1$ rather than by n. This adjustment is made so that the sample variance can be used to estimate the population variance without bias. Typically, the standard deviation is the preferred measure of variability for a sample variable because the units of the sample standard deviation are the same as those of the variable, whereas the units of the sample variance are in units squared. Furthermore, the sample variance and sample standard deviation use all of the observations in a data set, whereas the range and *IQR* use only a few select values.

Example 3.11

For the sample of five first-year grade point averages presented in Table 3.1, we found that the sample mean is 2.66. We can find the sample variance and sample standard deviation by first taking the difference between each observation and the mean, and then squaring this difference, as illustrated in Table 3.9.

TABLE 3.9

Difference and Difference Squared between Each Observation
and the Sample Mean for the Data Given in Table 3.1

x_i	$(x_i - \bar{x})$	$(x_i - \bar{x})^2$
2.56	$2.56 - 2.66 = -0.10$	$(-0.10)^2 = 0.0100$
3.21	$3.21 - 2.66 = 0.55$	$(0.55)^2 = 0.3025$
3.56	$3.56 - 2.66 = 0.90$	$(0.90)^2 = 0.8100$
2.10	$2.10 - 2.66 = -0.56$	$(-0.56)^2 = 0.3136$
1.87	$1.87 - 2.66 = -0.79$	$(-0.79)^2 = 0.6241$

Then to find the variance, the values in the third column in Table 3.9 are added up to get

$$\sum_{i=1}^{n}(x_i - \bar{x})^2 = 2.0602$$

This quantity is then divided by $n - 1$ as follows:

$$s^2 = \frac{1}{n-1}\sum_{i=1}^{n}(x_i - \bar{x})^2$$

$$= \frac{1}{5-1}\sum_{i=1}^{5}(x_i - \bar{x})^2$$

$$= \frac{1}{4}(0.0100 + 0.3025 + 0.8100 + 0.3136 + 0.6241)$$

$$= \frac{1}{4}(2.0602) = 0.5151$$

To find the sample standard deviation, we simply take the square root of the variance as follows:

$$s = \sqrt{s^2} = \sqrt{.5151} \approx 0.7177$$

Similarly, if we are given data for an entire population, then we could calculate the *population variance* and the *population standard deviation* using the following formulas:

$$\sigma^2 = \frac{1}{N}\sum_{i=1}^{N}(x_i - \mu)^2$$

$$\sigma = \sqrt{\frac{1}{N}\sum_{i=1}^{N}(x_i - \mu)^2}$$

where μ is the population mean, σ^2 is the population variance, σ is the population standard deviation, and N is the population size.

Example 3.12

To find the variance and standard deviation for the population data presented in Table 3.3, we would take the sum of the squared difference between each observation and the population mean, and then divide by the population size, which is $N = 10$, to obtain the population variance of $\sigma^2 = 33.01$ and the population standard deviation of $\sigma = 5.745$.

3.5   Using MINITAB to Calculate Descriptive Statistics

MINITAB can be used to calculate many types of descriptive statistics. Using the sample of five grade point averages given in Table 3.1, we can have MINITAB calculate the descriptive statistics of interest by first entering the data into a MINITAB worksheet and then selecting **Stat** from the top menu bar and then **Basic Statistics** and **Display Descriptive Statistics**, as presented in Figure 3.1.

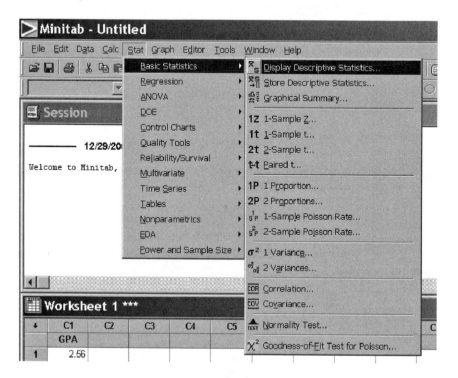

FIGURE 3.1
MINITAB commands to calculate descriptive statistics.

This gives the descriptive statistics dialog box that is presented in Figure 3.2.

To select the variable(s) that you wish to calculate the descriptive statistics for, simply highlight the variable(s) of interest and choose **Select**. This will place the variable(s) of interest in the **Variables** box.

Clicking on the **Statistics** tab in Figure 3.2 brings up the descriptive statistic dialog box, which allows you to select which specific descriptive statistics you want MINITAB to calculate for you, as can be seen in Figure 3.3.

Checking the boxes for the mean, standard deviation, variance, Q_1, median, Q_3, and the *IQR* provides the MINITAB printout given in Figure 3.4.

In many situations, it can be very difficult or even impossible to obtain the data for an entire population. However, even when it is not possible to obtain population-level data, we can make an inference about various population parameters by using the information from a representative sample that is obtained from the population of interest. For example, we can use the sample mean to estimate the population mean, and similarly, we can use the sample variance and sample standard deviation to estimate

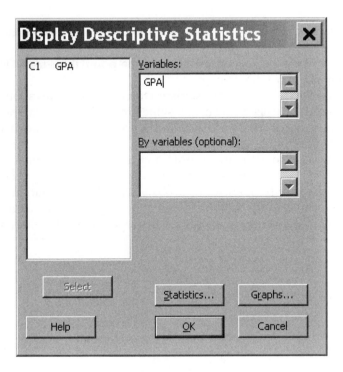

FIGURE 3.2
MINITAB dialog box for calculating descriptive statistics.

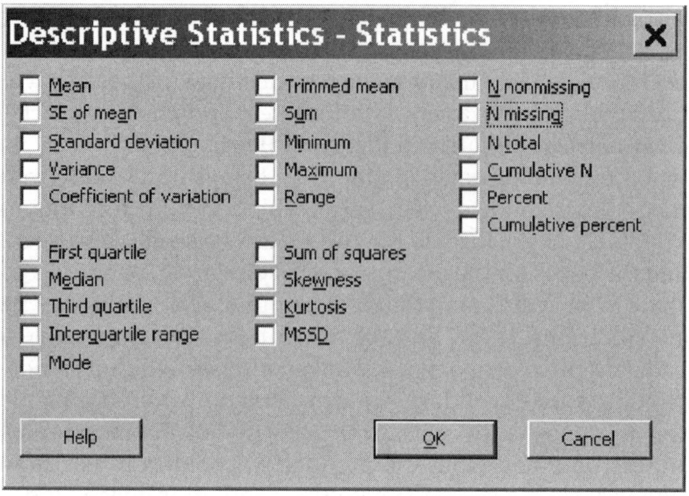

FIGURE 3.3
MINITAB dialog box for selecting descriptive statistics.

the population variance and the population standard deviation. Thus, $\bar{x}$, the sample mean, can be used in combination with s, the sample standard deviation, to provide a good estimate of the population mean μ and the population standard deviation σ.

But before we can use sample statistics to make inferences about population parameters, we need to consider some properties of the sample statistic that we are using. In particular, we need to consider what the distribution or shape of the sample statistic looks like in addition to how this distribution can be used to estimate an unknown population parameter. Similar to how a histogram can illustrate the distribution of a given variable along with its respective frequencies or probabilities, the distribution of a sample statistic can also display the values of a sample statistic along with its respective frequencies or probabilities. In order to describe distributions of

Descriptive Statistics: GPA

Variable	Mean	StDev	Variance	Q1	Median	Q3	IQR
GPA	2.660	0.718	0.515	1.985	2.560	3.385	1.400

FIGURE 3.4
MINITAB output of selected descriptive statistics.

sample statistics, we will first discuss random variables and their associated probability distributions.

3.6 Random Variables and Their Distributions

A *random variable* takes on different numerical values according to chance. A random variable can be discrete or continuous. A random variable is characterized as *discrete* if it can assume either a finite or a countably infinite number of values. For example, flipping a fair coin three times and counting the number of tails is an example of a discrete random variable because the values of the random variable, which is the number of tails in three coin flips, takes on a finite number of values, namely, 0, 1, 2, and 3, and these values are based on chance. An example of a discrete random variable, which is countably infinite, is the number of flips of a fair coin until the first head appears. This random variable is countably infinite because theoretically you could flip a coin indefinitely until the first head appears.

The probability pattern that is associated with the possible values of a random variable constitutes the *probability distribution* of the random variable. For instance, if the random variable is the number of tails counted after flipping a fair coin three times, then the probability pattern that associates each of the possible values of the random variable with its corresponding probability is presented in Table 3.10.

The way that the probability distribution in Table 3.10 was constructed was to first list the set of all possible outcomes of the random variable. For this example, this corresponds to flipping a fair coin three times as follows:

$$\{HHH, HHT, HTH, HTT, THH, THT, TTH, TTT\}$$

Since there are a total of eight possible outcomes, the probability of observing 0 tails in three coin flips is 1/8 (which corresponds to observing *HHH*),

TABLE 3.10

Probability Pattern Associated with the Values of the Random Variable That Consists of Counting the Number of Tails after Flipping a Fair Coin Three Times

Number of Tails X	Probability $p(x_i)$
$x_1 = 0$	$p(x_1) = 1/8$
$x_2 = 1$	$p(x_2) = 3/8$
$x_3 = 2$	$p(x_3) = 3/8$
$x_4 = 3$	$p(x_4) = 1/8$

the probability of observing one tail in three coin flips is 3/8 (which corresponds to observing *HHT, HTH,* or *THH*), the probability of observing two tails in three coin flips is 3/8 (which corresponds to observing *HTT, THT,* or *TTH*), and the probability of observing three tails in three coin flips is 1/8 (which corresponds to observing *TTT*).

Typically, we represent random variables by uppercase letters and specific outcomes of the random variable by lowercase letters. For example, in counting the number of tails from flipping a fair coin three times, the uppercase letter X would constitute the random variable itself, and lowercase letters would represent the specific values that the random variable X can take on, which in this case would be either $x_1 = 0$, $x_2 = 1$, $x_3 = 2$ or $x_4 = 3$.

There are two properties that must hold true for the probability distribution of a discrete random variable. First, the probabilities associated with each value of the random variable must be numbers between 0 and 1 inclusive, in other words:

$$0 \le p(x_i) \le 1$$

where $p(x_i)$ is the probability of observing the specific value x_i of the random variable X.

Second, all of the probabilities in the probability distribution for the random variable must sum to 1, which means that

$$\sum p(x_i) = 1$$

Random variables can also be summarized by calculating a measure of center and a measure of spread of the random variable.

We can find the mean, or expected value, of a random variable X as follows:

$$\mu_X = \sum x_i \, p(x_i)$$

where x_i is the *i*th observation of the random variable, and $p(x_i)$ is the probability that observation x_i will occur.

We can also find the standard deviation and variance of a random variable X as follows:

$$\sigma_X^2 = \sum (x_i - \mu_X)^2 \cdot p(x_i)$$

$$\sigma_X = \sqrt{\sigma_X^2}$$

Example 3.13

To find the mean, or expected value, of the random variable given in Table 3.10, we need to take the sum of the products of all of the possible outcomes of the random variable with their corresponding probabilities as follows:

$$\mu_X = \sum x_i \cdot p(x_i) = 0\left(\frac{1}{8}\right) + 1\left(\frac{3}{8}\right) + 2\left(\frac{3}{8}\right) + 3\left(\frac{1}{8}\right) = 1.50$$

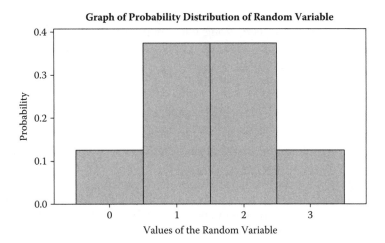

Graph of Probability Distribution of Random Variable

Values of the Random Variable

FIGURE 3.5
Graph of the probability distribution for the random variable in Table 3.10.

Thus, the mean, or expected value, of the random variable X is $\mu_x = 1.50$.

The variance of the random variable X is found by taking the sum of the squared difference between each value of the random variable and the mean of the random variable multiplied by the respective probability as follows:

$$\sigma_X^2 = \sum (x_i - \mu_X)^2 \cdot p(x_i) = (0 - 1.5)^2 \left(\frac{1}{8}\right) + (1 - 1.5)^2 \left(\frac{3}{8}\right) + (2 - 1.5)^2 \left(\frac{3}{8}\right)$$

$$+ (3 - 1.5)^2 \left(\frac{1}{8}\right) = 0.75$$

And the standard deviation of the random variable X is found by taking the square root of the variance as follows:

$$\sigma_X = \sqrt{\sigma_X^2} = \sqrt{0.75} \approx 0.866$$

The graph of the probability distribution of a random variable can be useful for visualizing the probability pattern of the random variable. The graph of a probability distribution can be found by plotting the possible values of the random variable on the x-axis versus the respective probabilities for each of the values of the random variable on the y-axis, as illustrated in Figure 3.5.

3.7 Sampling Distributions

When using sample statistics, such as the sample mean and the sample standard deviation, to estimate the mean and standard deviation of an unknown population, we need to consider the *probability distribution* of these sample statistics that we are using. In other words, we need to describe the possible values of the sample statistic along with their respective probabilities. The probability distribution of a sample statistic is referred to as the *sampling distribution* of the sample statistic.

The following example illustrates how a sampling distribution for the sample mean can be created.

Example 3.14

Consider the probability distribution for a random variable X given in Table 3.11.

We can graph this probability distribution by plotting the values of the random variable against their respective probabilities, as shown in Figure 3.6.

We can also calculate the mean and standard deviation of this probability distribution as follows:

$$\mu_X = \sum x_i \cdot p(x_i) = 1(0.25) + 2(0.25) + 3(0.25) + 4(0.25) = 2.50$$

$$\sigma_X = \sqrt{\sum (x_i - \mu_X)^2 p(x_i)}$$

$$= \sqrt{(1-2.5)^2(0.25) + (2-2.5)^2(0.25) + (3-2.5)^2(0.25) + (4-2.5)^2(0.25)}$$

$$= \sqrt{1.25} \approx 1.12$$

Now suppose that we can list every possible random sample of size $n = 2$ that could be drawn from the random variable that is given in Table 3.11. If we sample with replacement, which means that we allow for the second outcome to be the same as the first, then all such samples of size $n = 2$ are presented in Figure 3.7.

Assuming that each random sample of size $n = 2$ has the same probability of being selected, and because there are sixteen possible samples of size $n = 2$, the probability of drawing any one given sample of size $n = 2$ from Figure 3.7 would be $1/16 = 0.0625$.

We are now going to define a new random variable that describes the mean, or average, of all of the samples of size $n = 2$, where the mean values for each of the samples of size $n = 2$ are presented in Figure 3.8.

TABLE 3.11

Probability Distribution for the Random Variable X

X	$p(x_i)$
$x_1 = 1$	0.25
$x_2 = 2$	0.25
$x_3 = 3$	0.25
$x_4 = 4$	0.25

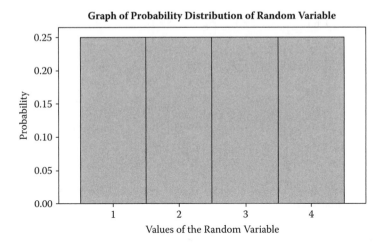

FIGURE 3.6
Graph of the probability distribution given in Table 3.11.

To create the probability distribution for the random variable that represents the sample mean (this probability distribution is referred to as the sampling distribution of the sample mean), we take each possible value of the random variable, which in this case is the sample mean for the samples of size $n = 2$, and assign to each of these values the appropriate probability. For instance, since there are four ways to get the sample mean of 2.5, as can be seen in Figure 3.8, and the probability of selecting each sample is equally likely, with a probability of .0625, the probability of drawing a sample of size $n = 2$ and getting a sample mean of 2.5 would be:

$$4(0.0625) = 0.2500$$

Similarly for the other possible values of the sample means for the samples of size $n = 2$, the associated probabilities are presented in Table 3.12.

$$\{1,1\} \quad \{2,1\} \quad \{3,1\} \quad \{4,1\}$$
$$\{1,2\} \quad \{2,2\} \quad \{3,2\} \quad \{4,2\}$$
$$\{1,3\} \quad \{2,3\} \quad \{3,3\} \quad \{4,3\}$$
$$\{1,4\} \quad \{2,4\} \quad \{3,4\} \quad \{4,4\}$$

FIGURE 3.7
All possible samples of size $n = 2$ for the random variable given in Table 3.11.

$$\frac{1+1}{2} = 1 \qquad \frac{2+1}{2} = 1.5 \qquad \frac{3+1}{2} = 2 \qquad \frac{4+1}{2} = 2.5$$

$$\frac{1+2}{2} = 1.5 \qquad \frac{2+2}{2} = 2 \qquad \frac{3+2}{2} = 2.5 \qquad \frac{4+2}{2} = 3$$

$$\frac{1+3}{2} = 2 \qquad \frac{2+3}{2} = 2.5 \qquad \frac{3+3}{2} = 3 \qquad \frac{4+3}{2} = 3.5$$

$$\frac{1+4}{2} = 2.5 \qquad \frac{2+4}{2} = 3 \qquad \frac{3+4}{2} = 3.5 \qquad \frac{4+4}{2} = 4$$

FIGURE 3.8

Mean values for all the samples of size $n = 2$ as given in Figure 3.7 from the probability distribution given in Table 3.11.

The sampling distribution of the sample mean in Table 3.12 illustrates the probability pattern for all of the sample means for the samples of size $n = 2$. Notice that all of the probabilities are between 0 and 1 inclusive, and the sum of the probabilities is equal to 1.

Graphing the sampling distribution of the sample mean requires plotting the possible values of the random variable (which in this case would be the values of the sample mean) on the x-axis, and the corresponding probability (expressed as a percent) on the y-axis. Figure 3.9 gives the graph of the sampling distribution of the sample mean from Table 3.12.

In general, to use the sampling distribution of a sample statistic to estimate a population parameter, we would want the sampling distribution to center about the true but unknown population parameter of interest. Notice in Figure 3.9 that the sampling distribution of the sample mean is centered about the true population mean of the original random variable as given in Figure 3.6 ($\mu_x = 2.5$). We also want the standard deviation of the sampling distribution (which is also referred to as the standard error) to be as small as possible.

TABLE 3.12

Sampling Distribution of the Sample Mean for Drawing a Sample of Size $n = 2$ from the Probability Distribution for the Random Variable Presented in Table 3.11

$\bar{x}$	$p(\bar{x})$
1	0.0625
1.5	0.1250
2	0.1875
2.5	0.2500
3	0.1875
3.5	0.1250
4	0.0625

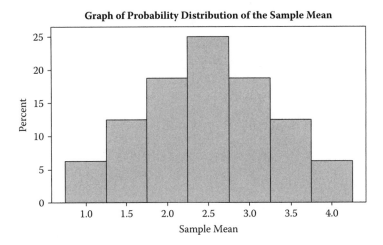

FIGURE 3.9
Graph of the sampling distribution of the sample mean from the probability distribution given in Table 3.12.

There is a famous theorem, called the central limit theorem, which states that for large samples (sample sizes larger than thirty), the sampling distribution of the sample mean will approximately have the shape of a normal, or bell-shaped, curve for *any* shape of the distribution of the underlying population that is being sampled from. It is because of this theorem that we will be able to make inferences about unknown population parameters of interest by using sample data. But before we begin to use sampling distributions to make inferences about population parameters of interest, we will first describe continuous random variables.

With a *continuous random variable*, between any two values of the random variable other values can exist. Another way to describe a continuous random variable is a random variable that consists of an uncountable number of values. For example, consider the time it takes to fly from Boston to New York City. If it takes at least 1 hour to fly from Boston to New York, we can call this smallest possible flight time t_1, where $t_1 = 1$ hour. It is impossible to precisely describe the next possible flight time, or flight time t_2, because it could be 1 hour and 2 minutes, 1 hour and 1 minute, 1 hour and 30 seconds, 1 hour and 5 seconds, etc. Thus, the flight time from Boston to New York City is a continuous random variable because between any two possible flight times other possible flight times can possibly exist. Consequently, there are an uncountable number of values that can exist between any two observed flight times.

Similar to discrete random variables, the relationship between a continuous random variable and its respective probability pattern can be displayed graphically. Perhaps the most common distribution of a continuous random variable is the *normal distribution*, which is the shape of a bell-shaped curve, as presented in Figure 3.10.

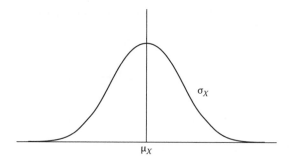

FIGURE 3.10
Normal distribution with a mean of μ_X and spread of σ_X.

A normal distribution can be characterized by its shape, its center, and its spread about the center. The center, or mean, of a random variable X that has the shape of a normal distribution is characterized by μ_X, and σ_X describes the spread or variability of the random variable about the mean, as illustrated in Figure 3.10. Notice that the normal distribution is symmetric about the mean μ_X. Figure 3.11 shows the comparison of two different normal distributions, one distribution that has a mean of 5 and a standard deviation of 10, and the other distribution that has a mean of −3 and a standard deviation of 5.

When a normal distribution represents the probability distribution of a continuous random variable, the area under the curve over an interval of possible values corresponds to the probability that the random variable will take on a value within the given interval. The total area under the normal

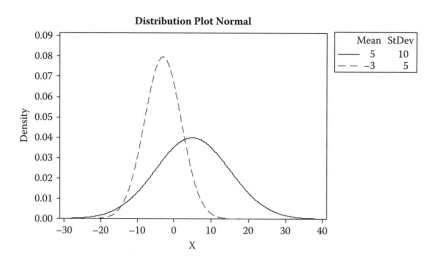

FIGURE 3.11
Comparison of two normal distributions, one with a mean of 5 and a standard deviation of 10, and the other with a mean of −3 and a standard deviation of 5.

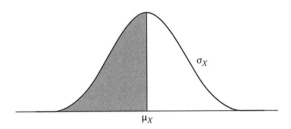

FIGURE 3.12
Shaded region that represents the area to the left of the mean that is equal to 0.50.

distribution must equal 1, and since the distribution is symmetric about the mean, the area to the left (or right) of the mean is equal to 0.50, as can be seen in Figure 3.12.

We denote $p(a < x < b)$ as the probability that a random variable X takes on a value between a and b. This probability corresponds to the shaded area under the curve, as illustrated in Figure 3.13.

Because the area under the normal curve represents the probability that a continuous random variable X will take on a value within a given range, the probability that the random variable X will take on a specific value is 0. This is because the area under the curve for any specific value of a normally distributed random variable is equal to 0.

There is one special normal distribution, called the *standard normal distribution*. The standard normal distribution is a normal distribution that has a mean of 0 and a standard deviation of 1, as illustrated in Figure 3.14.

The standard normal distribution is very important because it allows us to find probabilities for random variables (or areas under the curve) by using what is called a *standard normal table*, as given in Table 1 of Appendix A. This standard normal table presents the areas under the standard normal curve with respect to a given value of z. Thus, the shaded region in the table represents the probability that a random variable will take on a value between 0 and z.

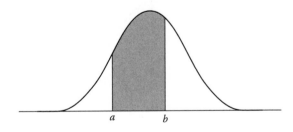

FIGURE 3.13
Shaded region that represents the probability that a normally distributed random variable X takes on values between a and b.

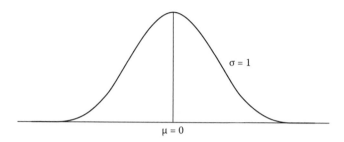

FIGURE 3.14
Standard normal distribution that has a mean of 0 and a standard deviation of 1.

Example 3.15

Suppose we want to find the area under the standard normal curve that represents the probability that a random variable takes on a value less than $z = 0.25$.

Because we want to find the probability that a random variable that has the standard normal distribution takes on a value less than $z = 0.25$, this corresponds to the area under the standard normal curve that is represented by the shaded region, as illustrated in Figure 3.15.

In order to find this probability, we can use the standard normal table (Table 1 of Appendix A) because the mean is equal to 0 and the standard deviation is equal to 1. Notice that this table is referenced to the mean of 0, where the shaded area represents the probability that a random variable will take on a value between 0 and z. Since the shaded area between 0 and $z = 0.25$ is 0.0987, the probability that z is less than 0.25 is equal to $0.5000 + 0.0987 = 0.5987$. We could also express this as $p(z < 0.25) = 0.5987$.

Example 3.16

To find the probability that a random variable with a standard normal distribution takes on a value between $z = -1.00$ and $z = 0.43$, we would use the standard normal tables to find the probability that is represented by the shaded region give in Figure 3.16.

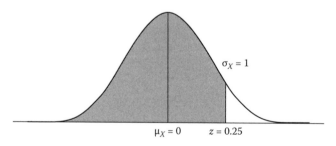

FIGURE 3.15
Standard normal distribution where the shaded area corresponds to the probability that the random variable takes on a value less than $z = 0.25$.

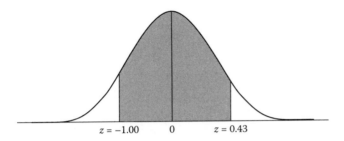

FIGURE 3.16
Area under the standard normal curve that represents the probability that a random variable with the standard normal distribution falls between $z = -1.00$ and $z = 0.43$.

To determine this probability, we would use the standard normal tables to find the probability that the random variable falls between $z = -1.00$ and $z = 0$, which is 0.3413, and the probability that the random variable falls between $z = 0$ and $z = 0.43$, which is 0.1664. Then adding these two probabilities gives

$$p(-1.00 < z < 0.430) = 0.5077$$

However, most normally distributed continuous random variables do not have a mean of 0 and a standard deviation of 1. Because there are an infinite number of possible means and standard deviations that describe different normally distributed random variables, we can *standardize* any normally distributed random variable in order to convert it to the standard normal distribution, which allows us to use the standard normal table to find probabilities (or the areas under the curve). By standardizing a normal distribution, we are taking a normal distribution with any given mean and any standard deviation and are converting it to a normal distribution, which has a mean of 0 and a standard deviation of 1.

For a nonstandard normal distribution, the shaded area under the curve, as illustrated in Figure 3.17, represents the probability that a random variable takes on a value less than x. By standardizing this normal distribution, we are

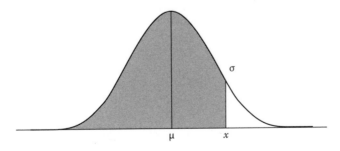

FIGURE 3.17
Nonstandard normal distribution where the shaded area represents the probability that a random variable X takes on a value less than x.

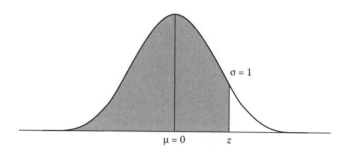

FIGURE 3.18
Standardized normal distribution from Figure 3.17.

essentially converting this nonstandard normal distribution to a standard normal distribution, which has a mean of 0 and a standard deviation of 1.

To transform any normal distribution into a standard normal distribution requires the use of the following transformation formula:

$$z = \frac{x - \mu}{\sigma}$$

where x is the specific value of the random variable under consideration, μ is the mean of the nonstandard normal distribution, and σ is the standard deviation of the nonstandard normal distribution.

Standardizing converts the nonstandard normal distribution in Figure 3.17 to a standard normal distribution by converting the x-value to a z-value and converting the mean to 0 and the standard deviation to 1, as illustrated in Figure 3.18.

Example 3.17

Suppose we want to find the probability that the random variable X takes on a value less than 150 if X has a normal distribution with a mean of 140 and a standard deviation of 40.

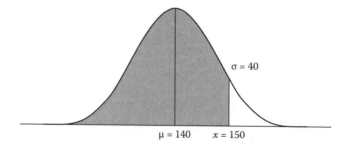

FIGURE 3.19
Nonstandard normal distribution to find the probability that a random variable with a mean of 140 and standard deviation of 40 takes on a value less than 150.

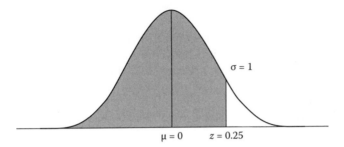

FIGURE 3.20
Standardized normal distribution from Figure 3.19.

The shaded region in Figure 3.19 illustrates the shaded area that corresponds to the probability that the random variable takes on a value less than 150 if X is a normally distributed random variable with a mean of 140 and a standard deviation of 40.

By transforming the given x-value of the random variable to a z-value we get

$$z = \frac{x - \mu}{\sigma} = \frac{150 - 140}{40} = 0.25$$

This now gives us the standard normal distribution such that $p(x < 150) = p(z < 0.25)$, which is illustrated in Figure 3.20.

Now we can use the standard normal tables (Table 1 of Appendix A) to find the area of the shaded region that corresponds to the probability that the random variable takes on a value less than 150. Notice that in referring to the standard normal table, the shaded area in the table corresponds to the probability that x falls between 0 and z. Thus, for our example, the shaded area that corresponds to the probability of interest can be found by adding the two areas that are shown in Figure 3.21.

Therefore, the probability that the random variable X is less than 150 is

$$p(x < 150) = p(z < 0.25) = 0.5000 + .0987 = 0.5987$$

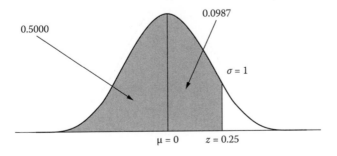

FIGURE 3.21
Probabilities under the standard normal curve corresponding to $p(z < 0.25)$.

Example 3.18

Suppose we want to find the probability that the random variable X takes on a value between 46 and 54 if X has a normal distribution with a mean of 52 and a standard deviation of 2. Figure 3.22 shows the shaded region that corresponds to the probability that the random variable will take on a value between 46 and 54.

The standardized (or z) values are found by transforming each of the x-values as follows:

$$z = \frac{46 - 52}{2} = -3$$

$$z = \frac{54 - 52}{2} = 1$$

This converts the nonstandard x-values of 46 and 54 to the standard z-values of −3 and 1, as illustrated in Figure 3.23.

Thus, to find the probability, we need to add the areas of the given shaded region in Figure 3.23 as follows:

$$p(46 < x < 54) = p(-3 < z < 1) = 0.4987 + 0.3413 = 0.8400$$

Exercises

1. The data given in Table 3.13 represent a sample of the daily rainfall (in inches) for a 1-week period during the summer for a small town in Connecticut.

    a. Calculate (by hand) the mean, median, mode, range, variance, and standard deviation for this set of data.

    b. Check your calculations using MINITAB. Which measure of center do you think best represents this set of data?

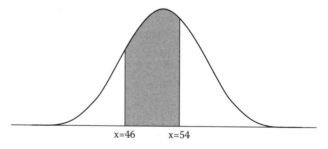

FIGURE 3.22
Shaded area that corresponds to the probability that the random variable X falls between 46 and 54 if X is normally distributed with a mean of 52 and a standard deviation of 2.

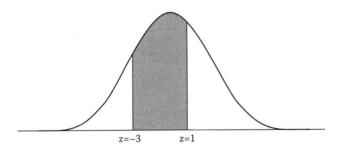

FIGURE 3.23
Standardized normal distribution for the random variable in Figure 3.22.

2. For the data set "Crime Rate by County," using MINITAB, find the mean, standard deviation, variance, and first, second, and third quartiles, and the IQR for the crime rate for the years 1991 and 1996.

3. The weighted mean, denoted as $\bar{x}_w$, is another measure of center in which weights can be associated with each of the individual observations. The formula for the weighted mean is as follows:

$$\bar{x}_w = \frac{\displaystyle\sum_{i=1}^{n} w_i \cdot x_i}{\displaystyle\sum_{i=1}^{n} w_i}$$

where w_i is the weight associated with observation x_i.

Suppose an instructor counts examinations twice as much as quizzes, and counts the final examination three times as much as quizzes. If a student earned an average examination grade of 83, an average quiz grade of 87, and a final examination grade of 91, using the weighted mean, what would the student's average grade be? What would the student's grade be using the sample mean, $\bar{x}$? Describe why there is a difference between these two values.

TABLE 3.13

Daily Rainfall (in inches) during a one-week period for a Small Town in Connecticut

1.25	0.20	0.00	0.00	0.00	0.10	2.45

TABLE 3.14

First-Year Salaries (in thousands of dollars) for a
Sample of Ten Graduates from a Certain University

25	34	26	95	45	52	34	51	29	38

4. Table 3.14 gives a sample of the first-year salaries (in thousands of dollars) for a sample of ten graduates from a certain university.

   a. Using MINITAB, calculate the mean and median of this set of data. Which of these two measures do you think best represents the center of the underlying data set? Why?

   b. Using MINITAB, calculate the standard deviation and the range of this set of data. Which of these two measures best represents the underlying data set? Why?

5. Table 3.15 gives the speeds (in miles per hour) for a random sample of twelve cars driving down a given section of highway that has a speed limit of 65 miles per hour.

   a. Calculate Q_1, Q_2 and Q_3 for these highway speeds by hand.

   b. Using MINITAB, draw a box plot for these highway speeds.

TABLE 3.15

Speed (in miles per hour) for a Random Sample of Twelve
Cars Driving Down a Highway

55	68	72	79	81	63	72	70	61	77	68	66

6. Often we may be interested in comparing the amount of variability between two different data sets. The *coefficient of variation* (COV) can be used to measure the relative variability between two different data sets. It is calculated by dividing the standard deviation by the mean and then multiplying by 100%, as can be seen in the following formula:

$$COV = \frac{s}{\bar{x}} \cdot 100\%$$

The data set that has the larger value of the coefficient of variation represents the data set that has more variability relative to the mean.

Table 3.16 gives the time (in minutes) that Phil and Barbara spend talking on their cell phones each day over the course of a 5-day period.

Using the coefficient of variation, which of these two individuals shows more variability in the time spent (in minutes) talking on his or her cell phone?

TABLE 3.16

Time (in minutes) That Phil and Barbara Spend
Talking on Their Cell Phones over a Five-Day Period

Day	Phil	Barbara
Monday	52	32
Tuesday	127	45
Wednesday	285	29
Thursday	6	38
Friday	29	36

7. The trimmed mean is also a useful measure of center. It is found by trimming off the smallest 5% of the observations and the largest 5% of the observations (rounded to the nearest integer), and then calculating the mean of the remaining values. The trimmed mean can be used if there are outliers at the extremes of the data set. For the data given in Table 3.5, which represents the number of hours that a sample of ten homeowners spend working on their lawn each week, find the trimmed mean.

8. The *kurtosis* of a data set represents a measure of how the distribution of a data set is peaked (in other words, how "sharp" the data set appears when you consider its distribution). The kurtosis can be used to gauge whether the distribution of a data set differs from the normal distribution. For instance, the closer the kurtosis value is to 0, the more normally distributed the data are; the closer the kurtosis value is to 1, the sharper the peak of the distribution; and a negative kurtosis value represents a flat distribution. The formula for kurtosis is as follows:

$$Kurtosis = \left[\frac{n(n+1)}{(n-1)(n-2)(n-3)} \right] \cdot \sum_{i=1}^{n} \left(\frac{x_i - \bar{x}}{s} \right)^4 - \frac{3(n-1)^2}{(n-2)(n-3)}$$

where n is the sample size, x_i are the individual observations, $\bar{x}$ is the sample mean, and s is the sample standard deviation for the given data.

a. For the "Commuting Distance" data set (Table 2.1), draw a histogram and calculate the value of the kurtosis statistic.

b. Does the measure of kurtosis make sense for shape of the distribution of the "Commuting Distance" data set?

9. The *skewness* of a data set can be used as a measure of how symmetrical a distribution is. If the skewness is positive, this indicates that more observations are below the mean than are above the mean, and if the skewness is negative, this indicates that there are more observations above the mean than are below the mean. The formula to calculate the skewness is as follows:

$$Skewness = \frac{n}{(n-1)(n-2)} \cdot \sum_{i=1}^{n} \left(\frac{x_i - \bar{x}}{s} \right)^3$$

a. For the data set "Commuting Distance" (Table 2.1), calculate the value of the skewness.

b. Does the measure of skewness make sense for the "Commuting Distance" data set?

10. For a standard normal distribution, find the following probabilities that correspond to the area under the standard normal curve:

a. $p(z < 1.25)$

b. $p(z < -0.49)$

c. $p(-1.37 < z < 1.16)$

d. $p(z > 1.25)$

e. $p(z < 1.03)$

11. Suppose you are given a random variable that has the standard normal distribution.

a. Find the value of z such that only 5% of all values of this random variable will be greater.

b. Find the value of z such that only 1% of all values of this random variable will be greater.

c. Find the value of z such that 95% of the standard normal distribution is between $-z$ and $+z$.

12. If a random variable X has a normal distribution with a mean of 20 and a standard deviation of 10, find the probability that the random variable X takes on a value greater than -10.

13. Grades that students receive on computerized examinations can often be approximated by a random variable that can be modeled using a normal distribution. Assume that the average grade on a statistics test is 73.25 points with a standard deviation of 12.58 points.

a. What is the probability that a grade selected at random will be greater than 80.50 points?

    b. What is the probability that a grade selected at random will be between 60.25 and 70.75 points?

    c. What is the probability that a grade selected at random will be exactly 90.23?

14. A random variable X has a normal distribution with $\mu = 70$ and $\sigma = 12$. Find each of the following probabilities:

    a. $p(X < 50)$

    b. $p(X > 70)$

    c. $p(40 < X < 60)$

    d. $p(120 < X < 160)$

    e. $p(60 < X < 95)$

    f. $p(X < 72.3)$

15. Justify whether or not each of the following tables represents the probability distribution of a random variable X:

x	$p(x)$
−1	0.26
−2	0.45
−3	0.22
−4	0.07

x	$p(x)$
0.21	0.20
0.22	0.20
0.23	0.20
0.24	0.20
0.25	0.20

x	$p(x)$
0.245	0.16
1.25	0.18
2.89	−0.12
3.47	0.39
0.18	0.39

16. Create a table that shows the probability distribution for the discrete random variable X given in Figure 3.24.

17. A cumulative probability distribution for a discrete random variable displays the probability that a random variable takes on a value less than or equal to some value of the random variable. For instance, consider the following probability distribution:

x	p(x)
1	0.25
2	0.25
3	0.25
4	0.25

A cumulative probability distribution would be as follows:

x	$p(X \leq x)$
1	0.25
2	0.50
3	0.75
4	1.00

So the probability that the random variable takes on a value less than or equal to 3 is $p(X \leq 3) = 0.25 + 0.25 + 0.25 = 0.75$.

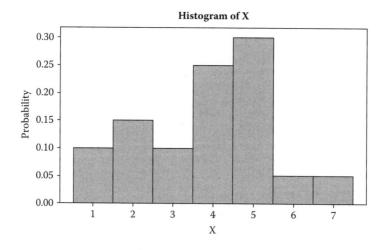

FIGURE 3.24
Probability distribution for the discrete random variable X.

Using the following discrete probability distribution, find the cumulative probability distribution:

x	$p(x)$
1	0.20
2	0.27
3	0.19
4	0.34

18. Find the mean, variance, and standard deviation for the following probability distribution:

x	$p(x)$
1	0.20
2	0.27
3	0.19
4	0.34

19. For the probability distribution in Problem 18, find the sampling distribution of the sample mean for all samples of size $n = 2$.

20. The data set "Automobiles" contains the make and model along with measures of their city and highway miles per gallon, and the amount of greenhouse gases they emit for a sample of 641 vehicles. Often with large data sets such as this one, you may want to calculate some descriptive measures for parts of the data set that are based on some variable. This type of calculation can easily be done in MINITAB by specifying the variables in the **Variables** box that you want to calculate the descriptive statistics for and the variables that are to be used to partition the data in the **By variables** box, as can be seen in Figure 3.25.

   a. Using MINITAB, find the mean and standard deviation of the highway miles per gallon for the six different models of automobiles (small car, family sedan, upscale sedan, luxury sedan, minivan, and SUV).

   b. Using MINITAB, find the mean and standard deviation of the highway miles per gallon for the forty-one different makes of automobiles (Acura, Audi, etc.).

   c. Using MINITAB, find the average annual fuel cost by make and by model of vehicle.

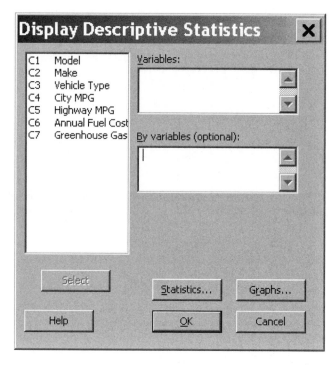

FIGURE 3.25
MINITAB dialog box to calculate the descriptive statistics by a given variable.

21. MINITAB can also be used to create a graphical summary of a given variable. The graphical summary can be found by selecting the **Stat** menu; then under **Basic Statistics** select **Graphical Summary**. There are four graphs that appear in the summary: a histogram with a normal curve superimposed, a box plot, and 95% confidence intervals for the mean and median, respectively.

   a. Using MINITAB, create a graphical summary for the annual fuel cost for the "Automobile" data set.

   b. Create a graphical summary for the annual fuel cost for SUVs.

   c. Create a graphical summary for the annual fuel cost for small cars.

4

Basic Statistical Inference

4.1 Introduction

Statistical inference involves making an inference or prediction about an unknown population parameter of interest from information that is obtained by using a sample statistic. In this chapter we will be describing two basic types of statistical inference: estimation and hypothesis testing.

In making statistical inferences using estimation, we are trying to obtain an estimate of a likely range of values for an unknown population parameter of interest. For example, we could use estimation techniques to infer the average amount spent on advertising for a population of car dealers. In making statistical inferences with hypothesis tests, we can test to see if an unknown population parameter differs enough from some hypothesized value. For instance, we may be interested in determining if the population of students who take a final examination receive an average score greater than 70%. For both estimation and hypothesis testing, we will be using data collected from a sample to make inferences or predictions about an unknown population parameter.

This chapter presents some basic statistical methods that can be used for estimation and hypothesis testing for a single mean, difference between two means, one-sample proportion, and two-sample proportion. This chapter also elaborates on the importance of a statistical power analysis.

4.2 Confidence Intervals

Estimation is concerned with using sample data to estimate a value or likely range of values for an unknown population parameter of interest. One such way to estimate an unknown population parameter is by using a sample statistic. When an estimate of a population parameter consists of only a single sample statistic, it is called a *point estimate*. For example, the sample mean $\bar{x}$ can be used as a point estimate for an unknown population mean μ.

We can also estimate a likely range of values for a population parameter of interest by calculating what is called a *confidence interval*. A confidence interval for a population mean can be used to estimate a range of values where we have some degree of confidence that this interval contains the true but unknown population mean. More specifically, a confidence interval for a population mean is determined by using the sample mean and then calculating upper and lower boundary points such that between these two boundary points we are confident that the true but unknown population mean lies with some probability.

The theory behind confidence intervals for a population mean relies on the fact that for large samples, the sampling distribution of the sample mean is approximately normally distributed and is centered about the true but unknown population mean with a standard deviation that can be estimated from the sample. For small samples, we need to assume that the population being sampled from is approximately normally distributed. Then by standardizing a given value of the sample mean this gives the following test statistic, which represents a random variable that follows the *t-distribution*:

$$T = \frac{\bar{x} - \mu}{\frac{s}{\sqrt{n}}}$$

where $\bar{x}$ is the sample mean, s is the sample standard deviation, n is the sample size, and μ is the unknown population mean.

The *t*-distribution resembles a standard normal distribution in that it is a bell-shaped curve that is centered at 0. However, the shape of the *t*-distribution depends on what are called the *degrees of freedom*. Degrees of freedom (often abbreviated as *df*) is a measure of the amount of information available in the sample that can be used to estimate a population parameter or parameters of interest. Typically, the degrees of freedom are found by taking the number of sample observations minus the number of parameters that are being estimated. For example, if we are using the *t*-distribution to compute a confidence interval for a population mean, then the degrees of freedom would be equal to $n - 1$ for a sample of size n because we are estimating a single parameter, namely, the population mean μ, and this requires that we subtract 1 from the sample size.

If the random variable T follows a *t*-distribution, then we can claim that the random variable T will fall between $-t_{\alpha/2}$ and $t_{\alpha/2}$ with probability $1 - \alpha$, as illustrated in Figure 4.1.

Because T is a random variable and if $-t_{\alpha/2} < T < t_{\alpha/2}$, then

$$-t_{\frac{\alpha}{2}} < \frac{\bar{x} - \mu}{\frac{s}{\sqrt{n}}} < t_{\frac{\alpha}{2}}$$

Multiplying through by $\frac{s}{\sqrt{n}}$ gives

$$-t_{\frac{\alpha}{2}}\left(\frac{s}{\sqrt{n}}\right) < \bar{x} - \mu < t_{\frac{\alpha}{2}}\left(\frac{s}{\sqrt{n}}\right)$$

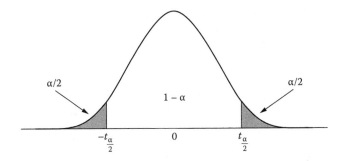

FIGURE 4.1
Distribution of the sample mean where the sample statistic T falls between $-t_{\alpha/2}$ and $t_{\alpha/2}$ with probability $1 - \alpha$.

Subtracting $\bar{x}$ from the inequality gives

$$-\bar{x} - t_{\frac{\alpha}{2}}\left(\frac{s}{\sqrt{n}}\right) < -\mu < -\bar{x} + t_{\frac{\alpha}{2}}\left(\frac{s}{\sqrt{n}}\right)$$

Multiplying by –1 requires changing the sign of the inequality as follows:

$$\bar{x} + t_{\frac{\alpha}{2}}\left(\frac{s}{\sqrt{n}}\right) > \mu > \bar{x} - t_{\frac{\alpha}{2}}\left(\frac{s}{\sqrt{n}}\right)$$

Correcting the direction of the inequality gives

$$\bar{x} - t_{\frac{\alpha}{2}}\left(\frac{s}{\sqrt{n}}\right) < \mu < \bar{x} + t_{\frac{\alpha}{2}}\left(\frac{s}{\sqrt{n}}\right)$$

Thus, the true but unknown population mean μ will fall between $\bar{x} - t_{\alpha/2}\left(\frac{s}{\sqrt{n}}\right)$ and $\bar{x} + t_{\alpha/2}\left(\frac{s}{\sqrt{n}}\right)$ with probability $1 - \alpha$.

We can also denote this confidence interval in the following manner:

$$\bar{x} \pm t_{\frac{\alpha}{2}} \cdot \frac{s}{\sqrt{n}} = \left(\bar{x} - t_{\frac{\alpha}{2}} \cdot \frac{s}{\sqrt{n}}, \bar{x} + t_{\frac{\alpha}{2}} \cdot \frac{s}{\sqrt{n}}\right)$$

This interval is called a $100(1 - \alpha)\%$ confidence interval. If we take repeated samples of size n from the population of interest and calculate confidence intervals for all such samples of the same size in the same manner, then the true but unknown population mean μ will be contained in approximately $100(1-\alpha)\%$ of all such intervals.

For instance, a $100(1-0.05)\% = 95\%$ confidence interval for a population mean μ obtained from a random sample of size n can be determined such that for all samples of size n, approximately 95% of all such intervals calculated in the same manner are expected to contain the population parameter μ. In other words, for repeated samples of size n, 95% of all such intervals of the form

$$\left(\bar{x}-t_{\frac{\alpha}{2}}\cdot\frac{s}{\sqrt{n}}, \bar{x}+t_{\frac{\alpha}{2}}\cdot\frac{s}{\sqrt{n}}\right),$$ where $\alpha = 0.05$ and $t_{0.025}$ is the value that describes the

upper $1-0.025 = 0.975$ percentage of the t-distribution with $n-1$ degrees of freedom, will contain the true but unknown population mean. Thus, we can say that we are 95% confident that any given interval will contain the true but unknown population mean. Such a confidence interval is illustrated in Figure 4.2.

In order to find the value of $t_{\alpha/2}$, we need to reference a t-table (see Table 2 of Appendix A). Notice that the t-table is different from the standard normal table in that the degrees of freedom are given in the first column, values of α are given in the top row, and the reference area is to the right of t. For instance, if we wanted to find $t_{0.025}$ for 17 degrees of freedom, we would locate the value of 0.025 in the top row of the t-table, and find 17 degrees of freedom in the left-most column of the t-table. This would give a value of $t_{0.025} = 2.110$.

Example 4.1

A random sample of eighteen car dealers found that this particular sample of dealers spent an average amount of $5,500 per month on the cost of adver-tising, with a sample standard deviation of $500. Suppose that we want to find a 95% confidence interval for the true but unknown average amount of money that the *population* of car dealers spends monthly on advertising.

Assuming that the population we are sampling from is normally distrib-uted, we can use the sample mean ($\bar{x} = 5500$) and the sample standard devia-tion ($s = 500$) to calculate a 95% confidence interval. We need to use the t-tables

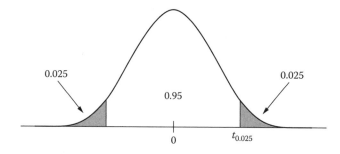

FIGURE 4.2
Distribution of the sample mean for calculating a 95% confidence interval for the true but unknown population mean.

(Table 2 of Appendix A) to find the value for $t_{\alpha/2} = t_{0.05/2} = t_{0.025}$. Since $n = 18$, we have $n - 1 = 17$ degrees of freedom; then the values for $t_{\alpha/2} = t_{0.05/2} = t_{0.025}$ that are found using the t-tables and these values describe the upper and lower boundary points of our confidence interval are as follows:

$$t_{\frac{\alpha}{2}} = t_{0.025} = 2.110$$

This is illustrated in Figure 4.3.

Therefore, a 95% confidence interval for the true but unknown population mean would be

$$\left(5500 \pm 2.110 \cdot \frac{500}{\sqrt{18}}\right) = (5500 \pm 248.67) = (5251.33,\ 5748.67)$$

Thus, if repeated samples of size 18 are drawn from the amount that the true population of car dealers spend per month on advertising, and the confidence intervals are calculated in a manner similar to that above, then 95% of all such intervals will contain the true but unknown population mean. In other words, we say that we are 95% confident that the true but unknown population mean amount that car dealers spend on advertising per month falls between $5251.33 and $5748.67.

Example 4.2

The final examination grades for a sample of twenty statistics students selected at random are presented in Table 4.1.

Assuming that the population we are sampling from is normally distributed, calculating the necessary descriptive statistics for this sample gives

$$\bar{x} = 76.65$$

$$s = 10.04$$

$$n = 20$$

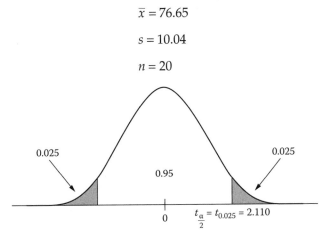

0.025

0.025

0.95

0

$t_{\frac{\alpha}{2}} = t_{0.025} = 2.110$

FIGURE 4.3
Distribution of the sample mean with the value of t that defines the upper boundary point for a 95% confidence interval with 17 degrees of freedom.

TABLE 4.1

Sample of Final Examination Grades for
a Random Sample of Twenty Students

71	93	91	86	75
73	86	82	76	57
84	89	67	62	72
77	68	65	75	84

Suppose we want to find a 99% confidence interval for the mean final examination grade for the population of all students taking the statistics examination at our site. Since $1 - \alpha = 0.99$, then $\alpha = 0.01$. So for a sample size of 20 with 19 degrees of freedom, the upper boundary point $t_{\alpha/2} = t_{0.01/2} = t_{0.005}$ would be described by the value of $t = 2.861$, as illustrated in Figure 4.4.

Based on the t-distribution with $20 - 1 = 19$ degrees of freedom, we have that $t_{\alpha/2} = t_{0.005} = 2.861$; therefore, a 99% confidence interval would be

$$76.65 \pm 2.861 \cdot \left(\frac{10.04}{\sqrt{20}} \right) = 76.65 \pm 6.42 = (70.23, 83.07)$$

This confidence interval suggests that if we repeat the process of taking samples size $n = 20$ from the population of all statistics students taking the final examination at our site and calculate the confidence intervals in the same manner, then 99% of all such intervals would contain the true but unknown population mean final examination grade. In other words, we can say that we are 99% confident that the population of statistics students will score on average between 70.23 and 83.07 points on the final examination.

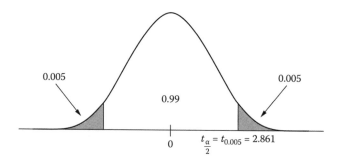

FIGURE 4.4
T-distribution with the value of t that defines the upper boundary point of a 99% confidence interval with 19 degrees of freedom.

4.3 Using MINITAB to Calculate Confidence Intervals for a Population Mean

MINITAB can be used to calculate confidence intervals for a population mean either by entering the raw data in a worksheet or by entering a collection of sample statistics.

To use MINITAB to calculate a 99% confidence interval for the population mean score on the statistics final examination for the data given in Table 4.1 using either the raw data or a collection of descriptive statistics, select **Basic Statistics** and **1-Sample *t*,** as illustrated in Figure 4.5.

To calculate a confidence interval, either the raw data or the summarized data can be entered into MINITAB, as can be seen from the one-sample *t*-test dialog box that is presented in Figure 4.6.

By clicking on the **Options** box we can input the appropriate confidence level, as indicated in Figure 4.7. We also need to make sure that the alternative is set to "not equal" whenever we are calculating a confidence interval

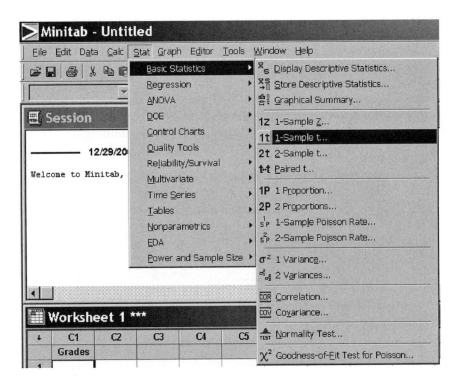

FIGURE 4.5
MINITAB commands to calculate a confidence interval for a population mean.

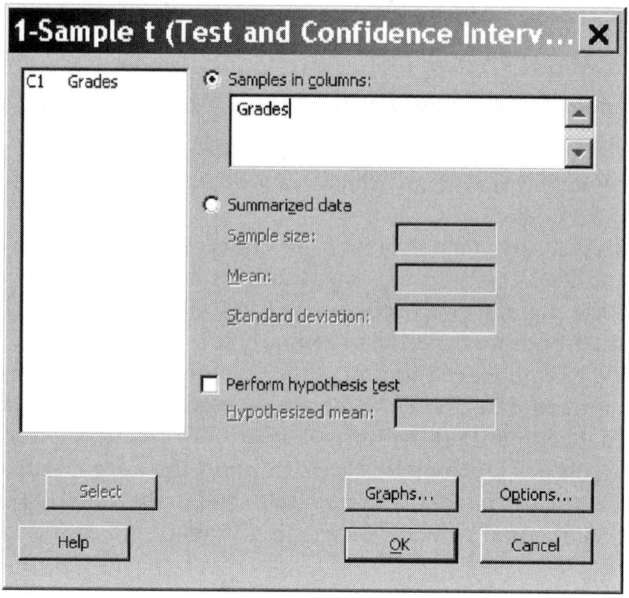

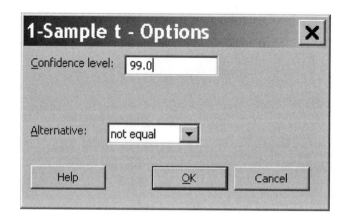

FIGURE 4.6
MINITAB dialog box for a confidence interval for a population mean.

(the reason for doing so will become more apparent when we cover hypothesis tests in the next section).

Then selecting **OK** gives the MINITAB printout illustrated in Figure 4.8. The highlighted portion in Figure 4.8 gives the 99% confidence interval for the true population mean score on the statistics final examination.

FIGURE 4.7
MINITAB options dialog box for calculating for a confidence interval for a population mean.

One-Sample T: Grades

```
Variable    N    Mean   StDev   SE Mean       99% CI
Grades     20   76.65   10.04      2.24   (70.23, 83.07)
```

FIGURE 4.8
MINITAB printout for a 99% confidence interval for the true population mean score on the statistics final examination.

We would get the same results if we entered the summary data as shown in Figure 4.9.

4.4 Hypothesis Testing: A One-Sample *t*-Test for a Population Mean

A confidence interval for a population mean describes an interval based on sample data where we are $(1 - \alpha) \cdot 100\%$ confident that the true but unknown population parameter is contained in this interval. A hypothesis test, which is also based on sample data, can tell us whether the true but unknown

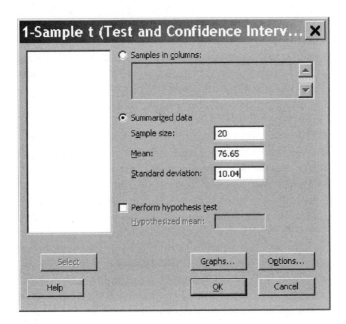

FIGURE 4.9
MINITAB dialog box for calculating a confidence interval using summarized data.

population parameter is different enough from some hypothesized value. For instance, we can conduct a hypothesis test about a population mean that would allow us to determine whether the true but unknown population mean is significantly different from some specific or hypothesized value of interest. Typically, the hypothesized value of the population mean we are testing is denoted by μ_0.

In order to perform a hypothesis test about a population mean we first have to set up a null hypothesis and an alternative hypothesis. The *alternative hypothesis* is established based on what we are investigating, and it is found by considering what it is that we are looking to accept that is different from a given hypothesized population value. The *null hypothesis* is the hypothesized population value that is set up for the purpose of being rejected. For instance, in using the data from Table 4.1, which provides the final examination grades for a sample of twenty statistics students, suppose we want to investigate whether the true but unknown population mean grade on the statistics final examination is different from 75. Since we want to know whether the true population mean is different from the hypothesized value of 75, we would state our alternative hypothesis as $H_A : \mu_0 \neq 75$ (this is the hypothesis that we are interested in testing), and the null hypothesis would then be the opposite of the alternative hypothesis, $H_0 : \mu_0 = 75$ (this is the hypothesis that we are interested in rejecting).

What we are trying to infer with hypothesis testing is whether it is likely that a given sample statistic comes from a population whose mean is significantly different from 75. Because we are using sample data to make an inference about an unknown population mean, we would expect there to be some amount of variability in the sample data that is due to sampling error. For instance, from our last example, recall that we calculated the sample mean to be 76.65. In performing a hypothesis test, what we are really interested in showing is whether a sample mean of 76.65 is reasonable and would likely be obtained from a population whose true mean is 75. We want to make sure that the difference between the sample mean and the hypothesized mean is not simply due to sampling variability.

Recall from Chapter 3 that for large samples ($n \geq 30$) the distribution of all possible sample means is approximately normally distributed and is centered about the true population mean. If $n < 30$ and the population we are sampling from is approximately normally distributed, then we can assume that the distribution of the sample mean will be also be normally distributed. One way to decide whether the sample we selected comes from a population with a mean of μ_0 is that we would expect a given sample mean to fall within two standard deviations of the true population mean with a high degree of certainty. Therefore, if the standardized value of the sample mean falls *more* than two standard deviations away from the true population mean, then we may believe that the population we are sampling from does not likely have a mean equal to μ_0. Thus, if the mean of a random sample is not likely to be drawn from a population that has a hypothesized mean of μ_0, we may infer that the mean of the true population we are sampling from is different from the hypothesized mean of μ_0.

The graph in Figure 4.10 illustrates the sampling distribution of the sample mean centered about the true population mean μ_0 that is specified in the null hypothesis. The small shaded area in Figure 4.10 represents the area of the sampling distribution of the sample mean where there is only a small chance that the sample mean would be drawn from a population with a hypothesized mean of μ_0. If it were the case that a given sample mean had only a very small chance of coming from a population with a mean of μ_0, then we would tend to believe that the population we are sampling from may not have a true mean of μ_0. This shaded region is called the *rejection region*, and it represents the probability that a given sample mean will be drawn from a population with a mean of μ_0. In order to accept the alternative hypothesis that the mean of the population we are sampling from is significantly different from the hypothesized mean of μ_0, we would expect a sample value to fall in either of these shaded regions.

Hypothesis testing is basically a decision process that is used to decide between the null and alternative hypotheses. If it were the case that a given sample mean was not likely to come from a population whose mean was μ_0, then we would accept the alternative hypothesis and reject the null hypothesis. For this particular example, if it were the case that we could accept the alternative hypothesis of $H_A : \mu_0 \neq 75$ and reject the null hypothesis of $H_0 : \mu_0 = 75$, then we would infer that the population we are sampling from has a mean that is significantly different from the hypothesized value of $\mu_0 = 75$.

In order to define the rejection region, we need to determine the area under the sampling distribution of the sample mean that corresponds to the collection of sample means that only have a small chance to coming from a population whose mean is μ_0. The *level of significance*, denoted as α, represents the shaded area of the distribution of the sample mean that describes the

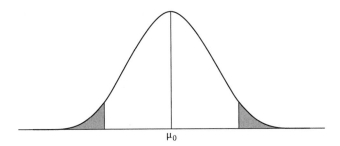

FIGURE 4.10
Distribution of the sample mean centered about the true population mean, μ_0, where the shaded area represents the area under the standard normal curve, which corresponds to the values of the sample mean that only have a small chance of coming from a population with a mean of μ_0.

probability of observing a sample mean that comes from a population with a mean of μ_0. The level of significance is established prior to beginning the hypothesis test. The level of significance corresponds to how much we are willing to risk that the true population we are sampling from actually does have a mean of μ_0, even though the sample mean obtained may be different enough from μ_0 to suggest otherwise. However, it could be the case that the population we are sampling from does have a mean of μ_0, even though it is very unlikely. We say that a *Type I error* is committed when we accept the alternative hypothesis and reject the null hypothesis when in fact the null hypothesis is actually true. In other words, a Type I error is committed when we claim that the mean of the population we are sampling from has a mean that is different enough from μ_0 (this is when we accept the alternative hypothesis), even though it is the case that the population we are sampling from really does have a mean of μ_0, and this can happen $\alpha\%$ of the time.

It is for this reason that typical values of α tend to be fairly small, such as $\alpha = 0.05$ or $\alpha = 0.01$. This is because if our sample mean does fall far enough away from the hypothesized mean into the rejection region, then for a significance level of either 5% or 1%, we have either a 5% or 1% chance that a given sample mean can actually be drawn from a population whose true mean is μ_0. The rejection region for testing a single population mean against some hypothesized value is determined by the level of significance, the sampling distribution of the test statistic, and the direction of the alternative hypothesis.

Another type of error that could be committed is called a *Type II error*. This occurs by failing to reject the null hypothesis when in fact the alternative hypothesis is actually true. We will talk more about Type II errors in Section 4.6, when we describe a power analysis.

There are three types of alternative hypotheses that can be used for testing whether the true population we are sampling from has a mean different from some hypothesized value μ_0. Although the null hypothesis can be described as the opposite of the alternative hypothesis, for mathematical simplicity, we will always state the null hypothesis as a simple equality. The three alternative hypotheses and their appropriate rejection regions for any given value of α are represented in Figures 4.11 to 4.13.

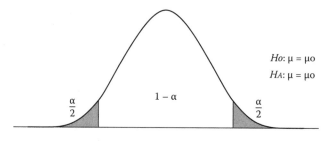

FIGURE 4.11
Two-tailed test.

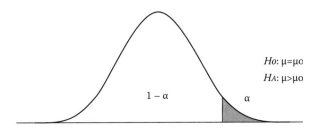

FIGURE 4.12
Right-tailed test.

The shaded rejection regions in Figures 4.11 to 4.13 represent that for a given hypothesized mean of μ_0 the probability is α that the population mean is consistent with what was stated in the alternative hypothesis. The area of the rejection region is part of the distribution of the test statistic under consideration, and for these regions, the values of the sample mean support what is stated in the alternative hypothesis.

After initially establishing the appropriate rejection region based on the level of significance and the direction of the alternative hypothesis, we can then determine whether the test statistic T, which is a random variable that represents the standardized value of our sample mean, $T = \dfrac{(\bar{x} - \mu_0)}{\frac{s}{\sqrt{n}}}$ (which has a t-distribution with $n - 1$ degrees of freedom), is consistent with what was specified in the null hypothesis.

Example 4.3

For the data given in Table 4.1 with a significance level of $\alpha = 0.05$ and a two-tailed hypothesis test ($H_0 : \mu_0 = 75,\ H_A : \mu_0 \neq 75$) for a sample size of $n = 20$ (with 19 degrees of freedom), we would define the rejection region as presented in Figure 4.14.

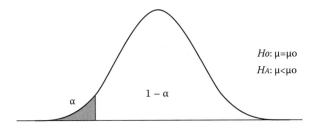

FIGURE 4.13
Left-tailed test.

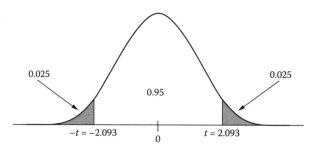

FIGURE 4.14
A two-tailed test of a population mean using a sample of size $n = 20$ with 19 degrees of freedom.

Notice in Figure 4.14 that the rejection region is defined by splitting the value of α between both tails. This is because the alternative hypothesis is specified as an inequality, which means that a given sample statistic could fall in either the right tail or the left tail, to be inconsistent with what is specified in the null hypothesis.

Since $n < 30$ and assuming that the population we are sampling from is normally distributed, by using the sample mean of $\bar{x} = 76.65$ and a sample standard deviation of $s = 10.04$ for a sample size of $n = 20$, the predetermined level of significance $\alpha = 0.05$, the test statistic for this sample mean would be calculated as follows:

$$T = \frac{\bar{x} - \mu_0}{\frac{s}{\sqrt{n}}} = \frac{76.65 - 75}{\frac{10.04}{\sqrt{20}}} \approx 0.73$$

If this test statistic falls into the rejection region in either the right or left tail, which corresponds to a value greater than $t = 2.093$ or less than $-t = -2.093$, then we could reject the null hypothesis (that the population

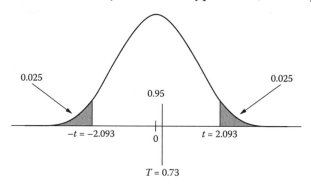

FIGURE 4.15
Value of the test statistic T in comparison to the values of t defining the rejection region for a level of significance of .05 with 19 degrees of freedom for a two-tailed test.

mean is equal to 75) and accept the alternative (that the population mean is not equal to 75). However, notice in Figure 4.15, the value of the test statistic T does not fall into the rejection region because it is not greater than $t = 2.093$ or less than $t = -2.093$. Thus, we cannot claim that the true population mean is significantly different from 75. To interpret this finding within the context of our example, we would claim that we do not have reason to believe that the mean of the population we are sampling is significantly different from 75.

4.5 Using MINITAB for a One-Sample *t*-Test

Using MINITAB to perform a one-sample *t*-test for the data given in Table 4.1 requires some of the exact same steps that are used when calculating confidence intervals for a population mean. One difference is that the mean being tested under the null hypothesis has to be provided, as illustrated in Figure 4.16.

In addition to providing the hypothesized mean, the confidence level and the direction of the alternative hypothesis have to be specified. This is done by selecting the **Options** tab, as can be seen in Figure 4.17.

Figure 4.18 provides the MINITAB printout for testing whether the true population mean of grades received on the final examination is significantly different from 75.

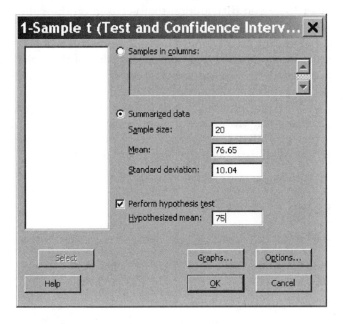

FIGURE 4.16
MINITAB dialog box for a one-sample *t*-test.

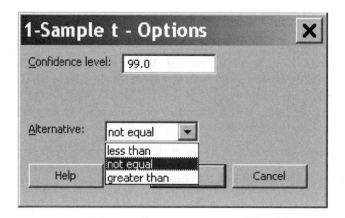

FIGURE 4.17
Options box to specify confidence level and alternative hypothesis for a one-sample *t*-test.

Notice that the highlighted portion of the MINITAB printout in Figure 4.18 provides the descriptive statistics along with the specification of null and alternative hypotheses and the value of the test statistic *T*.

However, the information in the MINITAB printout does not explicitly state whether to accept the alternative hypothesis and reject the null hypothesis. Instead, the printout provides what is called a *p-value*. The *p*-value describes the area under the given distribution that is based on the value of the test statistic. The *p*-value is the observed level of significance, and it describes the likelihood of observing an extreme value as it represents what is the smallest value of α that we can expect from the given value of the test statistic that will lead to rejecting the null hypothesis. We can use the *p*-value to assess whether we can accept the alternative hypothesis.

From the MINITAB printout in Figure 4.18, notice that the *p*-value is 0.4710. This *p*-value tells us that the minimum level of significance is 0.4710. In other words, a *p*-value of .4710 describes the smallest value of α that we can expect from our given test statistic, as illustrated in Figure 4.19.

One-Sample T

Test of mu = 75 vs not = 75

N	Mean	StDev	SE Mean	99% CI	T	P
20	76.65	10.04	2.25	(70.23, 83.07)	0.73	0.471

FIGURE 4.18
MINITAB printout for the one-sample *t*-test of whether the true population mean is different from 75 using the random sample of data provided in Table 4.1.

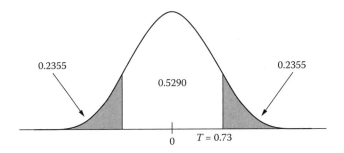

FIGURE 4.19
How the rejection region would be defined by the test statistic $T = 0.73$, which corresponds to a p-value of .4710.

The p-value of .4710 represents the probability of observing a sample mean that comes from a population with a mean of $\mu_0 = 75$, and there is a 47.10% chance of this actually happening. Clearly, this does not make for a very strong argument, and so for most practical considerations, that is why we are almost always concerned with significance levels that are less than 10%.

In general, when the p-value is less than our predetermined level of significance, α, then we can reject the null hypothesis and accept the alternative hypothesis because the value of our test statistic falls in the rejection region. If the p-value is greater than our predetermined level of significance, α, then this implies that the value of the test statistic does not fall in the rejection region. All that we can say is that we do not have enough evidence to reject the null hypothesis, and thus cannot accept the alternative hypothesis. For our example, a p-value of .471 is greater than the predetermined level of significance, which suggests that the test statistic T does not fall in the rejection region; and, therefore, the true mean of the population we are sampling from is not significantly different from the hypothesized mean of $\mu_0 = 75$.

We will now review confidence intervals and hypothesis tests about an unknown population mean by working through two additional examples.

Example 4.4

Suppose we are interested in finding a confidence interval for the population mean time, in hours, it takes a manufacturing plant to produce a certain type of part from start to finish ($\alpha = 0.05$).

The sample in Table 4.2 gives the time, in hours, that it takes for the plant to produce this specific type of part for a sample of size $n = 25$.

Recall that to find a 95% confidence interval for the population mean time it takes to manufacture this part, we would need to calculate the sample mean and the sample standard deviation.

TABLE 4.2

Sample of the Time in Hours That It Takes to Manufacture a
Certain Type of Part from Start to Finish at a Given Plant ($n = 25$)

55	53	46	37	52
42	45	55	50	70
45	65	50	51	30
37	45	42	55	46
60	61	30	32	46

The sample mean is calculated as follows:

$$\bar{x} = \frac{\sum_{i=1}^{n} x_i}{n} = \frac{\sum_{i=1}^{25} x_i}{25} = 48.00$$

The sample standard deviation is calculated as follows:

$$s = \sqrt{\frac{\sum_{i=1}^{n}(x_i - \bar{x})^2}{n-1}} \approx 10.30$$

Assuming that the population we are sampling from is approximately normally distributed, then a $100 \cdot (1 - \alpha)\% = 100 \cdot (1 - 0.05)\% = 95\%$ confidence interval for the true but unknown population mean time it takes to manufacture the part would be calculated as follows:

$$\left(\bar{x} - t_{\frac{\alpha}{2}} \frac{s}{\sqrt{n}}, \bar{x} + t_{\frac{\alpha}{2}} \frac{s}{\sqrt{n}} \right)$$

$$= \left(48 - 2.064 \cdot \frac{10.30}{\sqrt{25}}, 48 + 2.064 \cdot \frac{10.30}{\sqrt{25}} \right) = (43.75, 52.25)$$

where $t_{\alpha/2} = t_{0.025} = 2.064$ is the value from the t-distribution with $n - 1 = 25 - 1 = 24$ degrees of freedom.

This confidence interval suggests that if we repeatedly collect random samples of size 25 from the unknown population of the time it takes to manufacture the given part and calculate confidence intervals in the same manner, then 95% of all such intervals will contain the true but unknown population mean time it takes to manufacture this part. Thus, we would claim that we are 95% confident that the true but unknown mean population time it takes to manufacture the part from start to finish falls between 43.75 and 52.25 hours. A confidence interval provides a way to arrive at an estimate of a possible range of values for a true but unknown population mean by using data that was collected from a representative sample.

We can also conduct a hypothesis test if we are interested in testing some hypothesis about an unknown population mean by using sample data.

Example 4.5

Using the data from Table 4.2 for the time it takes to manufacture a specific type of part at a certain factory, suppose that we are interested in whether the average time it takes to manufacture this part is significantly less than 55 hours ($\alpha = 0.05$). In order to do this test, we would first set up the appropriate null and alternative hypotheses as follows:

$$H_0 : \mu_0 = 55$$
$$H_A : \mu_0 < 55$$

Based on the direction of the alternative hypothesis, a level of significance of $\alpha = 0.05$, and 24 degrees of freedom, the rejection region would be defined as presented in Figure 4.20.

In Figure 4.20, the rejection region is defined in the left tail because we are testing whether the mean of the true population we are sampling from is *less than* the hypothesized mean of 55 hours.

Assuming the population we are sampling from is approximately normally distributed, the test statistic would be calculated as follows:

$$T = \frac{\bar{x} - \mu_0}{\frac{s}{\sqrt{n}}} = \frac{48.00 - 55.00}{\frac{10.30}{\sqrt{25}}} = -3.40$$

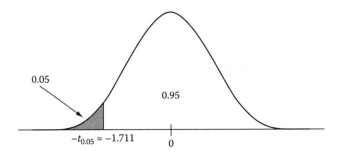

FIGURE 4.20
Rejection region for a left-tailed one-sample *t*-test with 24 degrees of freedom and a significance level of $\alpha = 0.05$.

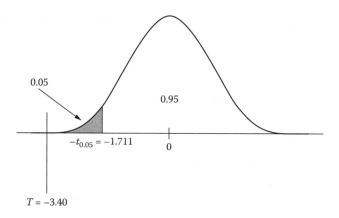

FIGURE 4.21
Rejection region and corresponding value of test statistic to determine if the population mean time to manufacture the part is significantly less than 55 hours.

Since the value of the test statistic falls into the rejection region ($T = -3.40 < t = -1.711$), as illustrated in Figure 4.21, we can accept the alternative hypothesis and reject the null hypothesis. This suggests that the true population mean time it takes to manufacture this specific part is significantly less than 55 hours.

Figure 4.22 provides the MINITAB printout for finding a 95% confidence interval for the true population mean time it takes to manufacture the part, and Figure 4.23 provides the MINITAB printout of the hypothesis test to determine whether or not the true population average time it takes to manufacture the part is less than 55 hours.

From the MINITAB printout in Figure 4.23, the value of the test statistic $T = -3.40$ corresponds to a p-value of .001. The p-value describes the area to the left of T that describes the minimum level of significance that we can expect from the given test statistic with the given alternative hypothesis. In other words, there is only a 0.1% chance of observing the given sample mean from a population whose mean is $\mu_0 = 55$. Figure 4.24 illustrates how the p-value describes the area that is to the left of T, which is clearly less than our predetermined level of significance, $\alpha = 0.05$. Because it is very unlikely that this sample mean comes from a population whose true mean is $\mu_0 = 55$, this leads to accepting the alternative hypothesis and rejecting the null hypothesis.

One-Sample T: Time (in hours)

```
Variable            N    Mean   StDev   SE Mean       95% CI
Time (in hours)    25   48.00   10.30      2.06   (43.75, 52.25)
```

FIGURE 4.22
MINITAB printout for the 95% confidence interval for the mean time it takes to manufacture the part.

One-Sample T: Time (in hours)

Test of mu = 55 vs < 55

					95% Upper		
Variable	N	Mean	StDev	SE Mean	Bound	T	P
Time (in hours)	25	48.00	10.30	2.06	51.53	-3.40	0.001

FIGURE 4.23
MINITAB printout for the hypothesis test ($\alpha = 0.05$) of whether, on average, it takes less than 55 hours to manufacture the part.

4.6 Power Analysis for a One-Sample *t*-Test

There are two types of conclusions that can be made when doing a one-sample hypothesis test about a population mean:

1. If the test statistic falls in the rejection region (this occurs when the *p*-value is less than the predetermined level of significance), we accept the alternative hypothesis and reject the null hypothesis and claim that there is enough evidence to suggest that we are sampling from a population whose mean is significantly different from what is specified in the null hypothesis.

2. If the test statistic does not fall in the rejection region (this occurs when the *p*-value is greater than the predetermined level of significance) and we do not reject the null hypothesis, we cannot claim that we are sampling from a population that has a mean that is significantly different from what is specified in the null hypothesis.

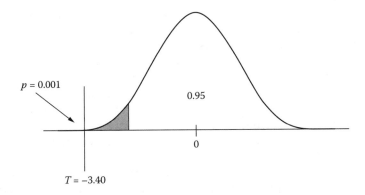

FIGURE 4.24
Illustration of the *p*-value being the area to the left of the test statistic *T*.

For the second conclusion to be true, we need to determine whether the population we are sampling from, in fact, does not have a true population mean that is different from what is specified in the null hypothesis. We need to decide if the true population mean is not different enough from a specific hypothesized value, and that it is not simply the case that there is not a large enough sample to be able to detect whether or not the true underlying population mean is different enough from the hypothesized value.

This is where the notion of a *power analysis* comes into play. Essentially, a power analysis provides you with the necessary sample size to detect a minimum difference for a given population mean. The *power of a test* is defined as the probability of finding a difference of a specific size provided that such a difference actually exists. The power of a test tends to range from 70% to 90%, with 80% being the typical power value. The power of a test is also described as the probability of *not* committing a Type II error (recall that a Type II error occurs by failing to reject the null hypothesis when in fact the alternative hypothesis is actually true).

A power analysis can be an important step in basic statistical inference because when you do not have enough evidence to reject the null hypothesis, and thus cannot claim that the population mean is different enough from some hypothesized value, you may want to maximize the probability of making the correct inference by deciding whether the collected sample is large enough to detect such a difference provided that such a difference were to exist.

4.7   Using MINITAB for a Power Analysis for a One-Sample *t*-Test

Consider the data given in Table 4.1, where we were interested in determining whether the true population mean grade on a statistics final examination is significantly different from 75. Recall that since we were interested in testing whether the population mean is different from 75, we stated our alternative hypothesis to be $H_A : \mu_0 \neq 75$, and the null hypothesis would then be the opposite of the alternative hypothesis, $H_0 : \mu_0 = 75$.

We found that we did not have enough evidence to suggest that the population mean grade on the examination was significantly different from 75. Even though we did not find a significant difference, we need to be sure that the sample that was collected was large enough to detect a difference if one were to exist.

Suppose we wanted to detect a minimum difference of 10 points on the examination score with 80% power. In other words, we want a mean difference of at least 10 points between the hypothesized mean and the standardized sample mean to be significant 80% of the time. We can use MINITAB to perform a power analysis to tell us how large of a sample would be needed to detect such a difference for a one-sample *t*-test. This can be done by clicking on **Stat**, then **Power** and **Sample Size**, and then **1-Sample *t***, as illustrated in Figure 4.25.

This brings up the dialog box that is presented in Figure 4.26, where we need to provide the difference we are interested in finding, the desired power, and an estimate of the standard deviation.

Notice in Figure 4.26 that we have to specify two of the three values of sample size, differences, and power values, along with an estimate of the standard deviation. By specifying a difference of 10 points, a power value of 80%, and the sample standard deviation of 10.04, this gives the MINITAB printout that appears in Figure 4.27.

As the highlighted portion of the MINITAB printout in Figure 4.27 suggests, a sample size of at least 10 would be needed to find a difference of at least 10 points between the true and hypothesized population means 80% of the time.

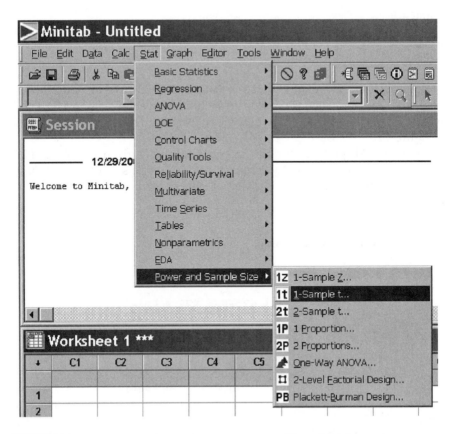

FIGURE 4.25
MINITAB commands for running a power analysis for a one-sample *t*-test.

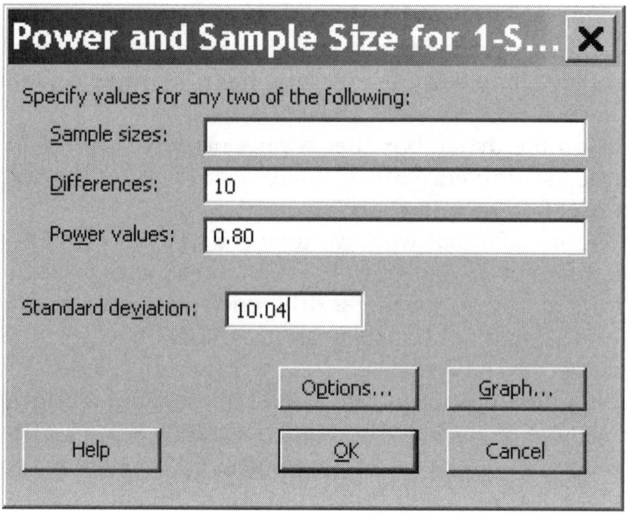

FIGURE 4.26
MINITAB dialog box for testing a difference of 10 points with a power level of 80% for a one-sample *t*-test.

Suppose that we wanted to detect a smaller difference of at least 3 points (also at 80% power). Then we could run a similar power analysis in MINITAB by specifying the difference, power level, and standard deviation as described in Figure 4.28.

The MINITAB printout for this power analysis is given in Figure 4.29. Notice from Figure 4.29 that we would need a sample size of at least 90 to detect if the population mean is different from the hypothesized mean by at least 3 points.

Power and Sample Size

```
1-Sample t Test

Testing mean = null (versus not = null)
Calculating power for mean = null + difference
Alpha = 0.05 Assumed standard deviation = 10.04

                   Sample  Target
   Difference       Size    Power   Actual Power
           10         10      0.8       0.800011
```

FIGURE 4.27
MINITAB printout for a power analysis for a one-sample *t*-test for a difference of size 10 and a power level of 80%.

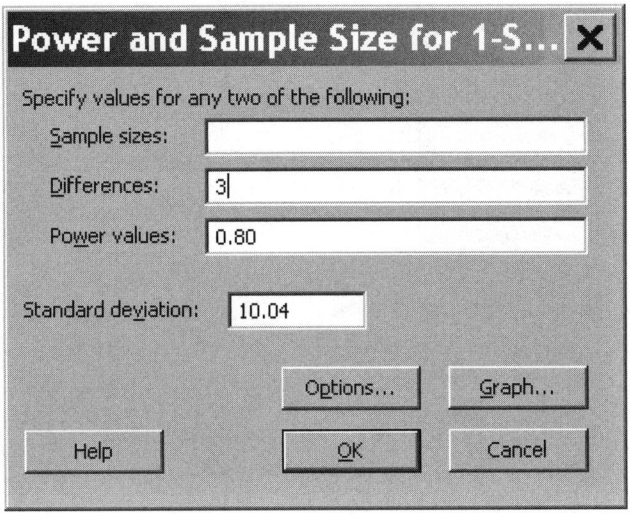

FIGURE 4.28
MINITAB dialog box for a power analysis for a one-sample *t*-test for a difference of 3 and a power level of 80%.

A larger difference is easier to detect, and therefore it does not require as large of a sample for a given power value. However, a small difference is much more difficult to find, which is why a much larger sample size would be needed to detect a small difference provided that such a difference were to exist.

We can also determine what the power would be for a given sample size and difference. For the sample given in Table 4.1, we have a sample of size 20.

Power and Sample Size

```
1-Sample t Test

Testing mean = null (versus not = null)
Calculating power for mean = null + difference
Alpha = 0.05 Assumed standard deviation = 10.04

               Sample  Target
Difference      Size   Power   Actual Power
     3           90     0.8      0.800676
```

FIGURE 4.29
MINITAB printout for a power analysis for a one-sample *t*-test of a difference of size 3 and a power level of 80%.

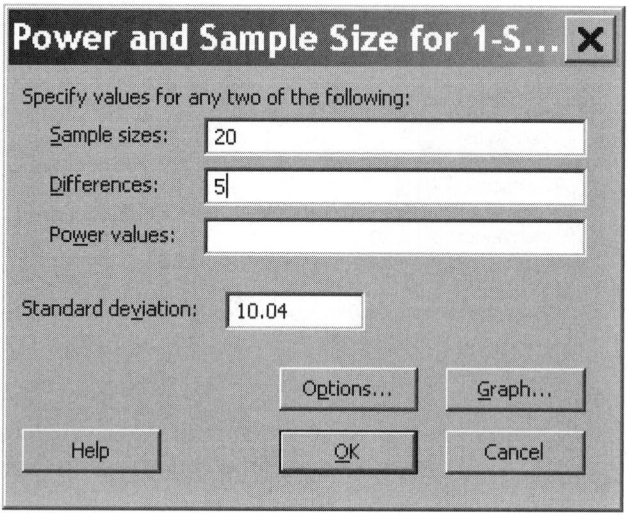

FIGURE 4.30
MINITAB dialog box for a power analysis for a one-sample *t*-test of a difference of size 5 with a sample of size *n* = 20.

Suppose we want to see what the power of our hypothesis test would be if we have a difference of at least 5. In other words, how likely is it that we will find a difference of at least 5 with a sample of size *n* = 20? Then we need to specify the sample size and the difference as presented in Figure 4.30.

The MINITAB printout for this power analysis is presented in Figure 4.31. As the highlighted printout in Figure 4.31 suggests, we only have (approximately) a 56% chance of detecting a difference of at least 5 with a sample of size *n* = 20 (provided that such a difference actually exists). Since this does

Power and Sample Size

```
1-Sample t Test

Testing mean = null (versus not = null)
Calculating power for mean = null + difference
Alpha = 0.05 Assumed standard deviation = 10.04

                Sample
Difference       Size    Power
        5          20    0.561180
```

FIGURE 4.31
MINITAB results for a power analysis for a one-sample *t*-test of a difference of size 5 with a sample of size *n* = 20.

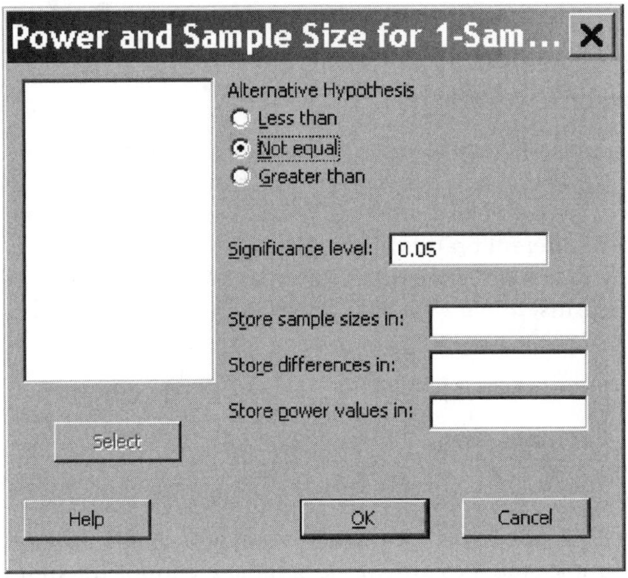

FIGURE 4.32
Options tab to select an alternative hypothesis and significance level.

not make for a very strong argument, we may want to consider increasing the sample size in order to detect such a difference with a higher probability. Similarly, we could also specify a given sample size and power value to have MINITAB determine what the difference would be.

Clicking on the **Options** tab in Figure 4.30 gives the options dialog box for a power analysis that is presented in Figure 4.32. This dialog box allows you to select the appropriate direction of the alternative hypothesis and the level of significance.

To summarize, in order to perform a power analysis, two of the three variables of sample size, minimum difference, and power level must be provided in order for MINITAB to calculate the other variable that is left blank.

In general, there are no hard-and-fast rules for conducting a power analysis, and the importance of using a power analysis varies for different fields of study, and most such analyses tend to be guided by generally accepted rules of thumb. Also, simply collecting a given sample of the appropriate size needed to detect a specified minimum difference does not guarantee that such a difference will actually be detected. All a power analysis can do is to provide you with some estimate as to the sample size needed to find such a difference if it were to exist. Furthermore, collecting the appropriate sample size for a given difference does not guarantee that any measures used to collect your data are valid.*

---

* For a nontechnical description of power analysis and validity, see Light et al. (1990).

4.8　Confidence Interval for the Difference between Two Means

We can also calculate confidence intervals and perform hypothesis tests for comparing the difference between means from two different populations. For instance, we could test whether there is a difference in the mean lifetimes for two different brands of cell phone batteries.

To find a $100 \cdot (1-\alpha)\%$ confidence interval for the difference between two population means, $\mu_1 - \mu_2$, we would use the following formula:

$$(\bar{x}_1 - \bar{x}_2) \pm t_{\frac{\alpha}{2}} \cdot s_p \cdot \sqrt{\frac{1}{n_1} + \frac{1}{n_2}}$$

where $s_p = \sqrt{\dfrac{(n_1 - 1)s_1^2 + (n_2 - 1)s_2^2}{n_1 + n_2 - 2}}$ is the pooled sample standard deviation and $t_{\alpha/2}$ is the upper $\alpha/2$ portion of the t-distribution with $n_1 + n_2 - 2$ degrees of freedom. The sample statistics are $\bar{x}_1$, the sample mean for the first population; $\bar{x}_2$, the sample mean for the second population; n_1, the sample size taken from the first population; n_2, the sample size taken from the second population; and s_1^2 and s_2^2, the sample variances for the first and second samples, respectively. The pooled standard deviation is used if the samples are drawn from two populations that both follow a normal distribution and if we assume that the two population variances are equal. However, for unequal variances we would need to use the following:

$$s = \sqrt{\frac{s_1^2}{n_1} + \frac{s_2^2}{n_2}}$$

Also, for unequal variances, the number of degrees of freedom would be found by rounding down the following to an integer:

$$d.f. = \frac{\left(\dfrac{s_1^2}{n_1} + \dfrac{s_2^2}{n_2}\right)^2}{\dfrac{\left(s_1^2/n_1\right)^2}{n_1 - 1} + \dfrac{\left(s_2^2/n_2\right)^2}{n_2 - 1}}$$

Then a $100 \cdot (1 - \alpha)\%$ confidence interval assuming unequal variances would be

$$(\bar{x}_1 - \bar{x}_2) \pm t_{\frac{\alpha}{2}} \cdot s$$

Example 4.6

Suppose we are interested in comparing the mean difference between the lifetimes of two different brands of cell phone batteries. If we collected a sample of fifteen lifetimes in hours from the two different brands of batteries, we could record the mean lifetime and standard deviation for each of the samples from the two different brands of batteries as follows:

Brand 1: $n_1 = 15$, $\bar{x}_1 = 74.2$ hours, $s_1 = 6.85$ hours

Brand 2: $n_2 = 15$, $\bar{x}_2 = 73.9$ hours, $s_2 = 7.05$ hours

Then a 95% confidence interval for the difference in the population mean lifetimes between the two different brands of batteries would be calculated, first finding the pooled standard deviation (assuming equal variances) as follows:

$$s_p = \sqrt{\frac{(n_1 - 1)s_1^2 + (n_2 - 1)s_2^2}{n_1 + n_2 - 2}}$$

$$= \sqrt{\frac{14(6.85)^2 + 14(7.05)^2}{15 + 15 - 2}}$$

$$\approx \sqrt{48.31} \approx 6.95$$

Then calculate the confidence interval:

$$(\bar{x}_1 - \bar{x}_2) \pm t_{\frac{\alpha}{2}} \cdot s_p \cdot \sqrt{\frac{1}{n_1} + \frac{1}{n_2}} = (74.2 - 73.9) \pm 2.048 \cdot (6.95) \cdot \sqrt{\frac{1}{15} + \frac{1}{15}}$$

Thus,

$$0.30 \pm 5.20 = (-4.90, 5.50)$$

Interpreting this confidence interval within the context of our problem suggests that we are 95% confident that the true but unknown population difference in the mean lifetimes between these two different brands of cell phone batteries is between −4.90 and 5.50 hours. Notice that the form of the confidence interval estimates the difference in the true population mean battery lifetimes between Brand 1 and Brand 2, namely, $\mu_1 - \mu_2$.

By looking at the range of numbers given in this confidence interval, we can expect that if we subtract the mean lifetime of Brand 2 from the mean lifetime of Brand 1, we would be 95% confident that the true but unknown difference in the population mean lifetimes between these two different brands of batteries would fall somewhere between −4.90 hours and 5.50 hours. Notice that this confidence interval covers the three cases where the difference between the population means for Brand 2 subtracted from

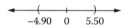

FIGURE 4.33
Ninety-five percent confidence interval for testing the mean difference between two different brands of cell phone batteries.

Brand 1 is 0, positive, or negative. This would suggest that there is no significant difference between the population mean battery lifetimes for these two different brands of cell phone batteries.

Another way to look at this interval would be to consider graphing the confidence interval on a number line, as illustrated in Figure 4.33.

This interval on the number line illustrates the total range of possible differences in the population mean lifetimes between the two different brands of cell phone batteries. Since this interval contains both positive and negative values, and thus the value of 0, we cannot infer that either brand of cell phone battery lasts longer than the other.

Similarly, if we were to assume unequal variances, then

$$s = \sqrt{\frac{s_1^2}{n_1} + \frac{s_2^2}{n_2}} = \sqrt{\frac{6.85^2}{15} + \frac{7.05^2}{15}} \approx 2.538$$

And there would be 27 degrees of freedom, which is found by rounding down the results of the following calculation to an integer:

$$d.f. = \frac{\left(\frac{s_1^2}{n_1} + \frac{s_2^2}{n_2}\right)^2}{\frac{\left(s_1^2/n_1\right)^2}{n_1 - 1} + \frac{\left(s_2^2/n_2\right)^2}{n_2 - 1}} = \frac{\left(\frac{6.85^2}{15} + \frac{7.05^2}{15}\right)^2}{\frac{\left(6.85^2/15\right)^2}{15 - 1} + \frac{\left(7.05^2/15\right)^2}{15 - 1}} \approx 27.98$$

Therefore, a 95% confidence interval would be

$$(\bar{x}_1 - \bar{x}_2) \pm t_{\frac{\alpha}{2}} \cdot s = (74.2 - 73.9) \pm 2.052\,(2.538) \approx (-4.91, 5.51)$$

where $t_{\alpha/2} = t_{0.025} = 2.052$ for 27 degrees of freedom.

4.9 Using MINITAB to Calculate a Confidence Interval for the Difference between Two Means

MINITAB can easily calculate the confidence interval for the difference between two population mean battery lifetimes, as illustrated in Figure 4.34.

This brings up the dialog box for a two-sample *t*-test and confidence interval, as presented in Figure 4.35.

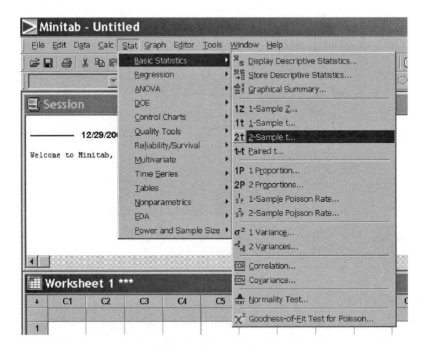

FIGURE 4.34
MINITAB commands for a two-sample confidence interval.

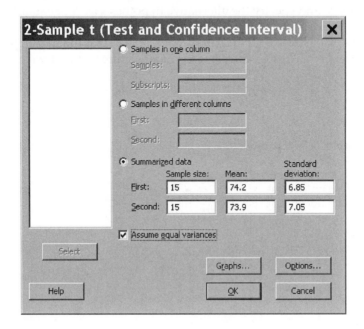

FIGURE 4.35
MINITAB dialog box for a two-sample confidence interval.

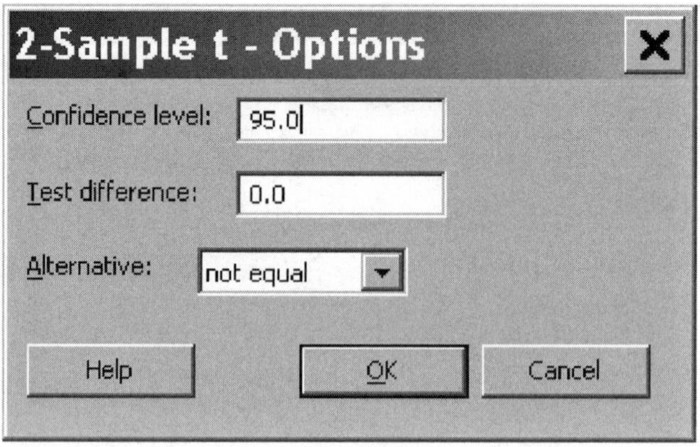

FIGURE 4.36
Options dialog box for a two-sample confidence interval.

Notice in Figure 4.35 that we can enter the data in either raw or summa-rized form. If we are assuming equal variances, then we need to check the box **Assume equal variances** (by not checking this box you will be assuming unequal variances). Also, under the **Options** tab, we need to select the level of confidence and specify the alternative hypothesis as "not equal," as illus-trated in Figure 4.36.

Figure 4.37 gives the MINITAB printout. Notice that the highlighted por-tion in Figure 4.37 gives the same confidence interval that we calculated by hand.

Two-Sample T-Test and CI

```
Sample   N   Mean   StDev   SE Mean
1        15  74.20  6.85    1.8
2        15  73.90  7.05    1.8

Difference = mu (1) - mu (2)
Estimate for difference: 0.30
95% CI for difference: (-4.90, 5.50)
T-Test of difference = 0 (vs not =): T-Value = 0.12  P-Value = 0.907  DF = 28
Both use Pooled StDev = 6.9507
```

FIGURE 4.37
MINITAB printout for the two-sample confidence interval for the difference in the population mean lifetimes for the two different brands of cell phone batteries.

4.10   Testing the Difference between Two Means

We can also conduct a hypothesis test to see if there is a significant difference between two population means by testing any of the following null and alternative hypotheses:

$$H_0 : \mu_1 = \mu_2$$

$$H_A : \mu_1 \neq \mu_2$$

$$H_0 : \mu_1 = \mu_2$$

$$H_A : \mu_1 > \mu_2$$

$$H_0 : \mu_1 = \mu_2$$

$$H_A : \mu_1 < \mu_2$$

Assuming equal variances, the test statistic for comparing two means follows a *t*-distribution with $n_1 + n_2 - 2$ degrees of freedom:

$$T = \frac{\bar{x}_1 - \bar{x}_2}{s_p\sqrt{\dfrac{1}{n_1} + \dfrac{1}{n_2}}}$$

where $s_p = \sqrt{\dfrac{(n_1 - 1)s_1^2 + (n_2 - 1)s_2^2}{n_1 + n_2 - 2}}$ is the pooled standard deviation, which is used when we assume equal variances, and this value serves as an estimate of the population standard deviation. If we are assuming unequal variances, then the test statistic would be

$$T = \frac{\bar{x}_1 - \bar{x}_2}{s}$$

where

$$s = \sqrt{\frac{s_1^2}{n_1} + \frac{s_2^2}{n_2}}$$

And the number of degrees of freedom would be found by rounding down the following to an integer:

$$d.f. = \frac{\left(\dfrac{s_1^2}{n_1} + \dfrac{s_2^2}{n_2}\right)^2}{\dfrac{\left(s_1^2/n_1\right)^2}{n_1 - 1} + \dfrac{\left(s_2^2/n_2\right)^2}{n_2 - 1}}$$

Example 4.7

Two types of drugs are used to treat migraine headaches, Drug 1 and Drug 2. A researcher wants to know if Drug 2 provides a faster relief time of migraine headaches than Drug 1. The researcher collected data as given in Table 4.3 ("Migraine Drugs"), which consists of the time it takes (in minutes) for a sample of seventeen patients using Drug 1 and fifteen patients using Drug 2 to begin feeling pain relief from their migraine headaches:

TABLE 4.3

Time (in minutes) for Patients to Feel Pain
Relief Using Either Drug 1 or Drug 2

Drug 1	Drug 2
22	26
18	28
31	29
26	24
38	23
25	25
29	28
31	28
31	27
28	29
26	31
19	30
45	22
31	27
27	28
16	
24	

The descriptive statistics for Drugs 1 and 2 obtained from Table 4.3 are as follows:

For Drug 1:

$$n_1 = 17$$

$$\bar{x}_1 = 27.47 \text{ minutes}$$

$$s_1 = 7.14 \text{ minutes}$$

For Drug 2:

$$n_2 = 15$$

$$\bar{x}_2 = 27.00 \text{ minutes}$$

$$s_2 = 2.56 \text{ minutes}$$

Notice that you do not necessarily need to have equal sample sizes from each population for a two-sample t-test.

We want to test whether the mean time it takes for Drug 2 to provide relief is faster than the mean time it takes for Drug 1 to provide relief ($\alpha = 0.01$). Thus, the appropriate null and alternative hypotheses would be

$$H_0 : \mu_1 = \mu_2$$

$$H_A : \mu_1 > \mu_2$$

The rejection region, which is based on the direction of the alternative hypothesis, would be in the right tail, and the value of t that defines the rejection region would be for $\alpha = 0.01$ and $17 + 15 - 2 = 30$ degrees of freedom, as illustrated in Figure 4.38.

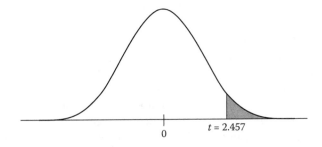

0          $t = 2.457$

FIGURE 4.38
Shaded rejection region defined by $\alpha = 0.01$ and $17 + 15 - 2 = 30$ degrees of freedom for a right-tailed, two-sample t-test.

If we are assuming equal variances, then the pooled standard deviation is calculated as follows:

$$S_p = \sqrt{\frac{(n_1 - 1)s_1^2 + (n_2 - 1)s_2^2}{n_1 + n_2 - 2}}$$

$$= \sqrt{\frac{16 \cdot (7.14)^2 + 14(2.56)^2}{17 + 15 - 2}} \approx \sqrt{30.25} \approx 5.50$$

Thus, the test statistic for the two-sample t-test would be as follows:

$$T = \frac{(\bar{x}_1 - \bar{x}_2)}{S_p\sqrt{\frac{1}{n_1} + \frac{1}{n_2}}} = \frac{27.47 - 27.00}{5.50\sqrt{\frac{1}{17} + \frac{1}{15}}} \approx 0.24$$

The value of the test statistic T does not fall into the defined rejection region ($t = 2.457$, as in Figure 4.38; this is because we are doing a right-tailed test and the value of our test statistic T is less than t). This suggests that we cannot claim that the population mean time it takes for Drug 2 to provide relief for migraine headaches is significantly less than the population mean time it takes for Drug 1 to provide pain relief.

4.11   Using MINITAB to Test the Difference between Two Means

Once the sample data or summarized data from each of the two populations are entered in MINITAB, a two-sample t-test can be done by selecting **Basic Statistics** and then **2-Sample t**, as illustrated in Figure 4.39.

Because in Example 4.7 we have the raw data in two different columns, we could use the option illustrated in Figure 4.40.

Note that similar to a one-sample t-test, we can also enter in the summarized data for each of the two samples. Also, the box for **Assume equal variances** needs to be checked if we are assuming equal variances. We also need to specify the level of significance and the appropriate direction of the alternative hypothesis. This can be done by selecting the **Options** tab in Figure 4.40 to give the options dialog box that is presented in Figure 4.41. We can also test whether there is a specific value of the difference between the two population means by specifying such a difference value in the **Options** box (see Exercise 12).

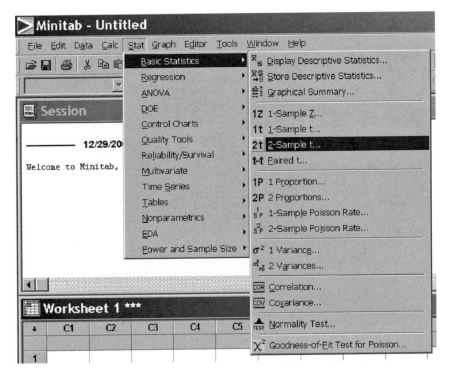

FIGURE 4.39
MINITAB commands to run a two-sample *t*-test.

The MINITAB printout for the two-sample *t*-test is presented in Figure 4.42.

Notice the highlighted value for the *T*-statistic in Figure 4.42 is the same as what we calculated by hand. Similarly, the highlighted *p*-value of .405 supports the same conclusion: we cannot claim that Drug 2 provides a faster relief time for migraine headaches than Drug 1. This is because the *p*-value is greater than our predetermined level of significance of $\alpha = 0.01$.

4.12 Using MINITAB to Create an Interval Plot

Another technique that can be used to compare the difference between two population means is to plot the confidence intervals for each population mean on a single graph and visually inspect whether the two confidence intervals overlap each other. To do this we can create what is called an *interval plot* using MINITAB. An interval plot is a plot of the confidence intervals for one or more population means on the same graph. Using the pain relief data from Table 4.3, the MINITAB commands to draw such an interval

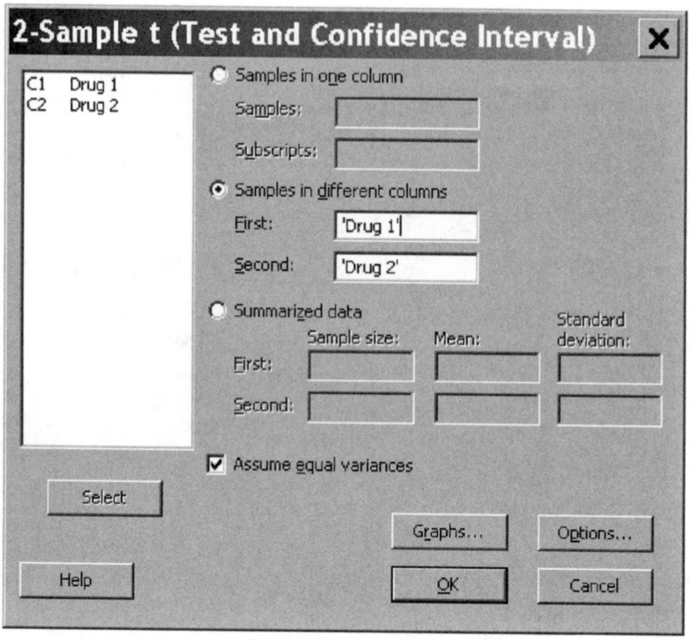

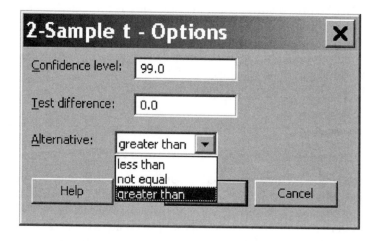

FIGURE 4.40
MINITAB dialog box to select either raw or summarized data options for a two-sample *t*-test.

FIGURE 4.41
MINITAB options box to select confidence level, difference, and direction of the alternative hypothesis.

Two-Sample T-Test and CI: Drug 1, Drug 2

```
Two-sample T for Drug 1 vs Drug 2

          N    Mean   StDev   SE Mean
Drug 1   17   27.47    7.14      1.7
Drug 2   15   27.00    2.56     0.66


Difference = mu (Drug 1) - mu (Drug 2)
Estimate for difference:  0.47
99% lower bound for difference:   -4.32
T-Test of difference = 0 (vs >):  T-Value = 0.24   P-Value = 0.405   DF = 30
Both use Pooled StDev = 5.5022
```

FIGURE 4.42
MINITAB printout for a two-sample *t*-test on whether there is a significant difference in the mean time to pain relief between the two different brands of drugs.

plot require selecting **Interval Plot** under the **Graphs** menu, as illustrated in Figure 4.43.

The dialog box for creating an interval plot is presented in Figure 4.44, which shows how we can select if we want to draw a simple interval plot for one variable or if we want to draw a multiple interval plot for two or more variables.

We can then select a simple plot of the **Multiple Y's** since we want to compare the confidence intervals for two different population means (namely, Drug 1 and Drug 2), as illustrated in Figure 4.45.

The interval plot for comparing the true population mean time to pain relief from migraines for Drugs 1 and 2 is given in Figure 4.46.

We can use the graph in Figure 4.46 to see that there does not appear to be a difference in the mean time to pain relief between the two different brands of drugs. This is because these two confidence intervals overlap each other. If a significant difference in the mean times were to exist, the confidence intervals would not overlap much.

4.13 Using MINITAB for a Power Analysis for a Two-Sample *t*-Test

Similar to a one-sample *t*-test, we can perform a power analysis for a two-sample *t*-test to determine what sample size is needed to find a minimum difference of a given size for a specified power value. Also similar to a one-sample *t*-test, two of the three values of sample size, difference, and power

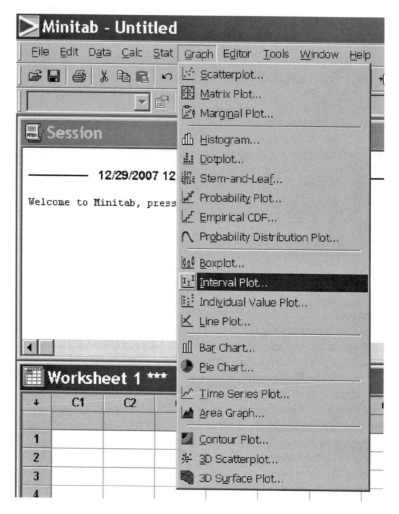

FIGURE 4.43
MINITAB commands to draw an interval plot.

need to be specified. We can use the pooled standard deviation as an estimate of the population standard deviation if we assume equal variances, or we can use *s* if we assume unequal variances. The dialog box for a two-sample *t*-test using MINITAB is illustrated in Figure 4.47.

Example 4.8

Suppose that we want to find the sample size needed for Drug 2 to be at least 10 minutes faster in providing pain relief than Drug 1 at 80% power.

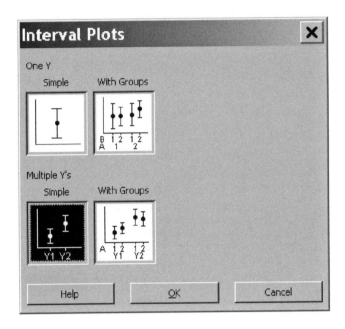

FIGURE 4.44
MINITAB dialog box for selecting the type of interval plot.

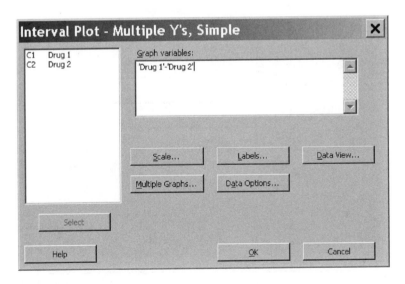

FIGURE 4.45
MINITAB dialog box for an interval plot of multiple populations.

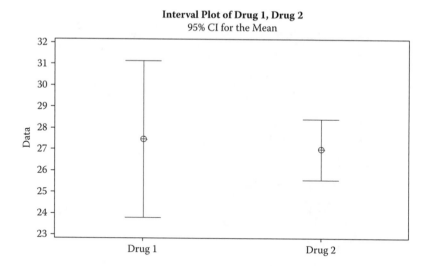

FIGURE 4.46
Interval plot illustrating the confidence intervals for the population mean time to pain relief for Drug 1 and Drug 2.

Using MINITAB, we would need to specify such a difference, along with the power and the pooled standard deviation, as illustrated in Figure 4.48.

By selecting the **Options** tab, illustrated in Figure 4.49, this allows us to specify a given alternative hypothesis and also a level of significance.

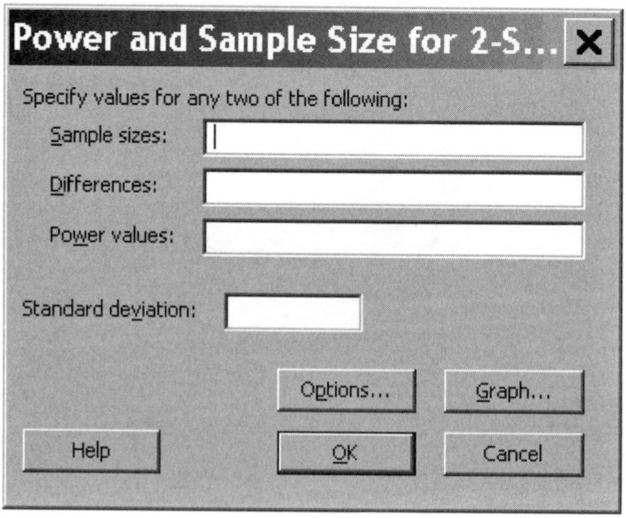

FIGURE 4.47
MINITAB dialog box for a power analysis for a two-sample *t*-test.

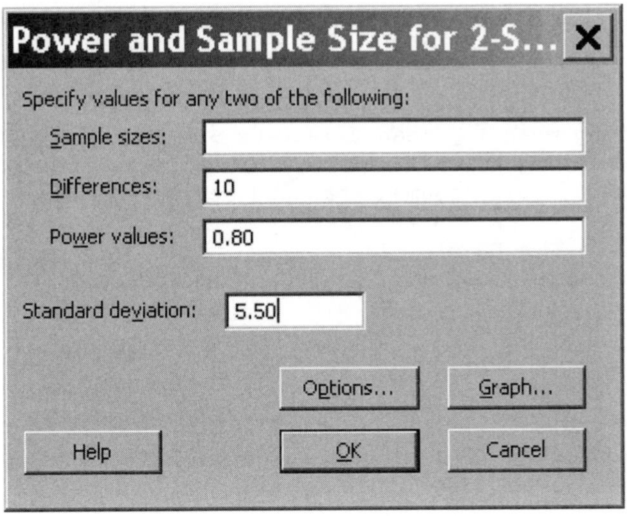

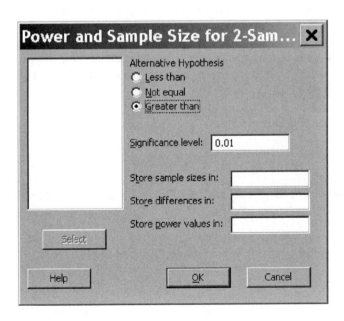

FIGURE 4.48
MINITAB dialog box for a power analysis for a two-sample *t*-test for a difference of 10 minutes and a power value of 80%.

FIGURE 4.49
MINITAB options box for a power analysis for a two-sample *t*-test.

Power and Sample Size

```
2-Sample t Test

Testing mean 1 = mean 2 (versus >)
Calculating power for mean 1 = mean 2 + difference
Alpha = 0.01   Assumed standard deviation = 5.5

                   Sample  Target
Difference          Size   Power  Actual Power
        10             8    0.8        0.828960

The sample size is for each group.
```

FIGURE 4.50
MINITAB power analysis for a two-sample *t*-test to detect a difference of 10 minutes at 80% power.

Specifying a one-tailed test and a significance level of .01, the MINITAB printout for the power analysis is provided in Figure 4.50.

As the highlighted portion in the MINITAB printout in Figure 4.50 suggests, in order to find a difference of at least 10 minutes between the mean time to relief for the two different brands of pain relief drugs, a minimum sample size of at least 8 would be needed from each population.

If we wanted to detect a smaller effect, say a difference of at least 3 minutes, then we would need a sample size of at least 69 from each group, as illustrated in the MINITAB printout presented in Figure 4.51.

Recall that it is much easier to detect a larger difference, and thus a smaller sample is needed, whereas it takes a much larger sample to be able to detect a smaller difference.

Power and Sample Size

```
2-Sample t Test

Testing mean 1 = mean 2 (versus >)
Calculating power for mean 1 = mean 2 + difference
Alpha = 0.01   Assumed standard deviation = 5.5

            Sample  Target
Difference   Size   Power  Actual Power
         3     69    0.8        0.801089

The sample size is for each group.
```

FIGURE 4.51
MINITAB power analysis for a two-sample *t*-test to detect a difference of 3 minutes at 80% power.

4.14 Confidence Intervals and Hypothesis Tests for Proportions

Many applied problems in education and science deal with percentages or proportions. Similar to the work we have done thus far with means, we can calculate confidence intervals to obtain a likely range of values for an unknown population proportion, and we can perform hypothesis tests to infer whether or not the true but unknown population proportion is significantly different from some hypothesized value.

We can calculate a sample proportion by taking the ratio of the number in the sample that meet the given characteristic of interest and dividing it by the sample size as follows:

$$\hat{p} = \frac{x}{n},$$

where x is the number in the sample that meet the given characteristic, and n is the sample size. For instance, if there are twenty students in a class and sixteen are males, then the proportion of males is

$$\hat{p} = \frac{x}{n} = \frac{16}{20} = 0.80 = 80\%$$

A confidence interval for a population proportion takes the following form:

$$\hat{p} \pm z_{\frac{\alpha}{2}} \cdot \sqrt{\frac{\hat{p} \cdot (1 - \hat{p})}{n}},$$

where $\hat{p}$ is the sample proportion, n is the sample size, and $z_{\alpha/2}$ is the value that defines the upper $\alpha/2$ portion of the standard normal distribution.

For large samples ($n \geq 30$), the sampling distribution of the sample proportion $\hat{p}$ is normally distributed and is centered about the true but unknown population proportion p.

Example 4.9

Suppose that in a taste test given to a random sample of 350 shoppers at a local grocery store, 210 of these shoppers preferred the taste of a new brand of soft drink over a competitor's brand. To find a 95% confidence interval for the true but unknown population proportion of shoppers who would prefer this new brand of soft drink, we would have to find the sample proportion

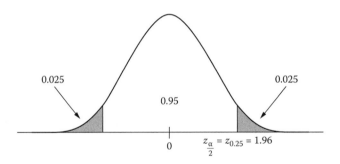

FIGURE 4.52
Distribution of the sample proportion for a 95% confidence interval.

and the value of z that defines the upper 2.5% of the standard normal distribution, as illustrated in Figure 4.52.

The sample proportion is found by taking the number who prefer the taste of the new soft drink and dividing it by the total number of shoppers who participated in the test. This can be calculated as follows:

$$\hat{p} = \frac{210}{350} = 0.60$$

Thus, a 95% confidence interval would be

$$\hat{p} \pm z_{\frac{\alpha}{2}} \cdot \sqrt{\frac{\hat{p} \cdot (1-\hat{p})}{n}} = 0.60 \pm 1.96 \cdot \sqrt{\frac{0.60(1-0.60)}{350}} = 0.60 \pm 0.051 \approx (0.549, 0.651)$$

Interpreting this confidence interval suggests that we are 95% confident that between 54.9% and 65.1% of the true population of shoppers would prefer this particular brand of soft drink over the competitor's.

We can also perform hypothesis tests to compare the true population proportion against some hypothesized value. For large samples ($n \geq 30$), we can use the following test statistic:

$$Z = \frac{x - n \cdot p_0}{\sqrt{n \cdot p_0(1-p_0)}},$$

where x is the number in the sample who meet the given characteristic of interest, n is the sample size, and p_0 is the population proportion being tested under the null hypothesis. This test statistic for the sample proportion follows the standard normal distribution, and it is centered about the true but unknown population proportion p_0, as illustrated in Figure 4.53.

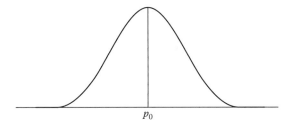

FIGURE 4.53
Distribution of the sample proportion if $H_0 = p_0$ were true.

Example 4.10

Suppose a soft drink company wants to test whether at least 55% of all shoppers would prefer their brand of soft drink over the competitor's. A random sample of 350 shoppers found that 210 preferred the particular brand of soft drink over the competitor's. To test whether or not the true population proportion of shoppers that would prefer their particular brand of soft drink is significantly greater than 55% ($\alpha = 0.05$), we would set up the appropriate null and alternative hypotheses as follows:

$$H_0 : p_0 = 0.55$$

$$H_A : p_0 > 0.55$$

Figure 4.54 shows the appropriate rejection region that is defined by the value of $z_{0.05} = 1.645$.

The test statistic would then be calculated as follows:

$$Z = \frac{x - n \cdot p_0}{\sqrt{n \cdot p_0(1 - p_0)}} = \frac{210 - (350)(0.55)}{\sqrt{(350)(0.55)(1 - 0.55)}} = 1.88$$

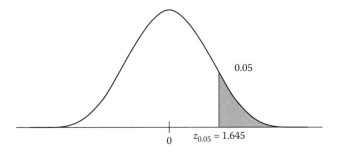

FIGURE 4.54
Rejection region for testing if the true population proportion of shoppers who prefer the brand of soft drink is significantly greater than 55%.

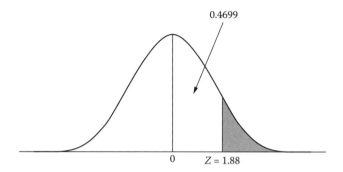

FIGURE 4.55
Area between 0 and the test statistic $Z = 1.88$.

Since this test statistic falls in the rejection region, as illustrated in Figure 4.54, we can accept the alternative hypothesis and reject the null hypothesis. Thus, the true but unknown population mean proportion of shoppers who would prefer this particular brand of soft drink is significantly greater than 55%.

Because the test statistic for a sample proportion follows a normal distribution, we can go one step further and actually calculate the exact p-value. Recall that the p-value is the minimum level of significance that is determined based on where the test statistic falls on the given sampling distribution, and it describes the probability of observing a sample proportion that comes from a population with a true proportion of p_0.

Since the value of the test statistic for the population proportion follows a standard normal distribution, we can use the standard normal tables (Table 1 in Appendix A) to actually calculate the p-value. This is done by finding the area to the right of the test statistic Z. Notice in Figure 4.55 that the area between 0 and $Z = 1.88$ is 0.4699.

Thus, the p-value is the area that lies to the right of the test statistic Z, which is

$$0.5000 - 0.4699 = 0.0301$$

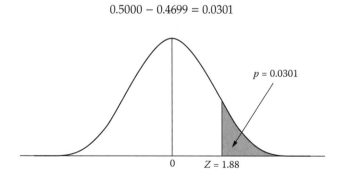

FIGURE 4.56
Area to the right of the test statistic $Z = 1.88$ corresponds to a p-value of .0301.

Figure 4.56 shows that the test statistic $Z = 1.88$ gives the area to the right of Z equal to 0.0301. This area corresponds to the p-value of .0301, which supports the conclusion that the true population proportion of shoppers who prefer the given brand of soft drink over the competitor's is significantly greater than 55% because the p-value is less than the predetermined level of significance of .05.

It is fairly easy to calculate an exact p-value by hand for a test statistic that follows a normal distribution. For a right-tailed test, the p-value represents the area to the right of the test statistic on the standard normal curve. For a left-tailed test, the p-value represents the area to the left of the test statistic on a standard normal curve. For a two-tailed test, the p-value represents twice the area to the right of a positive test statistic on a standard normal curve or twice the area to the left of a negative test statistic on a standard normal curve. However, it is much more difficult to find the exact p-value by hand for a test statistic that follows the t-distribution, or any other distribution for that matter. This is because the t-tables and tables for many other distributions only provide some values of the area, and thus cannot be used to determine an exact value, whereas the standard normal table can be used to find the exact p-value.

4.15 Using MINITAB for a One-Sample Proportion

MINITAB can also be used to calculate confidence intervals and to perform hypothesis tests for a one-sample proportion. To calculate a confidence interval, under the **Basic Statistics** option under the **Stat** command on the menu bar you can select **1 Proportion** to test one population proportion, as illustrated in Figure 4.57.

The dialog box for a one-sample proportion confidence interval and hypothesis test is presented in Figure 4.58, where we can enter either the actual sample data or the summarized data.

Then under the **Options** tab you can select the confidence level and specify that the alternative has to be set to "not equal" to find a confidence interval, as can be seen in Figure 4.59. Also, notice that the box that specifies to conduct the hypothesis test and calculate confidence intervals based on the normal distribution is checked, and this is done so that MINITAB will use the same formulas that were introduced in the calculations.

This gives the MINITAB printout in Figure 4.60.

Similarly, we could use MINITAB to run a hypothesis test by specifying the proportion being tested under the null hypothesis, the direction of the alternative hypothesis, and the level of significance, as can be seen in Figures 4.61 and 4.62.

The MINITAB printout is given in Figure 4.63. Notice in Figure 4.63 that the highlighted p-value of .0301 represents the area to the right of the test statistic $Z = 1.88$. Thus, it is very unlikely that the given sample would be drawn from a population with a true proportion of $p_0 = 0.55$. Since this p-value is

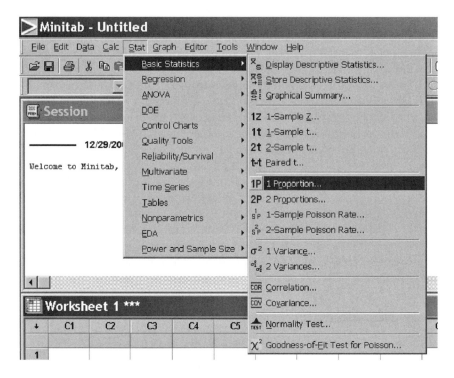

FIGURE 4.57
MINITAB commands to calculate a confidence interval and conduct hypothesis tests for a one-sample proportion.

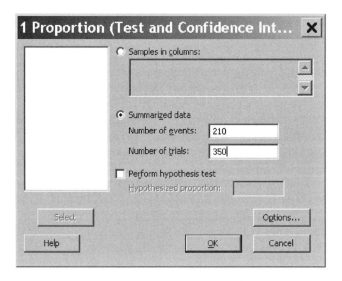

FIGURE 4.58
MINITAB dialog box for a one-sample proportion.

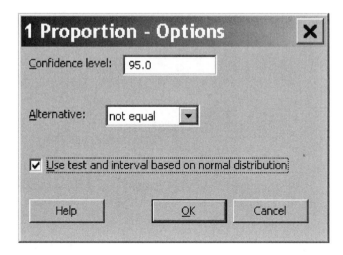

FIGURE 4.59
MINITAB options box for a one-sample proportion.

less than the predetermined level of significance, we can claim that the true but unknown population proportion is significantly greater than 55%.

4.16   Power Analysis for a One-Sample Proportion

Similar to all of the power analyses we have done thus far, we can use MINITAB to perform a power analysis for a one-sample proportion. The MINITAB dialog box for a one-sample proportion is given in Figure 4.64.

Notice that Figure 4.64 requires two of the three values for sample size, alternative values for p, or the power level, along with the proportion being tested under the null hypothesis. For our last example, suppose we want to know the sample size that would be needed to detect a population proportion of 60% with a power level of 80%. Note that 60% is 5% more than the hypothesized proportion of 55%. We would specify the 60% as the alternative value of p, and specify the value we are testing under the null hypothesis

Test and CI for One Proportion

```
Sample    X    N   Sample p            95% CI
1        210  350  0.600000   (0.548676, 0.651324)

Using the normal approximation.
```

FIGURE 4.60
MINITAB printout for the confidence interval for a one-sample proportion of those shoppers that prefer the particular brand of soft drink over the competitor's.

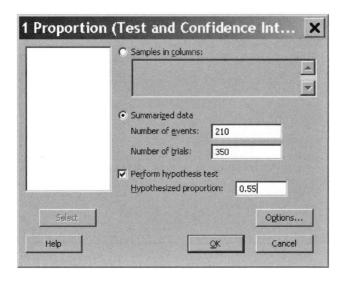

FIGURE 4.61
MINITAB dialog box to perform a hypothesis test for a hypothesized proportion of 55%.

as the hypothesized value of p. Essentially, we are trying to find the sample size needed where a difference of at least 5% of the hypothesized proportion test out as significant at the given power level. Because the direction of the alternative hypothesis was "greater than," we need to specify the significance level and the appropriate direction of the alternative hypothesis, as can be seen in the **Options** box in Figure 4.65.

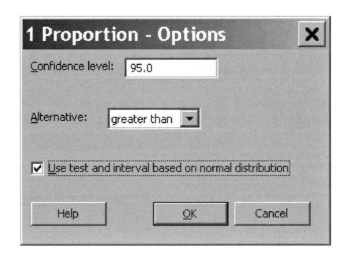

FIGURE 4.62
MINITAB options box to test if the true population proportion is greater than 55% ($\alpha = 0.05$).

Test and CI for One Proportion

```
Test of p = 0.55 vs p > 0.55

                           95% Lower
Sample    X    N   Sample p     Bound   Z-Value   P-Value
1       210   350  0.600000   0.556928    1.88     0.0301
```

Using the normal approximation.

FIGURE 4.63
MINITAB printout for the one-sample test of a population proportion of whether at least 55% of shoppers prefer the particular brand of soft drink over the competitor's.

This gives the MINITAB printout in Figure 4.66. Thus, as Figure 4.66 suggests, a sample size of at least 606 is needed to detect at least a 5% difference at 80% power.

4.17 Differences between Two Proportions

We can also calculate confidence intervals and perform hypothesis tests of whether or not there is a difference in the proportions between two different populations. Similar to a one-sample proportion for large samples ($n \geq 30$),

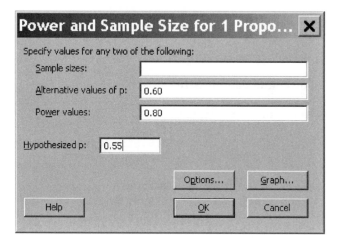

FIGURE 4.64
MINITAB dialog box for a power analysis for a one-sample proportion.

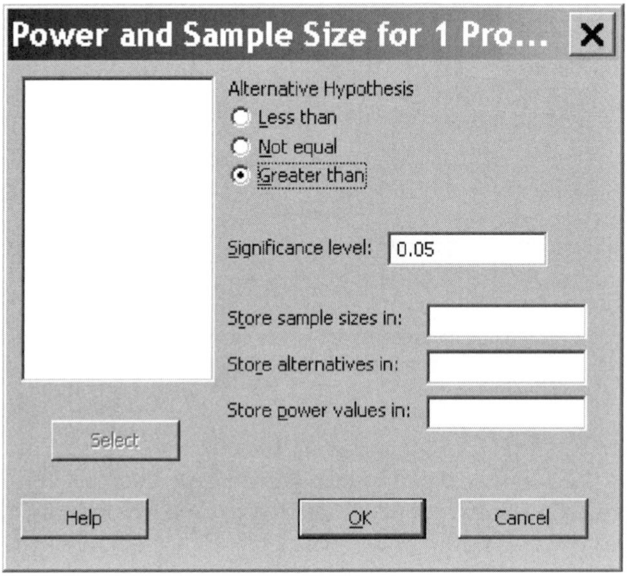

FIGURE 4.65
MINITAB options box for a one-sample proportion.

the sampling distribution for the difference between two proportions is normally distributed and is centered about the true difference between the two population proportions.

The formula for the confidence interval for the difference between the two population proportions is as follows:

$$(\hat{p}_1 - \hat{p}_2) \pm z_{\frac{\alpha}{2}} \cdot \sqrt{\frac{\hat{p}_1(1-\hat{p}_1)}{n_1} + \frac{\hat{p}_2(1-\hat{p}_2)}{n_2}},$$

Power and Sample Size

```
Test for One Proportion

Testing proportion = 0.55 (versus > 0.55)
Alpha = 0.05

Alternative   Sample    Target
Proportion    Size      Power    Actual Power
       0.6     606        0.8         0.800137
```

FIGURE 4.66
MINITAB printout for a power analysis to determine what size sample is needed to detect a difference of 5% over the hypothesized proportion of 55% with 80% power.

where $\hat{p}_1$ is the sample proportion for the first sample, $\hat{p}_2$ is the sample proportion for the second sample, n_1 is the sample size for the first sample, n_2 is the sample size for the second sample, and $z_{\alpha/2}$ is the value that defines the upper $\alpha/2$ portion of the standard normal distribution.

Example 4.11

The data set in Table 4.4 gives a comparison of the number of students who planned to graduate in the spring versus the number of students who actually graduated in the spring semester at two competing colleges.

We want to find a 95% confidence interval for the true population difference in the actual graduation rates between the two colleges.

From Table 4.4, the descriptive data would be calculated as follows:

$$\hat{p}_1 = \frac{218}{275} = 0.7927, \, n_1 = 275$$

$$\hat{p}_2 = \frac{207}{289} = 0.7163, \, n_2 = 289$$

Thus, a 95% confidence interval for the difference between the actual graduation rates for the population of students at College A and the population of students at College B would be as follows:

$$(0.7927 - 0.7163) \pm 1.96 \cdot \sqrt{\frac{0.7927\,(0.2073)}{275} + \frac{0.7163\,(0.2837)}{289}}$$

$$\approx 0.0764 \pm 0.0707 \approx (0.0057, 0.1471)$$

This confidence interval suggests that we are 95% confident that the difference in the population proportion of students who actually graduate from the two competing schools is between 0.57% and 14.71%. Because the confidence interval is testing the difference in the graduation rates of College B subtracted from College A, this confidence interval would also suggest that College A has a graduation rate that is 0.57% to 14.71% higher than College B's.

TABLE 4.4

Comparison of the Number of Students Who Plan to Graduate versus the Number of Students Who Actually Graduate for Two Competing Colleges

	College A	College B
Number of students who plan to graduate in the spring	275	289
Number of students who actually graduated in the spring	218	207

Similarly, we could also perform a hypothesis test to see if College A graduates a significantly different proportion of students than College B. The test statistic is as follows:

$$Z = \frac{(\hat{p}_1 - \hat{p}_2)}{\sqrt{\dfrac{\hat{p}_1(1-\hat{p}_1)}{n_1} + \dfrac{\hat{p}_2(1-\hat{p}_2)}{n_2}}}$$

Example 4.12

Using the data in Table 4.4, suppose we want to know if there is a difference in the proportion of students that College A graduates compared to College B ($\alpha = 0.05$). Therefore, the appropriate null and alternative hypotheses would be

$$H_0 : p_1 = p_2$$

$$H_A : p_1 \neq p_2$$

The rejection region for the test statistic would be established by the level of significance and the direction of the alternative hypothesis, as shown in Figure 4.67.

The value of the test statistic is then calculated as follows:

$$Z = \frac{(\hat{p}_1 - \hat{p}_2)}{\sqrt{\dfrac{\hat{p}_1(1-\hat{p}_1)}{n_1} + \dfrac{\hat{p}_2(1-\hat{p}_2)}{n_2}}} = \frac{(0.7927 - 0.7163)}{\sqrt{\dfrac{0.7927(0.2073)}{275} + \dfrac{0.7163(0.2837)}{289}}} \approx 2.12$$

By comparing the value of this test statistic to the value of z that defines the rejection region, as illustrated in Figure 4.67, we can see that the test statistic does fall in the rejection region because it is greater than $z = 1.96$. Therefore,

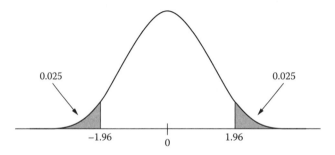

FIGURE 4.67
Rejection region for testing whether there is a difference in the proportion of students that College A graduates compared to College B ($\alpha = 0.05$).

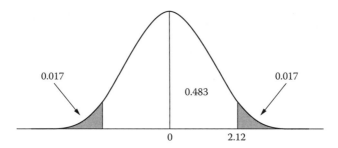

FIGURE 4.68
P-value representing twice the area to the right of the corresponding value of the test statistic
Z = 2.12.

we can claim that there is a significant difference in the proportion of students who graduate from College A compared to the proportion of students who graduate from College B.

Because the sampling distribution for the two-sample proportion follows a normal distribution, we can also calculate the *p*-value by finding the area that is defined by the value of the test statistic, as presented in Figure 4.68. Because this is a two-sided test, the *p*-value is found by taking twice the area that is to the right of the test statistic Z. Therefore, since the area between 0 and 2.12 is 0.4830, the *p*-value for this test will be 2(0.5000 − 0.4830) = 0.034.

4.18 Using MINITAB for Two-Sample Proportion Confidence Intervals and Hypothesis Tests

To use MINITAB for a two-sample proportion using the data in Table 4.4, under the **Stat** menu, select **Basic Statistics**, and then **2 Proportions,** as illustrated in Figure 4.69.

This gives the dialog box presented in Figure 4.70, where you can enter the raw data or the summarized data to represent the number of events and trials for each of the two samples.

To calculate a confidence interval, we select the **Options** tab to bring up the dialog box in Figure 4.71. This gives the MINITAB printout in Figure 4.72.

Notice that the highlighted portion of Figure 4.72 provides the 95% confidence interval for the difference in the graduation proportions for College A and College B, similar to what we have calculated.

If we were interested in testing whether College A graduated a larger proportion of students than College B ($\alpha = 0.01$), we could test the following null and alternative hypotheses:

$$H_0 : p_1 = p_2$$

$$H_A : p_1 > p_2$$

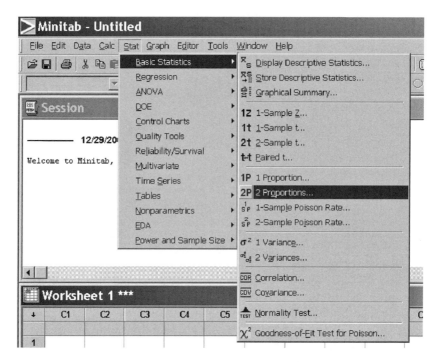

FIGURE 4.69
MINITAB commands for comparing two proportions.

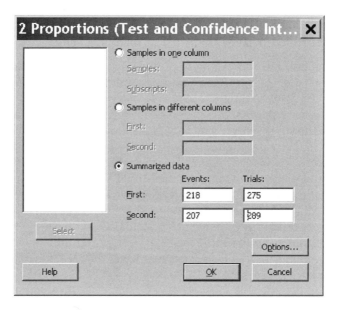

FIGURE 4.70
MINITAB dialog box for a two-sample proportion.

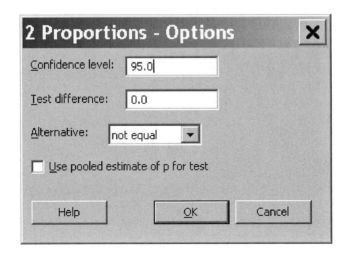

FIGURE 4.71
MINITAB options box for two-sample proportions.

Adjusting for these factors in MINITAB, as illustrated in Figure 4.73, gives the MINITAB printout in Figure 4.74.

4.19 Power Analysis for a Two-Sample Proportion

Example 4.13

Suppose we want to find a difference of 5% in the proportion of graduates between College A and College B, and also suppose that the estimated proportion of graduates for the population of all colleges in general is 75%.

Test and CI for Two Proportions

```
Sample    X    N   Sample p
1        218  275  0.792727
2        207  289  0.716263

Difference = p (1) - p (2)
Estimate for difference:   0.0764643
95% CI for difference:   (0.00577742, 0.147151)
Test for difference = 0 (vs not = 0):   Z = 2.12   P-Value = 0.034

Fisher's exact test: P-Value = 0.040
```

FIGURE 4.72
MINITAB printout for two-sample proportions for the college data in Table 4.4.

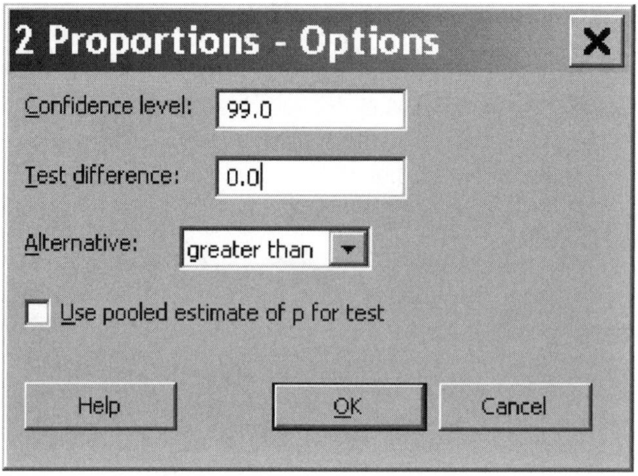

FIGURE 4.73
MINITAB options box specifying the level of significance as .01 and the direction of the alternative hypothesis as "greater than."

To find the sample sizes needed to achieve 80% power that can detect a 5% difference in the proportion of students graduating from College A and College B, you would enter in the respective values, as illustrated in Figure 4.75.

Figure 4.76 gives the MINITAB printout for the given power analysis. Thus, as Figure 4.76 illustrates, we would need a sample size of at least 1,251 to detect a 5% difference in the mean graduation rate between the two colleges for a proportion of 0.70, and a sample size of 1,094 for a proportion of 0.80.

Test and CI for Two Proportions

```
Sample    X    N   Sample p
1        218  275  0.792727
2        207  289  0.716263

Difference = p (1) - p (2)
Estimate for difference:  0.0764643
99% lower bound for difference:  -0.00743636
Test for difference = 0 (vs > 0):  Z = 2.12  P-Value = 0.017

Fisher's exact test: P-Value = 0.022
```

FIGURE 4.74
MINITAB printout for the hypothesis test with a level of significance of .01 and an alternative hypothesis of "greater than."

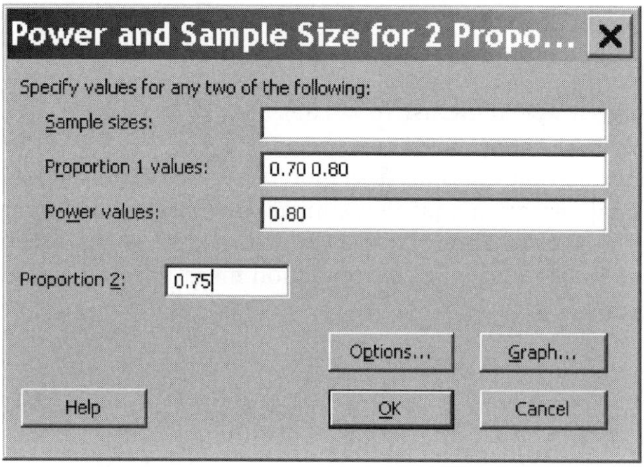

FIGURE 4.75
Power analysis to find a 5% difference from a hypothesized population proportion of 75% with 80% power.

Notice that in this example, we had to specify what we believe the true population proportion is and then add and subtract the difference we were interested in testing, as illustrated in Figures 4.75 and 4.76.

Power and Sample Size

```
Test for Two Proportions

Testing proportion 1 = proportion 2 (versus not =)
Calculating power for proportion 2 = 0.75
Alpha = 0.05

               Sample  Target
Proportion 1    Size    Power   Actual Power
        0.7     1251     0.8       0.800090
        0.8     1094     0.8       0.800095

The sample size is for each group.
```

FIGURE 4.76
MINITAB output for the power analysis to find a 5% difference from a hypothesized population proportion of 75% with 80% power.

Exercises

Unless otherwise specified, use ($\alpha = 0.05$).

1. If the population standard deviation σ is known, confidence intervals can be calculated in a fashion similar to what we have done, but by using the standard normal distribution instead of the t-distribution. A confidence interval for a population mean when σ is known is of the form

$$\left(\bar{x} - z_{\frac{\alpha}{2}} \cdot \frac{\sigma}{\sqrt{n}}, \, \bar{x} + z_{\frac{\alpha}{2}} \cdot \frac{\sigma}{\sqrt{n}} \right),$$

where $z_{\alpha/2}$ represents the upper $\alpha/2$ portion of the standard normal distribution (no degrees of freedom are needed when using the z-distribution), $\bar{x}$ is the sample mean, and σ is the population standard deviation.

   a. Using the above formula, find a 95% confidence interval for the true population mean for a sample of size 52 that has a population standard deviation of 4.56 and a sample mean of 24.81.

   b. You can also find this confidence interval using a one-sample z-test in MINITAB, as illustrated in Figure 4.77. Check your answer to part (a) using MINITAB.

2. If the population standard deviation σ is known, hypothesis tests can also be performed using the standard normal distribution with the following test statistic:

$$Z = \frac{\bar{x} - \mu_0}{\dfrac{\sigma}{\sqrt{n}}}$$

   a. For the sample statistics given in Exercise 1, test (by hand) whether or not the population mean is significantly different than 20 ($\alpha = 0.05$).

   b. Check your results using MINITAB (see Figure 4.77).

3. Recall that the p-value represents the minimum level of significance that you can expect from the value of the test statistic. For the hypothesis test in Exercise 2, calculate the p-value.

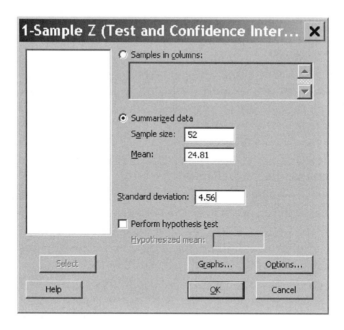

FIGURE 4.77
MINITAB dialog box to calculate a confidence interval.

4. a. Find a 99% confidence interval for the average time it takes to manufacture the given part using the data provided in Table 4.2.

b. Test the hypothesis that the population average time it takes to manufacture the part is significantly different from 50 hours ($\alpha = 0.05$).

c. For the data in Table 4.2, determine the power that you can expect for a minimum difference of 2 hours.

5. a. For the "Crime Rate by County" data set, using MINITAB find a 99% confidence interval for the difference between the mean population crime rates for 1991 and 1996.

b. For the given sample from part (a), for a power of 80%, what is the minimum difference that could be found provided that such a difference actually exists?

6. a. Using the "Crime Rate by County" data set, test whether or not the average number of violent crimes reported has significantly increased from 1991 to 1996 ($\alpha = 0.05$).

b. What sample size would be needed to find a mean increase of at least 500 violent crimes reported between 1991 and 1996 with 80% power?

7. A weight loss program claims that its clients have lost, on average, more than 40 pounds over the course of a 4-month period. A random sample of thirty participants of the weight loss program shows an average loss of 44 pounds in a 4-month period, with a standard deviation of $s = 11$ pounds.

a. For a significance level of $\alpha = 0.05$, test whether you would support the program's claims.

b. What sample size is needed to achieve 80% power in order to find a mean difference of at least 3 pounds over a 4-month period?

c. For the sample of size $n = 30$ participants and a power of 80%, determine the minimum mean difference in weight loss that can be detected provided that such a difference exists.

8. Table 4.5 gives the speeds (in miles per hour) for a random sample of twelve cars driving down a highway that has a speed limit of 65 miles per hour. Determine if the mean speed of the population of drivers who use this highway is faster than the speed limit of 65 miles per hour ($\alpha = 0.01$).

9. a. A random sample of $n = 19$ observations produced a sample mean of $\bar{x} = 69.7$ and a sample variance of $s^2 = 4.7$. Test the null hypothesis $H_0 : \mu = 70$ against the alternative, $H_A : \mu \neq 70$. Use $\alpha = 0.02$.

b. Given the sample statistics in part (a), find a 99% confidence interval for the true population mean.

10. a. Suppose you are testing a single population mean and have collected a sample of size $n = 28$. Find the value of t such that 5% of all values of t will be greater.

b. Find the value of t such that 10% of all values of t will be greater.

c. Find the value of t such that 95% of the distribution is between $-t$ and $+t$.

11. Using the "Crime Rate by County" data, find two 98% confidence intervals for the true population crime rate for both 1991 and 1996.

TABLE 4.5

Speed (in miles per hour) for a Random Sample of Cars Driving on a Highway with a Speed Limit of 65 Miles per Hour

55	68	72	79	81	63	72	70	61	77	68	66

12. We can also test whether two populations' mean values differ by some constant, C. The procedure is similar to testing whether two population means are different from each other; however, the possible null and alternative hypotheses are framed somewhat differently in that we need to include the difference we are testing for:

$$H_0 : \mu_1 - \mu_2 = C$$
$$H_A : \mu_1 - \mu_2 \neq C$$

$$H_0 : \mu_1 - \mu_2 = C$$
$$H_A : \mu_1 - \mu_2 > C$$

$$H_0 : \mu_1 - \mu_2 = C$$
$$H_A : \mu_1 - \mu_2 < C$$

Also (assuming equal variances) the test statistic below incorporates the constant:

$$T = \frac{(\bar{x}_1 - \bar{x}_2) - C}{S_p \sqrt{\dfrac{1}{n_1} + \dfrac{1}{n_2}}}$$

The data set "Rental Car Agencies" in Table 4.6 presents a sample of the wait times (in minutes) for renting a car from two competing car rental agencies at a local airport.

Assuming equal variances, you want to determine if Car Agency B is at least 5 minutes faster in renting their cars than Car Agency A ($\alpha = 0.05$).

a. Set up the appropriate null and alternative hypotheses.

b. Find the value of the test statistic.

c. Interpret your findings. Justify if you can claim that Car Agency B is at least 5 minutes faster in renting a car than Car Agency A.

d. At 80% power, how large of a sample would be needed to find whether Car Agency B is at least 5 minutes faster in renting their cars than Car Agency A provided that such a difference exists?

TABLE 4.6

Wait Times (in minutes) for the Time It Takes to
Rent a Car from Two Competing Car Agencies

Car Agency A	Car Agency B
25	19
18	20
20	21
12	18
19	19
12	17
19	20
22	21
21	22
20	17
21	15
25	8
28	12
24	19
28	20
24	22
35	23
15	25
8	27
16	
20	

13. The service department at a car dealership serviced 528 cars over the course of 1 week. These customers were asked to complete a survey regarding their experience with the service department. Of those 528 customers who serviced their cars during the course of this week, 484 said they were very satisfied with the service they received through the service department.

    a. Find a 95% confidence interval for the true population proportion of customers who would respond that they are very satisfied with the service department at this dealership.

    b. Test to determine if more than 90% of the population of customers will respond that they are very satisfied with the service department at this dealership ($\alpha = 0.05$).

14. A recent study suggests that more than 80% of all high school mathematics teachers use calculators in their classroom. To test this theory, a national sample of 1,286 high school mathematics teachers were

asked if they used calculators in their classroom, and 1,274 responded that they did.

a. Based on the information collected from this sample, do you agree with the results of this study that suggest that more than 80% of all high school mathematics teachers use calculators in their classroom ($\alpha = 0.05$)?

b. What size sample would be needed to detect at least a 5% difference at 70% power?

15. A recent report found that a higher percentage of college graduates earn more per year than high school graduates. Table 4.7 gives a summary of the number of respondents who are either high school graduates or college graduates and whether their yearly salary is more or less than $40,000.

a. Find the proportion of high school graduates that make more than $40K per year.

b. Find the proportion of college graduates that make more than $40K per year.

c. Test whether the proportion of college graduates who make more than $40K per year is significantly higher than the percentage of high school graduates who make more than $40K per year ($\alpha = 0.05$).

16. A manufacturing company manufactures parts to a given set of specifications. The parts are to be manufactured with a mean diameter of 15 cm. In order to see if their machines are working properly, the quality control manager collected a random sample of parts, as presented in Table 4.8.

a. Find the mean and standard deviation of the given sample.

b. Set up the appropriate test to determine whether or not the company is manufacturing the parts to the given specification for the mean diameter ($\alpha = 0.05$).

TABLE 4.7

Salary Comparison for a Sample of 19,561 High School and College Graduates

	Less Than $40K/Year	More Than $40K per Year	Total
High school graduate	4,287	4,988	9,275
College graduate	5,041	5,245	10,286
Total	9,328	10,233	19,561

TABLE 4.8

Diameter (in centimeters) of a Sample of Twenty-Four Parts

15.8	16.1	15.7	14.9	14.5	14.8
15.6	15.3	16.1	15.9	15.7	15.2
14.9	13.8	17.1	15.4	15.8	15.2
16.0	14.8	14.9	15.0	15.2	15.7

    c. Construct a 99% confidence interval of the population mean diameter for the parts.

17. The data set "Automobiles" contains the type, make, and model of different vehicles sold in the United States along with measures of their city and highway miles per gallon, and the amount of greenhouse gases they emit for a random sample of 641 vehicles.

    a. Using MINITAB, find a 95% confidence interval for the mean annual fuel cost for each of the six different types of vehicles. You may want to have MINITAB unstack the columns by using the **Unstack Columns** command under the **Data** tab. To unstack the data based on the annual fuel cost, enter the annual fuel cost in the **Unstack the data in** box and enter the types of vehicles in the **Using subscripts in** box.

    b. Test whether there is a difference in the mean annual fuel costs between minivans and SUVs ($\alpha = 0.05$).

    c. Test whether there is a difference in the mean annual fuel costs between Honda and Toyota ($\alpha = 0.05$).

    d. Test whether there is a difference in the mean annual fuel costs between Chevrolet and Ford ($\alpha = 0.05$).

    e. Find the minimum difference in the mean annual fuel cost between minivans and SUVs that can be found with the given sample at 80% power.

18. A random sample of 1,463 households across the United States found that 938 of these households watch television on more that one night per week. Find a 99% confidence interval for the true population proportion of households in the United States that watch television on more than one night per week.

19. A random sample of 1,463 households across the United States found that the average amount of television watched per week was 15.36 hours, with a standard deviation of 4.27 hours. Find a 99% confidence interval for the true population mean number of hours that households in the United States watch television.

20. You may have heard that most car accidents happen within 25 miles of home. The data set "Car Accidents" presented in Table 4.9 gives a random sample of the mileage away from home for thirty-six car accidents in a given state.

a. Find the mean and standard deviation of this data set.

b. Test the claim that on average, accidents happen within 25 miles of home ($\alpha = 0.05$).

c. How large of a sample size would be needed to find a difference of at least 5 miles at 80% power?

21. The methods of inference that we used to compare two means require that the samples collected are independent of each other. When the assumption of independence does not hold true, such as when we are doing a before and after comparison on the same subject, we can use what is called a *paired t-test*. We can calculate confidence intervals and perform hypothesis tests of the population mean difference between paired observations. A paired *t*-test is the same as testing whether the difference between any paired comparisons is different from some hypothesized value.

The test statistic is similar to a one-sample *t*-test:

$$T = \frac{\bar{d} - \mu_0}{\frac{s_d}{\sqrt{n}}},$$

where $\bar{d}$ is the mean of the paired sample difference, μ_0 is the hypothesized mean difference, s_d is the standard deviation of the difference, and n is the sample size.

The data set "Weight Loss" presented in Table 4.10 gives two measures of weight for a sample of twenty-four individuals: the

TABLE 4.9

Distance (in miles) Away from Home for a Sample of Thirty-Six Car Accidents in a Given State

2	18	27	47	29	24
18	46	38	36	12	8
26	38	15	57	26	21
37	18	5	9	12	28
28	17	4	7	26	29
46	20	15	37	29	19

individual's weight before beginning a weight loss program and the individual's weight after participating in the program for 4 weeks.

a. Find the mean and standard deviation for the weight before the program and for the weight after beginning the program.

b. Create a column in MINITAB that consists of the difference between the weight before the program and the weight after the program (in other words, create a new column that is the result of Weight Before – Weight After).

c. Using MINITAB, run a paired *t*-test (select **Stat**, then **Basic Statistics**, and then **Paired** *t*) to see if the population who participate in the weight loss program lose a significant amount of weight ($\alpha = 0.05$).

d. Using MINITAB, run a one-sample *t*-test on the difference between the before and after weights that was calculated in part

TABLE 4.10

Sample of Before and After Weights (in pounds) for Twenty-four Individuals

Weight Before	Weight After
184	184
191	187
207	209
176	174
155	147
189	183
254	238
218	210
170	168
154	151
148	145
137	135
167	158
129	125
174	170
225	219
218	219
175	171
182	175
194	193
177	177
209	207
176	170
164	163

(b) to determine whether this difference is significantly greater than 0, and show that these results are the same as if you ran a paired *t*-test in part (c) to determine whether the difference between the weight before the program and the weight after the program is significant.

22. We can also calculate confidence intervals for paired data by using the following formula:

$$\left(\bar{d} - t_{\frac{\alpha}{2}} \cdot \frac{s_d}{n}, \bar{d} + t_{\frac{\alpha}{2}} \cdot \frac{s_d}{n} \right),$$

where $\bar{d}$ is the mean of the paired sample difference, s_d is the standard deviation of the difference, $t_{\alpha/2}$ is the value that defines the upper boundary point for $n - 1$ degrees of freedom, and n is the sample size.

The data set "Business Establishments" given in Table 4.11 consists of a sample of the number of businesses for the years 1997 and 2002 by industry (http://www.census.gov/econ/census02/data/comparative/USCS.HTM).

TABLE 4.11

Number of Businesses by Year Based on Type of Industry

Industry	Number of Businesses 2002	Number of Businesses 1997
Mining	24,284	25,000
Construction	709,279	656,448
Manufacturing	350,728	362,829
Retail trade	1,110,983	1,118,447
Transportation and warehousing	199,618	178,025
Finance and insurance	440,268	395,203
Real estate and rental and leasing	322,805	288,273
Management of companies and enterprises	49,340	47,319
Administrative, support, waste management, remediation services	273,407	276,393
Health care and social assistance	709,133	645,853
Arts, entertainment, and recreation	110,324	99,099
Other services (except public administration)	524,879	519,715

a. Using MINITAB, find a 95% confidence interval for the paired difference.

b. Using MINITAB, find a 98% confidence interval for the paired difference.

c. Comment on the difference between the two confidence intervals found in parts (a) and (b).

d. Test whether there was a significant difference in the mean number of businesses between 1997 and 2002 ($\alpha = 0.05$).

e. Test whether there was a significant difference in the mean number of businesses between 1997 and 2002 ($\alpha = 0.01$).

f. Determine if, on average, there are at least 30,000 more businesses in 2002 than in 1997 ($\alpha = 0.05$).

g. Comment on the difference between the two tests done in parts (d) and (e).

23. Explain why conducting a two-tailed hypothesis test allows you to make inferences for a one-sided test. (*Hint*: Describe how the rejection region for a two-tailed test also includes the rejection region for a one-tailed test.)

Reference

Light, R., Singer, J., and Willett, J. 1990. *By design: Planning research on higher education.* Cambridge, MA: Harvard University Press.

5

Simple Linear Regression

5.1 Introduction

In the last chapter, we described some basic techniques for making inferences about different population parameters using sample statistics. However, there may be occasions when we are interested in determining whether two or more variables are related to each other. For instance, suppose we want to determine whether there is a relationship between a student's high school mathematics ability and how well he or she does during his or her first year of college. One way to quantify such a relationship could be to look at whether the score a student receives on the mathematics portion of the SAT, which students take when they are in high school, is related to his or her first-year college grade point average.

Consider Table 5.1, which presents the data set "SAT–GPA," which consists of the scores on the mathematics portion of the SAT examination and the corresponding first-year grade point averages for a random sample of ten students at a local university.

The range of the possible scores on the SAT mathematics examination is from 200 to 800 points, and the range of possible first-year grade point averages is from 0.00 to 4.00 points.

We want to determine whether the score received on the SAT mathematics examination (SATM) is related to the first-year grade point average (GPA). We also may want to further quantify this relationship by estimating how strong this relationship is, and we may want to develop some type of a model that can be used to describe the relationship between GPA and SATM that can also be used to predict GPA using SATM.

We will be using *simple regression analysis* to study the relationship between two variables. The variable that we are interested in modeling or predicting (in this case first-year GPA) is called the *dependent* or *response variable*. The variable that is used to predict the dependent or response variable (which in this case is SATM) is called the *independent* or *predictor variable*. *Simple linear regression analysis* is a statistical technique that can be used to develop a model that estimates the linear relationship between a single continuous response variable (or *y*-variable) and a single continuous predictor variable (or *x*-variable).

Before we begin a detailed discussion of simple linear regression analysis it is important to understand that it is very difficult to infer a *cause-and-effect*

TABLE 5.1

SAT Mathematics Examination Scores and Corresponding First-Year
Grade Point Averages for a Random Sample of Ten College Students

Observation Number	SAT Math Score (SATM) x	First-Year Grade Point Average (GPA) y
1	750	3.67
2	460	1.28
3	580	2.65
4	600	3.25
5	500	3.14
6	430	2.82
7	590	2.75
8	480	2.00
9	380	1.87
10	620	3.46

relationship between two variables. In most situations, we can only infer that
there is an association or relationship between two variables. It is extremely
difficult to claim that the variation in y is *caused* by the variable x because
there could be numerous factors other than the variable x alone that may
influence y.

For instance, in developing a model to predict first-year grade point
averages, although the score received on the mathematics portion of the
SAT examination may have an influence on the first-year grade point average,
numerous other factors, such as motivation, study habits, number of credits attempted, etc., could also impact a student's first-year grade point
average. Thus, it does not make sense to infer that one single variable,
such as the score received on the SAT mathematics examination, is the
only variable that has an influence on first-year grade point averages. And
because it may not be possible to isolate all such factors that influence the
given response variable, simple linear regression is typically only used to
describe an association or relationship between the response and predictor
variables.

5.2 Simple Linear Regression Model

Recall in Chapter 2 that we described a scatter plot as a way to visualize the
relationship between two variables. We can create a scatter plot to visualize
the relationship between SATM and GPA, and we can also use a marginal
plot to graph the scatter plot along with the histograms or box plots of each
of the two variables simultaneously. To create a scatter plot, we simply plot
the ordered pairs (x, y) on the Cartesian plane, as illustrated in Figure 5.1.

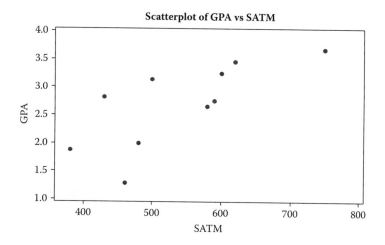

FIGURE 5.1
Scatter plot of first-year grade point average (GPA) versus the score received on the mathematics portion of the SAT examination (SATM) for the data given in Table 5.1.

The marginal plot with histograms for the response and predictor variables is presented in Figure 5.2.

By visually examining the trend in the scatter plot in Figure 5.1 and the marginal plot in Figure 5.2, we can decide whether we believe there is a relationship between SATM and GPA. We can also try to describe the pattern

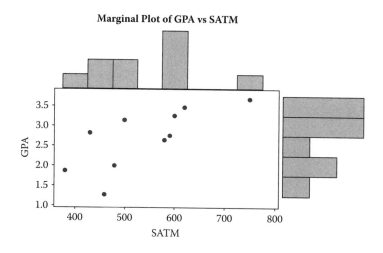

FIGURE 5.2
Marginal plot of first-year grade point average (GPA) versus the score received on the mathematics portion of the SAT examination (SATM) for the data given in Table 5.1.

of this relationship. The scatter plot in Figure 5.1 and the marginal plot in Figure 5.2 show that a positive trend appears to exist between SATM and GPA, where an increase in SATM seems to suggest an increase in GPA (in other words, as x increases, y increases). We can also describe the relationship between SATM and GPA as one that could be reasonably modeled by using a straight line. Using a straight line to approximate the relationship between two variables provides a simple way to model such a relationship because in using a linear model, the calculations are relatively simple and the interpretations are straightforward.

In order to use a straight line to model the relationship between two variables, we will begin by choosing an arbitrary line to model the relationship between our two variables and describe how to assess the fit of this line to the data. Consider Figure 5.3, which illustrates the line drawn on the scatter plot that connects the two most extreme data points: (380, 1.87) and (750, 3.67).

This line in Figure 5.3 is one of many different lines that could be used to model the relationship between x and y. Notice that for any given scatter diagram, we could conceivably find numerous straight lines that could be used to represent the relationship between the two variables. Once we have found a line to approximate the relationship between our two variables, we can then use the equation of this line to model the relationship between SATM and GPA and predict GPA based on SATM.

In order to determine the equation of the line in Figure 5.3, we will first review the slope-intercept form of a linear equation. You may recall that the slope-intercept form for the equation of a straight line is $y = mx + b$, where

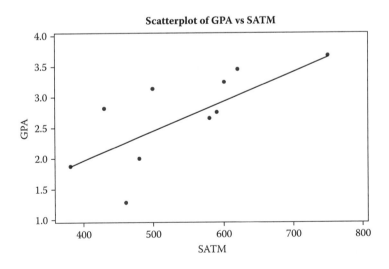

FIGURE 5.3
Scatter plot of GPA versus SATM that includes the line that connects the two most extreme data points: (380, 1.87) and (750, 3.67).

m is the slope of the line and b is the y-intercept (the y-intercept is the point where the line crosses the y-axis). Using the line drawn on the scatter plot in Figure 5.3, we can find the equation of this particular line by finding the slope and the y-intercept, as follows:

To find the slope, we simply find the quotient of the difference in the y-values and the difference in the x-values, as follows:

$$m = \frac{y_2 - y_1}{x_2 - x_1} = \frac{3.67 - 1.87}{750 - 380} = \frac{1.80}{370} = 0.0048648649$$

Now using the point (380, 1.87) and the slope $m \approx 0.0049$, we can find the y-intercept by plugging these values into the slope-intercept form of the equation of a straight line, as follows:

$$1.87 = (0.0048648649) \cdot (380) + b$$

$$b = 0.0213513514$$

Then the equation of the line that connects these two extreme points, as drawn in Figure 5.3, is $y = 0.0048648649x + 0.0213513514$.

This line can be used to approximate the relationship between x (SATM) and y (GPA). Notice that the calculations for the slope and the y-intercept were carried out to several decimal places, and this was done in order to avoid round off error in future calculations.

In statistics, we use a slightly different notation to represent the slope-intercept form of a linear equation. We will use β_0 to represent the y-intercept and β_1 to represent the slope. The primary objective in simple linear regression analysis is to determine the estimates of the true but unknown population parameters β_0 and β_1 for the line that describes the data, and determine how well this line fits the data.

The true but unknown population linear equation that we are interested in estimating is of the form $y = \beta_0 + \beta_1 x + \varepsilon$, where β_0 represents the y-intercept of the true but unknown population linear equation, β_1 represents the slope of the true but unknown population linear equation, and ε is the error component, which represents all other factors that may affect y as well as purely random disturbances. The line that we will be using to estimate this population line is of the form $\hat{y} = \hat{\beta}_0 + \hat{\beta}_1 x$, where the "hat" ($\wedge$) symbol is used to denote the estimates of the slope and y-intercept of the true but unknown population linear equation. Thus, our estimate of the true population regression line, which connects the two most extreme points, is as follows:

$$\hat{y} = 0.0213513514 + 0.0048648649x$$

One way to assess the usefulness of any line that describes the relationship between the predictor (or x) variable and the response (or y) variable is to look at the vertical distance between the *y*-values for each of the observations and the estimated line, as illustrated in Figure 5.4. We can measure the vertical distance between each of our observed (or sample) *y*-values and the line we are considering by simply taking the difference between the actual *y*-value for each observation in the sample and the value of the estimated or fitted *y*-value for each given data point.

For any given data point in the sample (x_i, y_i) the observed *y*-value is y_i, and the estimated (or predicted) value of *y* is denoted as $\hat{y}_i$. The value $\hat{y}_i$ can be found by substituting the value of x_i into the equation of the estimated line that we are considering. The difference between the observed value y_i and the estimated value $\hat{y}_i$ is called the *i*th *residual*. The *i*th residual is found by taking the difference between the observed and estimated values for the *i*th observation as follows:

$$\hat{\varepsilon}_i = (y_i - \hat{y}_i) = y_i - (\hat{\beta}_0 + \hat{\beta}_1 x_i)$$

Table 5.2 presents the estimated values for the line drawn in Figure 5.3 along with the difference between the observed and estimated values for each of the observations in the sample.

Notice that in doing the calculations in Table 5.2, the decimal values were carried out to several decimal places in order to avoid round-off error.

For instance, the residual for observation 10 is the vertical distance between the observed value and the estimated value, which can be calculated as follows:

$$y_{10} - \hat{y}_{10}$$

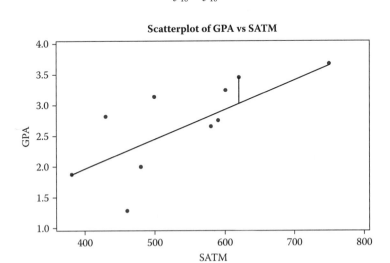

Scatterplot of GPA vs SATM

FIGURE 5.4
Vertical distance between the observed value of (3.46, 620) and the line that connects the two most extreme points.

TABLE 5.2

Estimated Values $(\hat{y}_i)$ and Residuals $(\hat{\varepsilon}_i = y_i - \hat{y})$ for the Line That Connects the Points (380, 1.87) and (750, 3.67), as Illustrated in Figure 5.3

Observation Number	Observed Value (y_i)	Independent Variable (x_i)	Estimated Value $(\hat{y}_i)$	Residuals $\hat{\varepsilon}_i = y_i - \hat{y}_i$
1	3.67	750	3.670000026	−0.000000026
2	1.28	460	2.259189205	−0.979189205
3	2.65	580	2.842972993	−0.192972993
4	3.25	600	2.940270291	0.309729709
5	3.14	500	2.453783801	0.686216199
6	2.82	430	2.113243258	0.706756742
7	2.75	590	2.891621642	−0.141621642
8	2.00	480	2.356486503	−0.356486503
9	1.87	380	1.870000013	−0.000000013
10	3.46	620	3.037567589	0.422432411

To find $\hat{y}_{10}$ we need to substitute the value $x_{10} = 620$ into the estimated equation we are considering, as follows:

$$\hat{y}_{10} = 0.0213513514 + 0.0048648649(620) \approx 3.037567589$$

$$y_{10} - \hat{y}_{10} = 3.46 - 3.037567589 = +0.422432411$$

The residual term for the tenth observation, denoted as $\hat{\varepsilon}_{10}$, can also be illustrated as representing the magnitude of the vertical distance between the observed y-value of the data point $(x_{10} = 620, y_{10} = 3.46)$ and the estimated value $(\hat{y}_{10})$, as presented in Figure 5.4. This residual is positive because the point lies above the line.

Now suppose that we find the residuals for each of the ten observations in the sample, and then find the sum of all such residual terms as follows:

$$\sum_{i=1}^{10} \hat{\varepsilon}_i = 0.454864675$$

And if we square the value for each residual and then take the sum, then the sum of the squares of all of the residual terms would be

$$\sum_{i=1}^{10} \hat{\varepsilon}_i^2 = 2.387968790$$

These two summations tell us that the sum of all the residual terms is approximately 0.45, and the sum of all the residual terms squared is approximately 2.39. These summations can be used as a measure to assess how well the line that we have chosen fits the underlying set of data.

But how do we know that the line we have chosen is the line that *best* represents the linear relationship between x and y? Clearly, there are many other lines that could be used to represent the relationship between SATM and GPA, and some of these may represent a much better fit of the data than the line we have chosen (which is the line that connects the two most extreme points). There is one line in particular that is very interesting, and it is called the *line of best fit*. This is the line that best fits the given set of data.

The line of best fit is a unique line such that the sum of the residuals is equal to 0 and the sum of the squares of the residuals is as small as possible. Essentially, this unique line is the line that best fits the data because for the sum of the residuals to equal 0, this implies that the vertical distance between the observations and the line of best fit is approximately the same for those observed values that lie above the line of best fit as for those observed values that lie below the line of best fit. Furthermore, the sum of the squares of the vertical distances between the observed values and the line of best fit (this is also called the sum of the squares of the residual errors and is denoted by SSE) is a minimum. This value being a minimum suggests that the observed values lie closer to the line of best fit than they do to any other line. In other words, for a line to be the line of best bit, the following two conditions need to be met:

1. $\displaystyle\sum_{i=1}^{n} \hat{\varepsilon}_i = 0$

2. $SSE = \displaystyle\sum_{i=1}^{n} \hat{\varepsilon}_i^2 = \sum_{i=1}^{n} (y_i - \hat{y})^2 = \sum (\text{Observed value} - \text{Estimated value})^2$

   is a minimum.

This line of best fit is also referred to as the *least squares line*, or the *regression line*.

By minimizing the sum of the squares of the residual terms using sample data, we can calculate the estimates of the y-intercept $(\hat{\beta}_0)$ and slope of the line of best fit $(\hat{\beta}_1)$. These estimates can then be used to approximate the true but unknown population parameters β_0 and β_1. By calculating these values, we can estimate the true but unknown population linear equation by using the line of best fit as follows:

$$\hat{y} = \hat{\beta}_0 + \hat{\beta}_1 x$$

Notice that since we are using estimated values of the population parameters, we use the "hat" (^) symbol to denote an estimated parameter. The equation $\hat{y}$ gives the estimated (or predicted) values of the true but unknown population linear equation, and the values $\hat{\beta}_0$ and $\hat{\beta}_1$ are estimated values of the true but unknown population parameters β_0 and β_1.

Thus, by assuming that y is linearly related to x, we can use the equation of the line of best fit to estimate the true but unknown population linear equation:

$$y = \beta_0 + \beta_1 x + \varepsilon$$

where y denotes the population response (or dependent) variable, x corresponds to the population predictor (or independent) variable, ε represents all unknown variables, as well as random disturbances, that may impact y, and β_0 and β_1 represent the true but unknown linear equation parameters.

Just as when making basic statistical inferences about a population mean or population proportion, discussed in Chapter 4, we can also make inferences about the true but unknown population linear equation. In other words, we can use the estimated line of best fit, which is based on sample data, to make inferences about the true but unknown population linear equation.

5.3 Model Assumptions

Before we can use the line of best fit to make inferences about the true population linear equation, we first need to consider several model assumptions. These model assumptions are very important because they need to hold true in order to draw meaningful inferences about the unknown population linear equation using the estimated equation. If these assumptions do not hold reasonably true, then any inferences we make using the estimated regression equation about the unknown population equation may be suspect.

The model assumptions for a simple linear regression analysis rely on the unknown population error component, ε, being independent, normally distributed, and with constant variance, σ^2. Also, the specification of the regression model stipulates a linear relationship between x and y. These assumptions are described in more detail below:

1. *The errors associated with any two observed values are independent of each other.*

    This essentially means that for any two observations we collect, the error associated with any one observation does not depend on the error of any other observation.

    This assumption could be violated if we take repeated observations on the same individual over time. For instance, suppose we are interested in determining whether a relationship exists between the score received on the SAT examination and yearly grade point averages. If the data were collected such that we are looking at the yearly grade point averages for the same individuals over the course of

4 years, then it is likely that the error terms for a regression analysis would not be independent of each other.

2. *The error component is normally distributed.*

This assumption stipulates that the errors are essentially random and, thus, allow us to make inferences about the population linear equation by creating confidence intervals and performing hypothesis tests using the estimated line of best fit.

3. *The error component has constant variance.*

This assumption stipulates that the variance of the error term remains constant for all values of the independent variable x. This allows us to use simplified calculations to determine the estimated values of the population regression parameters (described in the next section).

4. *The functional form of the relationship between x and y can be established.**

For instance, we may want to assume that the population model, $y = \beta_0 + \beta_1 x + \varepsilon$, is linear with respect to the relationship between x and y. However, if the true but unknown model equation is not linear, and we are using a linear equation, then this assumption would be violated. This could be the case if x were quadratic or any other nonlinear power.

5.4 Finding the Equation of the Line of Best Fit

Assuming that all of the model assumptions described in the last section hold true, we can now obtain the estimates of the population parameters β_0 and β_1 using the following formulas:

$$\hat{\beta}_1 = \frac{S_{xy}}{S_{xx}}$$

$$\hat{\beta}_0 = \bar{y} - \hat{\beta}_1 \cdot \bar{x}$$

These equations provide the values of the estimated parameters that represent the slope and the intercept for the estimated line of best fit, where the

---

* For all of the models used throughout this text, we will always assume that the regression equations are linear in the beta parameters. For instance, equations such as $y = \beta_0 + \beta_1 x^3 + \varepsilon$ can be considered, but not equations such as $y = \beta_0 + \beta_1^3 x + \varepsilon$.

estimated line of best fit is a unique line such that the sum of the residuals is 0 and the sum of the squared residuals is a minimum.

The following statistics are needed in order to calculate the estimated parameters for the line of best fit:

$$\bar{x} = \frac{\Sigma x_i}{n}, \quad \bar{y} = \frac{\Sigma y_i}{n},$$

$$S_{xx} = \Sigma(x_i - \bar{x})^2$$

$$S_{yy} = \Sigma(y_i - \bar{y})^2$$

$$S_{xy} = \Sigma(x_i - \bar{x})(y_i - \bar{y})$$

Where the statistics $\bar{x}$ and $\bar{y}$ are the sample means for the x- and y-variables, respectively, from the sample of ordered pairs of size n; the values S_{xx} and S_{yy} are the sums of the squared differences of the individual observations from the means for x and y, respectively; and S_{xy} is the sum of the cross-products of the difference between each observation and the mean for both x and y. Note that S_{yy} is not used to calculate the line of best fit, but it will be used in future calculations.

Example 5.1

Using the data given in Table 5.1, these statistics are calculated as follows (see Exercise 2):

$$\bar{x} = 539.000$$

$$\bar{y} = 2.689$$

$$S_{xx} = 107490.000$$

$$S_{yy} = 5.228$$

$$S_{xy} = 555.490$$

Given these statistics, we can then obtain the estimates of $\hat{\beta}_0$ and $\hat{\beta}_1$ the true but unknown population parameters β_0 and β_1, as follows:

$$\hat{\beta}_1 = \frac{S_{xy}}{S_{xx}} = \frac{555.490}{107490.00} = 0.00517$$

$$\hat{\beta}_0 = \bar{y} - \hat{\beta}_1 \cdot \bar{x} = 2.689 - (0.00517) \cdot 539 = -0.09545$$

Once we find these estimates, the line of best fit is

$$\hat{y} = \hat{\beta}_0 + \hat{\beta}_1 x = -0.09545 + 0.00517 \cdot x$$

$$\hat{y} \approx -0.0955 + 0.0052x$$

This is the line that best fits our data because it is the unique line such that the sum of the residuals is 0 and the sum of the squared residuals is a minimum. The individual residual values are calculated as follows:

$$\hat{\varepsilon}_i = y_i - (\hat{\beta}_0 + \hat{\beta}_1 x_i), \quad i = 1, 2, \ldots, n$$

Table 5.3 gives the values of the observed and estimated values, and the residuals (notice again that we did not round off in order to avoid errors in future calculations).

Notice that the values for the residuals are both positive and negative, and the sum of the residuals is approximately 0 (due to a slight rounding error):

$$\sum \hat{\varepsilon}_i \approx 0.000000001$$

Furthermore, the sum of the squares of the residuals will be the minimum:

$$SSE = \sum_{i=1}^{10} \hat{\varepsilon}_i^2 = 2.357412$$

In comparing $\sum \hat{\varepsilon}_i$ and $\sum \hat{\varepsilon}_i^2$ for the line of best fit to those for the line that connected the two most extreme points on the scatter plot, the line of best fit has the sum of the residuals approximately equal to 0, and the sum of squares of the residuals is a minimum, and thus it is smaller than what was obtained using the line that connected the two most extreme data values.

TABLE 5.3

Estimated Values ($\hat{y}_i$) and Residuals ($\hat{\varepsilon}_i = y_i - \hat{y}$) for the Line of Best Fit

Observation Number	Observed Value (y_i)	Independent Variable (x_i)	Estimated Value ($\hat{y}_i$)	Residuals $\hat{\varepsilon}_i = y_i - \hat{y}_i$
1	3.67	750	3.779412038	-0.109412038
2	1.28	460	2.280741464	-1.000741464
3	2.65	580	2.900881012	-0.250881012
4	3.25	600	3.004237603	0.245762397
5	3.14	500	2.487454647	0.652545353
6	2.82	430	2.125706577	0.694293423
7	2.75	590	2.952559308	-0.202559308
8	2.00	480	2.384098056	-0.384098056
9	1.87	380	1.867315099	0.002684901
10	3.46	620	3.107594195	0.352405805

In fact, if you compare the sum of the residuals and the sum of the squares of the residuals to any other line, you may find that the sum of the residuals is greater than 0 or the sum of the squares of the residuals will be greater than for the line of best fit (see Exercise 1).

The sum of the squares of the residuals, SSE, is also an interesting statistic because it can be used as an estimate of the variance of the unknown error component (σ^2):

$$SSE = \sum_{i=1}^{n} \hat{\varepsilon}_i^2 = S_{yy} - \frac{S_{xy}^2}{S_{xx}}$$

To obtain an estimate of σ^2, we first find the value of SSE and then divide it by $n - 2$ (we subtract 2 because 2 degrees of freedom are used to estimate two parameters, β_0 and β_1).

Thus, an estimate of the variance of the unknown error component σ^2 is

$$S^2 = \frac{SSE}{n-2}$$

where S^2 is called the *mean square error*.

The *root mean square error* is the square root of the mean square error as follows:

$$S = \sqrt{S^2}$$

Example 5.2

To find S^2 for the data presented in Table 5.3, since $SSE = 2.357412$, then

$$S^2 = \frac{SSE}{n-2} = \frac{2.357412}{10-2} = 0.294677$$

We can also find the root mean square error by taking the square root of the mean square error as follows:

$$S = \sqrt{S^2} = \sqrt{641.146} = 0.54284$$

5.5 Using MINITAB for Simple Linear Regression

We can use MINITAB to draw a scatter plot with the line of best fit and to perform all of the calculations that we have done thus far. In order to use MINITAB for a simple linear regression analysis, the data have to be entered in two columns, one column for the x-variable and one column for the y-variable.

To draw the scatter plot with the line of best fit, select **Regression** from the **Stat** menu and then select **Fitted Line Plot**, as illustrated in Figure 5.5.

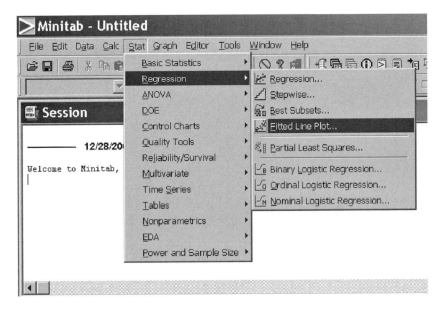

FIGURE 5.5
MINITAB commands to create a fitted line plot.

Executing these commands gives the dialog box in Figure 5.6.

Providing the appropriate response and predictor variables along with the type of relationship (which we are assuming is a linear relationship) and then hitting **OK** gives the scatter plot, which includes the graph and the equation of the line of best fit, as illustrated in Figure 5.7.

To run a regression analysis to obtain the estimated regression parameters, click on **Stat** on the top menu bar, and select **Regression** as in Figure 5.8 to open the regression dialog box that is presented in Figure 5.9.

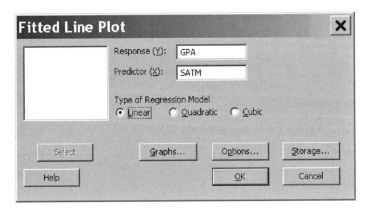

FIGURE 5.6
Dialog box for creating a fitted line plot.

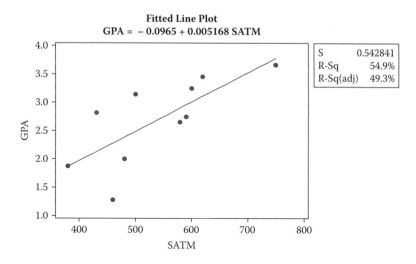

FIGURE 5.7
Fitted line plot describing the relationship between first-year grade point average (GPA) and the score received on the mathematics portion of the SAT examination (SAT).

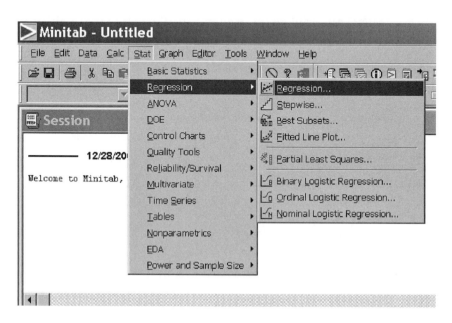

FIGURE 5.8
MINITAB commands to run a regression analysis.

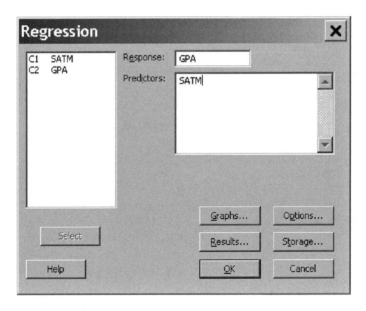

FIGURE 5.9
Regression dialog box.

By selecting GPA as the response (*y*) variable and SATM as the predictor (*x*) variable, selecting **OK** will give the MINITAB printout presented in Figure 5.10.

As the MINITAB printout in Figure 5.10 illustrates, the highlighted portion represents the line of best fit as GPA = −0.096 + 0.00517*SATM,

Regression Analysis: GPA versus SATM

```
The regression equation is
GPA = - 0.096 + 0.00517 SATM

Predictor      Coef     SE Coef        T       P
Constant     -0.0965     0.9088    -0.11    0.918
SATM        0.005168    0.001656     3.12    0.014

S = 0.542841   R-Sq = 54.9%    R-Sq(adj) = 49.3%

Analysis of Variance

Source             DF       SS       MS       F      P
Regression          1   2.8707   2.8707    9.74  0.014
Residual Error      8   2.3574   0.2947
Total               9   5.2281
```

FIGURE 5.10
MINITAB printout of a simple regression analysis for the data in Table 5.1.

($\hat{y} = -0.096 + 0.00517x$). Also notice that the other highlighted areas in Figure 5.10 provide the estimates of the intercept ($\hat{\beta}_0$) and slope ($\hat{\beta}_1$) parameters along with the root mean square error, $S = 0.54284$. As we begin to conduct further analyses, we will describe in more detail some of the other information that is given in a MINITAB printout for a regression analysis.

The line of best fit, $\hat{y} = \hat{\beta}_0 + \hat{\beta}_1 x$, is only an estimate of the true unknown population line, $y = \beta_0 + \beta_1 x + \varepsilon$, and we may want to use this estimated line of best fit along with the residuals that represent an estimate of the error component to find confidence intervals and perform hypothesis tests about the true but unknown population parameters, β_0 and β_1.

5.6 Regression Inference

Similar to when we calculated confidence intervals and performed hypothesis tests about an unknown population mean, we can also calculate confidence intervals and perform hypothesis tests about the unknown population parameters β_0 and β_1. Just as in using the sample mean to estimate the population mean, we will use the estimated (or sample) values $\hat{\beta}_0$ and $\hat{\beta}_1$ to estimate the true but unknown population parameters β_0 and β_1, and we will also use the residuals $\hat{\varepsilon}$ to estimate the true but unknown population error component ε.

In order to calculate confidence intervals and perform hypothesis tests on the population parameters, we first need to obtain an estimate of the standard error for each of the estimated parameters. Such standard errors for the estimates of β_0 and β_1 can be calculated as follows:

$$SE(\hat{\beta}_1) = \frac{S}{\sqrt{S_{xx}}}, \qquad SE(\hat{\beta}_0) = S\sqrt{\frac{1}{n} + \frac{\bar{x}^2}{S_{xx}}}$$

Where

$$S = \sqrt{\frac{SSE}{n-2}}$$

Example 5.3

We can use the above equations to calculate the standard error for each of the estimated regression parameters for the data from Table 5.3, as follows:

$$SE(\hat{\beta}_1) = \frac{0.54284}{\sqrt{107490}} = 0.00166$$

$$SE(\hat{\beta}_0) = (0.54284)\sqrt{\frac{1}{10} + \frac{(539)^2}{107490}} = 0.90879$$

Notice that the standard error for each of the estimated regression parameters appears on the MINITAB printout in Figure 5.10 as the standard error for the constant term (which is the intercept) and for the slope (SATM).

5.7   Inferences about the Population Regression Parameters

If all of our regression model assumptions hold true, then we can make inferences about the slope of the true but unknown population line by calculating confidence intervals and testing whether or not the population slope parameter β_1 is significantly different from some hypothesized value.

Similar to finding confidence intervals for the population mean, we can also find confidence intervals for the population parameter β_1, which can be estimated by using the sample statistic $\hat{\beta}_1$, which has the t-distribution with $n - 2$ degrees of freedom, as described below:

$$\left(\hat{\beta}_1 - t_{\frac{\alpha}{2}} \cdot \frac{S}{\sqrt{S_{xx}}}, \hat{\beta}_1 + t_{\frac{\alpha}{2}} \cdot \frac{S}{\sqrt{S_{xx}}} \right)$$

where $\hat{\beta}_1$ is the estimate of the true population slope parameter β_1, which falls between $-t_{\alpha/2}$ and $t_{\alpha/2}$ for a given value of α, S is the root mean square error, and S_{xx} is the sum of the squared differences between the individual x-values and the mean of x.

To test whether the true population slope parameter is significantly different from some hypothesized value, which is denoted as $\beta_{1(0)}$, we would use the following test statistic:

$$T = \frac{\hat{\beta}_1 - \beta_{1(0)}}{\dfrac{S}{\sqrt{S_{xx}}}}, \qquad df = n - 2$$

where $\hat{\beta}_1$ is the estimated value of the slope parameter, S is the root mean square error, and S_{xx} is the sum of the squared differences between the observed values of x and the mean of x.

Provided that the regression model assumptions hold true, the sampling distribution of the sample statistic $\hat{\beta}_1$ has approximately the t-distribution

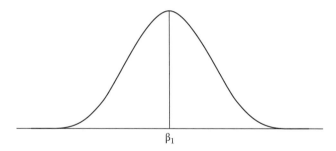

FIGURE 5.11
Sampling distribution of the sample statistic $\hat{\beta}_1$.

centered about the true population slope parameter β_1, with $n-2$ degrees of freedom, as illustrated in Figure 5.11.

To determine if the predictor variable x has a significant impact on the dependent variable y, we need to look at the line of best fit and determine whether or not we can infer that the slope of the true population line is significantly different from 0. In order to do this, we can test the null hypothesis against any one of the following three alternative hypotheses:

$$H_0 : \beta_1 = 0$$

$$H_A : \beta_1 \neq 0$$

$$H_0 : \beta_1 = 0$$

$$H_A : \beta_1 < 0$$

$$H_0 : \beta_1 = 0$$

$$H_A : \beta_1 > 0$$

To test the population slope parameter against the value of $\beta_{1(0)} = 0$ for any of the above hypotheses, the test statistic would be

$$T = \frac{\hat{\beta}_1 - 0}{\frac{S}{\sqrt{S_{xx}}}}, \qquad d.f. = n - 2$$

Example 5.4

Suppose we are interested in testing whether or not our true population slope parameter β_1 is *significantly different from 0* for the SAT–GPA data given in

Table 5.1 ($\alpha = 0.05$). The appropriate null and alternative hypotheses would be as follows:

$$H_0 : \beta_1 = 0$$

$$H_A : \beta_1 \neq 0$$

Since we are testing whether the true population slope parameter is significantly different from 0, we would use a two-sided test because for the population slope parameter to be different from 0 implies that it can be either less than 0 or greater than 0. Thus, we would use a two-sided test because of the specification of the alternative hypothesis.

Then for $10 - 2 = 8$ degrees of freedom, if the test statistic falls in the rejection region as defined by $t = 2.306$ and $-t = -2.306$, we can reject the null hypothesis and claim that our true but unknown population parameter β_1 is significantly different from 0, as illustrated in Figure 5.12.

Then based on our sample data we have the following information, which can be used to calculate the test statistic:

$$\hat{\beta}_1 = 0.00517 \quad \text{(parameter estimate)}$$

$$S^2 = \frac{SSE}{n-2} = \frac{2.357412}{10-2} = 0.294677, \quad S = \sqrt{0.294677} = 0.54284$$

$$S_{xx} = 107490.0000$$

$$SE(\hat{\beta}_1) = \frac{S}{\sqrt{S_{xx}}} = \frac{0.54284}{\sqrt{107490.0000}} = 0.001656 \quad \text{(standard error)}$$

$$T = \frac{0.00517}{0.001656} = 3.12$$

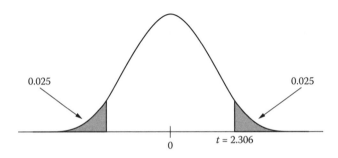

FIGURE 5.12
Rejection region for testing if the true population slope parameter β_1 is significantly different from 0 for 8 degrees of freedom.

Since $T = 3.12 > 2.306$ and thus falls in the rejection region as described in Figure 5.12, we can reject the null hypothesis and accept the alternative hypothesis. Thus, we can infer that our true population slope parameter β_1 is significantly different from 0.

To interpret this finding within the context of our example, we claim to have strong evidence to suggest that higher SAT math examination scores from high school are associated with higher first-year grade point averages in college. More specifically, we can state that for a 100-point increase in the score received on the mathematics portion of the SAT examination, the first-year GPA will increase by approximately 0.517 points. Notice that we did not claim that higher SAT math examination scores *cause* higher first-year grade point averages in college; we only infer that higher SAT math scores are associated with higher first-year grade point averages, and we can provide an estimate of the magnitude of such an effect.

We can also calculate a 95% confidence interval for the true population slope parameter β_1 with $n - 2$ degrees of freedom, as follows:

$$\left(\hat{\beta}_1 - t_{\frac{\alpha}{2}} \cdot \frac{S}{\sqrt{S_{xx}}} , \hat{\beta}_1 + t_{\frac{\alpha}{2}} \cdot \frac{S}{\sqrt{S_{xx}}} \right)$$

$$= \left(0.0052 - (2.306) \cdot \frac{0.54284}{\sqrt{107490}} , 0.0052 + (2.306) \cdot \frac{0.54284}{\sqrt{107490}} \right)$$

$$= (0.0014, 0.0090)$$

Thus, if we take repeated samples of size 10 and calculate confidence intervals in a similar manner, then 95% of these intervals will contain the true but unknown population parameter β_1. Another way to interpret this interval would be to say that we are 95% confident that for a 100-point increase in the SAT math examination score, the true but unknown population first-year grade point average will increase from between 0.14 and 0.90 points.

5.8 Using MINITAB to Test the Population Slope Parameter

The MINITAB printout obtained in Figure 5.10 contains information regarding the hypothesis test that we just used to determine whether the true but unknown population slope parameter β_1 is significantly different from 0. Notice the section that is highlighted on the MINITAB printout given in Figure 5.13.

Note that the calculations for the standard error for β_1 and the value of the test statistic T are what we obtained when we did the calculations by hand. Also notice the p-value of .014. Recall that the p-value gives the observed level of significance that is based on the test statistic, which in this case is less than our predetermined level of significance of $\alpha = 0.05$. Therefore, based on the

Regression Analysis: GPA versus SATM

```
The regression equation is
GPA = - 0.096 + 0.00517 SATM

Predictor        Coef     SE Coef         T        P
Constant       -0.0965      0.9088     -0.11    0.918
SATM          0.005168     0.001656      3.12    0.014


S = 0.542841     R-Sq = 54.9%     R-Sq(adj) = 49.3%


Analysis of Variance

Source             DF        SS         MS       F        P
Regression          1    2.8707     2.8707    9.74    0.014
Residual Error      8    2.3574     0.2947
Total               9    5.2281
```

FIGURE 5.13
MINITAB printout that highlights the estimate of the true population regression slope parameter for the data given in Table 5.1.

p-value, we can reject the null hypothesis and accept the alternative hypothesis, which suggests that the true population slope parameter β_1 is significantly different from 0. Of course, this inference relies on the linear regression model assumptions having been reasonably met. Chapter 6 will describe in more detail some different techniques and methods that can be used to check the regression model assumptions.

If we do not have enough evidence to reject the null hypothesis and accept the alternative hypothesis that the true population slope parameter is not significantly different and from 0 (in other words, if our test statistic does not fall in the rejection region and that happens when the *p*-value is greater than the predetermined level of significance), then we may infer that *y* is not related to *x*. This can be a misleading inference for two reasons. First, when determining the estimated model equation, we are only using a sample of *x*- and *y*-values. It could be the case that the given sample is not large enough for *x* to impact *y*. Second, we are assuming that all of the underlying model assumptions for the regression analysis hold true, and this includes the assumption that the true but unknown population model can be represented using a linear equation. However, if the true underlying population model is not reasonably represented by a linear equation, but yet the estimated model equation is linear, then it could be that *x* does have an influence on *y*, but that the relationship between *x* and *y* is not necessarily linear.

Similar to making inferences about the true population slope parameter β_1, we can also make inferences about the *intercept* of the true population equation β_0. Inferences about the intercept of the population equation are also based on the sampling distribution of the sample statistic $\hat{\beta}_0$ having the *t*-distribution centered about the true population parameter β_0 with $n-2$ degrees of freedom by using the following test statistic:

$$T = \frac{\hat{\beta}_0 - \beta_{0(0)}}{S\sqrt{\dfrac{1}{n} + \dfrac{\overline{x}^2}{S_{xx}}}}, \qquad d.f. = n-2$$

where $\hat{\beta}_0$ is the estimate of the true population intercept parameter, $\beta_{0(0)}$ is the specified value being tested under the null hypothesis, n is the sample size, S is the root mean square error, and S_{xx} is the sum of the squared differences between the observed values of x and the mean of x.

Similar to obtaining confidence intervals for the slope of the true population line, we can also calculate confidence intervals for the value of the true population intercept value β_0.

A $100(1-\alpha)\%$ confidence interval for β_0 with $n-2$ degrees of freedom is

$$\left(\hat{\beta}_0 - t_{\frac{\alpha}{2}} \cdot S\sqrt{\frac{1}{n} + \frac{\overline{x}^2}{S_{xx}}}, \hat{\beta}_0 + t_{\frac{\alpha}{2}} \cdot S\sqrt{\frac{1}{n} + \frac{\overline{x}^2}{S_{xx}}} \right)$$

However, for many situations (such as with the last example), it may not make sense to calculate confidence intervals and perform hypothesis tests about the population intercept. This is because in many circumstances the x-values are not collected around 0. Recall that the y-intercept of a line is where the line crosses the y-axis, and this is where $x = 0$. One concern for our analysis is that a score of 0 is not possible for the SAT mathematics examination because the scores on this examination range from 200 to 800. Thus, for data that are not collected around the value of 0, conducting a hypothesis test or finding a confidence interval about the population intercept parameter β_0 may not be of interest.

5.9 Confidence Intervals for the Mean Response for a Specific Value of the Predictor Variable

We can use the line of best fit to estimate the mean, or average, response value that corresponds to a specific value of the predictor variable (in other words, a specific x-value). For example, suppose that we want to obtain an estimate of the mean first-year grade point average for all of the students who have achieved a given score on the mathematics portion of the SAT examination. Using the line of best fit, the response at a specific value of the predictor variable, x_s, can be estimated as follows:

$$\hat{y}_s = \hat{\beta}_0 + \hat{\beta}_1 x_s$$

Then, a $100(1-\alpha)\%$ confidence interval for the population mean response for a specific value of x (x_s) is given by

$$(\hat{\beta}_0 + \hat{\beta}_1 x_s) \pm t_{\frac{\alpha}{2}} \cdot S \cdot \sqrt{\frac{1}{n} + \frac{(x_s - \bar{x})^2}{S_{xx}}}, \text{ with } n-2 \text{ degrees of freedom}$$

Example 5.5

Suppose we want to calculate a 95% confidence interval for the true but unknown population mean first-year grade point average for all students who score 550 on the mathematics portion of the SAT examination. The estimated regression line is $\hat{y} = -0.096 + 0.00517x$, so the first-year grade point average for the specific value of $x_s = 550$ would be calculated as follows:

$$\hat{y}_s = -0.096 + 0.00517(550) = 2.748$$

Then the estimated standard error for the confidence interval would be calculated as follows:

$$S\sqrt{\frac{1}{n} + \frac{(x_s - \bar{x})^2}{S_{xx}}} = 0.54284 \cdot \sqrt{\frac{1}{10} + \frac{(550 - 539)^2}{107490}} \approx 0.1726$$

So a 95% confidence interval for the mean first-year grade point average for all students who score 550 on the mathematics portion of the SAT exam would be:

$$= 2.748 \pm (2.306)(0.1726)$$

$$= 2.748 \pm 0.3980$$

$$\Rightarrow \quad (2.35, 3.15)$$

Therefore, if we take repeated samples of size $n = 10$ from the true but unknown population and calculate confidence intervals in a similar manner, then 95% of these intervals will contain the true but unknown population mean response value. Thus, we are 95% confident that for all students who score 550 on the SAT math examination, the population mean first-year grade point average will fall between 2.35 and 3.15 points.

5.10 Prediction Intervals for a Response for a Specific Value of the Predictor Variable

We just saw that we can obtain confidence intervals that can be used for estimating the population mean first-year grade point average for all students who scored 550 on the mathematics portion of the SAT examination.

However, there may be instances when we want to determine what the predicted response would be for a single specified value of the predictor variable. For instance, we may be interested in determining what the predicted first-year grade point average would be for a single student who has achieved a specific score on the mathematics portion of the SAT examination.

A prediction interval is very similar to a confidence interval for a specific value of the predictor, except the standard error of the prediction interval will be larger than a confidence interval because using a single observation to predict a response generates more uncertainty than including all of the observations at a specified value.

The calculations needed to create a prediction interval are similar to those for finding a confidence interval to estimate the mean response for a specific predictor value (x_s), except the estimated standard error is calculated as follows:

$$S \cdot \sqrt{1 + \frac{1}{n} + \frac{(x_s - \bar{x})^2}{S_{xx}}}$$

Therefore, a $100(1-\alpha)\%$ prediction interval for predicting a single response for a specified value of x (x_s) is given by

$$(\hat{\beta}_0 + \hat{\beta}_1 \cdot x_s) \pm t_{\frac{\alpha}{2}} \cdot S \cdot \sqrt{1 + \frac{1}{n} + \frac{(x_s - \bar{x})^2}{S_{xx}}}, \text{ with } n - 2 \text{ degrees of freedom}$$

Example 5.6

Suppose we want to predict (with 95% confidence) what the first-year grade point average would be for an individual student who scores 550 on the SAT math examination.

The first-year grade point average would be

$$\hat{y} = -0.096 + 0.00517(550) = 2.748$$

Then the prediction interval for the first-year grade point average for a single student who scores a 550 on the SAT math examination would be

$$2.748 \pm (2.306)(0.54284)\sqrt{1 + \frac{1}{10} + \frac{(550 - 539)^2}{107490}} = 2.748 \pm 1.314 \implies (1.43, \ 4.06)$$

If we take repeated samples of size $n = 10$ from the true population and calculate prediction intervals in a similar manner, then 95% of these intervals will contain the true but unknown population mean. Therefore, we are 95% confident that *an individual* student who scores 550 on the SAT mathematics examination would achieve a first-year grade point average that falls between 1.43 and 4.06 points.

Notice that the prediction interval is much wider than the confidence interval for the same specified value of the predictor (in this case a SAT mathematics examination score of 550). Also, the upper limit of the prediction interval actually

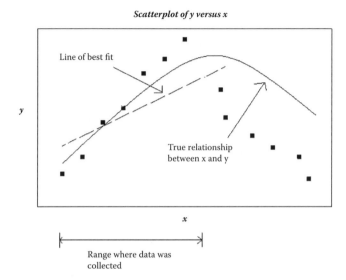

FIGURE 5.14
Graph of the true relationship being nonlinear outside the range of values used to obtain the line of best fit.

falls out of the range for the response variable (since GPA ranges from 0.00 to 4.00 points). This can happen when calculating prediction intervals because there is a large amount of uncertainty in estimating a single response using a single observation versus estimating the average, or mean, response by using the entire set of observations, as is the case with confidence intervals. Such a large standard error could generate a prediction interval where some values in the interval may fall outside of the range of the possible values for the given variable.

One thing to keep in mind when making inferences with confidence and prediction intervals is that it is not usually a good idea to make inferences when the specified value of interest, x_s, is too far removed from the range of the sample data that was used to obtain the line of best fit. Not only could using a value out of range cause the confidence and prediction intervals to become very wide, but the relationship may not be linear at more extreme values. The graph in Figure 5.14 shows a model that has a different functional form outside the range where the data was collected.

5.11 Using MINITAB to Find Confidence and Prediction Intervals

To use MINITAB to create confidence and prediction intervals for the data given in Table 5.1, use the **Regression** tab under the **Stat** menu and select **Fitted Line Plot**, as illustrated in Figure 5.15.

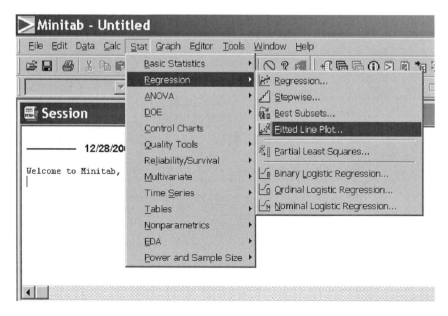

FIGURE 5.15
MINITAB commands to draw the fitted line plot.

Then provide the appropriate response and predictor variables, as illustrated in Figure 5.16.

Selecting the **Options** tab on the **Fitted Line Plot** dialog box and checking the boxes to display confidence and prediction intervals, as shown in Figure 5.17, gives the fitted line plot, including the confidence and prediction intervals illustrated in Figure 5.18.

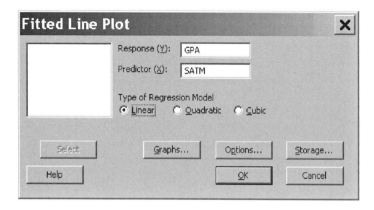

FIGURE 5.16
Fitted line plot dialog box.

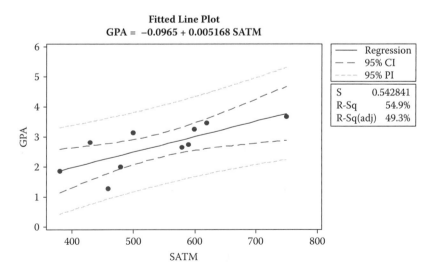

FIGURE 5.17
Options box to display confidence and prediction intervals on the fitted line plot.

The graph in Figure 5.18 provides the fitted line plot along with confidence and prediction intervals for the entire range of *x*-values that were used in estimating the line of best fit.

We can also find an estimate of the prediction and confidence intervals for any given *x*-value by using the **Crosshairs** command, which is found

Fitted Line Plot
GPA = −0.0965 + 0.005168 SATM

——	Regression
− −	95% CI
- - - -	95% PI

S	0.542841
R-Sq	54.9%
R-Sq(adj)	49.3%

FIGURE 5.18
Fitted line plot that illustrates confidence and prediction intervals.

by right-clicking on the graph and selecting **Crosshairs**, as shown in Figure 5.19.

Then by using the mouse to move the crosshair lines around the fitted line plot, you can line up the desired *x*-value to get an estimate of where the crosshairs intersect with the confidence or prediction interval. Figure 5.20 illustrates that the box at the top-left corner of the graph gives an estimate of the values of the *x*- and *y*-coordinates. Note that the value obtained for the confidence or prediction interval will not be exactly the same as if it were calculated by hand since using the crosshairs only gives values that are approximate.

Exact confidence and prediction intervals can also be found by selecting the **Options** tab in the **Regression** dialog box and entering the specific *x*-value of interest and the desired confidence level, as illustrated in Figure 5.21.

This gives the exact confidence and prediction intervals in addition to the information obtained in a MINITAB regression printout, as illustrated in

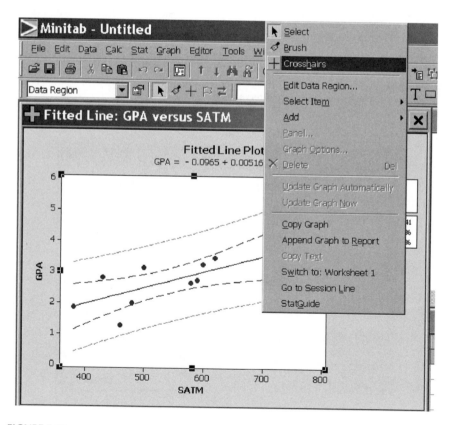

FIGURE 5.19
MINITAB commands to use crosshairs to estimate the confidence and prediction intervals for a specified *x*-value.

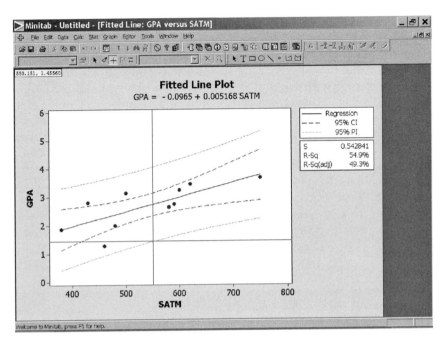

FIGURE 5.20
Crosshairs to estimate the prediction interval at $x_s = 550$.

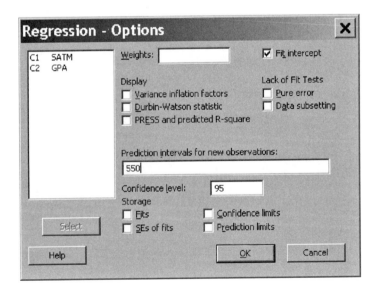

FIGURE 5.21
Regression options tab to find exact confidence and prediction intervals for a specified x-value of 550.

Figure 5.22. The confidence and prediction intervals for a score of 550 on the SAT mathematics examination are highlighted in Figure 5.22.

Example 5.7

Does spending more time studying result in higher exam grades? To answer this question, a survey was given to a random sample of thirty-two students participating in an introductory statistics course, which asked them how many

Regression Analysis: GPA versus SATM

```
The regression equation is
GPA = - 0.096 + 0.00517 SATM

Predictor       Coef    SE Coef      T        P
Constant     -0.0965     0.9088   -0.11    0.918
SATM        0.005168    0.001656    3.12    0.014

S = 0.542841    R-Sq = 54.9%    R-Sq(adj) = 49.3%

Analysis of Variance

Source           DF       SS       MS       F       P
Regression        1    2.8707   2.8707    9.74    0.014
Residual Error    8    2.3574   0.2947
Total             9    5.2281

Unusual Observations

Obs    SATM     GPA     Fit   SE Fit   Residual   St Resid
2       460   1.280   2.281    0.216     -1.001     -2.01R

R denotes an observation with a large standardized residual.

Predicted Values for New Observations

New
Obs    Fit   SE Fit      95%  CI          95%  PI
1    2.746    0.173   (2.348, 3.144)   (1.432,  4.059)

Values of Predictors for New Observations

New
Obs    SATM
1       550
```

FIGURE 5.22
MINITAB printout with exact 95% confidence and prediction intervals for a specified SAT math score of 550.

hours they spent studying for a statistics examination during the week before they took the exam. The data set "Time Studying," provided in Table 5.4, gives the number of hours the students reported they spent studying for the examination along with the score they received on the examination.

Because we want to know if more time spent studying is related to higher examination grades, we would let the predictor (or x) variable be the number

TABLE 5.4

Number of Hours Studying and Examination Score for a Random Sample of Thirty-Two Students

Observation	Hours Studying	Score Received on Exam
1	3	62
2	5	55
3	10	75
4	5	67
5	18	84
6	20	89
7	21	91
8	17	82
9	9	77
10	10	73
11	18	86
12	9	69
13	7	67
14	15	72
15	27	91
16	9	50
17	10	70
18	16	77
19	21	84
20	8	74
21	15	76
22	9	62
23	10	73
24	16	82
25	5	78
26	17	85
27	4	62
28	8	68
29	2	64
30	9	72
31	20	97
32	17	91

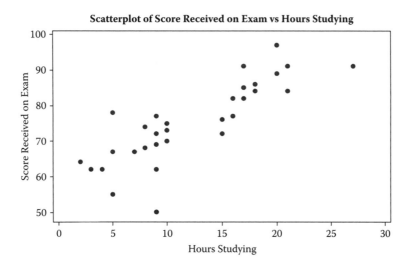

FIGURE 5.23
Scatter plot showing the relationship between the number of hours studying for an examination and the score received on the examination.

of hours studying and the response (or y) variable be the grade received on the examination.

We can first begin by drawing a scatter plot to see what the relationship between x and y looks like. This scatter plot is given in Figure 5.23.

The scatter plot illustrates that there appears to be a positive trend between the number of hours studying and the score received on the examination. Also, it appears that this relationship is one that could be reasonably approximated by using a straight line.

For a significance level of .01, we can use MINITAB to run the regression analysis for us to generate the printout given in Figure 5.24.

From the MINITAB printout, we can see that the estimated line of best fit is given by the following equation:

$$\hat{y} \approx 57.2 + 1.47x$$

This equation can be used to estimate the true population equation:

$$y = \beta_0 + \beta_1 x + \varepsilon$$

Thus, we can use $\hat{\beta}_0 = 57.207$ as an estimate of β_0 and $\hat{\beta}_1 = 1.4728$ as an estimate of β_1.

Regression Analysis: Score Received on Exam versus Hours Studying

```
The regression equation is
Score Received on Exam = 57.2 + 1.47 Hours Studying

Predictor              Coef  SE Coef       T        P
Constant             57.207    2.490   22.98    0.000
Hours Studying       1.4728   0.1822    8.08    0.000


S = 6.36683     R-Sq = 68.5%    R-Sq(adj) = 67.5%


Analysis of Variance

Source               DF        SS       MS       F        P
Regression            1    2648.1   2648.1   65.33    0.000
Residual Error       30    1216.1     40.5
Total                31    3864.2


Unusual Observations

                        Score
             Hours   Received
Obs       Studying    on Exam     Fit  SE Fit  Residual   St Resid
15            27.0      91.00   96.97    2.92     -5.97     -1.06 X
16             9.0      50.00   70.46    1.27    -20.46     -3.28R
25             5.0      78.00   64.57    1.73     13.43      2.19R


R denotes an observation with a large standardized residual.
X denotes an observation whose X value gives it large leverage.
```

FIGURE 5.24
MINITAB printout of the regression analysis for the data in Table 5.4.

We want to know if the population parameter β_1 is significantly different from 0. By looking at the MINITAB printout in Figure 5.24, we can see that the p-value for β_1 is less than our predetermined level of significance of .01. Then we can accept the alternative and reject the null hypothesis, which suggests that our true population value of β_1 is significantly different from 0.

We can interpret this result within the context of our problem as follows: For every additional 10 hours spent studying during the week prior to the examination, the grade on the examination will increase by approximately 14.7 points.

We can also calculate confidence and prediction intervals and use MINITAB to draw a fitted line plot with confidence and prediction intervals, as illustrated in Figure 5.25.

Suppose that we wanted to know the mean examination score for all students who study for 12 hours the week before the examination. Using MINITAB, we could get the value of the 99% confidence interval of (71.78, 77.98) for the true

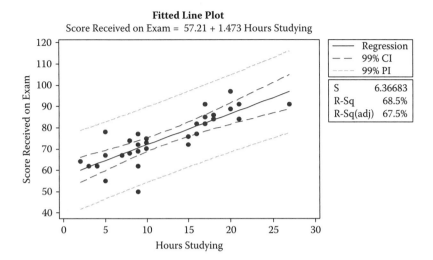

FIGURE 5.25
MINITAB fitted line plot with 99% confidence and 99% prediction intervals for the data in Table 5.4.

population mean examination score for all students who studied for 12 hours the week of the examination, as given in Figure 5.26.

Similarly, we could find the prediction interval for the examination score for a single student who studied for 12 hours during the week of the examination in the highlighted portion of the MINITAB printout in Figure 5.26. The prediction interval of (57.10, 92.66) suggests that we are 99% confident that an individual student who studies for 12 hours during the week of the examination will score between 57.10 and 92.66 on the examination.

Predicted Values for New Observations

```
New
Obs     Fit   SE Fit      99%  CI            99%  PI
  1   74.88    1.13   (71.78,  77.98)   (57.10,  92.66)

Values of Predictors for New Observations

New        Hours
Obs      Studying
  1         12.0
```

FIGURE 5.26
Confidence and prediction intervals for the specific value of the predictor of 12 hours.

Exercises

1. a. Using the data set in Table 5.1, which provides a set of ordered pairs of SAT math scores and first-year grade point averages, use the points (590, 2.75) and (480, 2.00) to calculate the line that connects these points, and calculate the sum of the residuals and the sum of the squares of the residuals for the line that connects these two points.

   b. How do these calculations compare to those that were found using the line of best fit?

2. Using the data from Table 5.1, verify that $S_{xx} = 107490.000$, $S_{yy} = 5.228$, and $S_{xy} = 555.490$.

3. Suppose we collected a set of ordered pairs (x, y) that generated the following set of summary statistics:

   $n = 15$, $\bar{x} = 50.33$, $\bar{y} = 150.33$

   $S_{xx} = 617.33$, $S_{xy} = 2624.33$, $S_{yy} = 16489.30$

   Using the above summary statistics, find the equation for the line of best fit. Find estimates of the regression parameters β_0 and β_1.

4. For the "Simple Linear" data set in Table 5.5, find S_{xx}, S_{yy}, and S_{xy}, and using these values, find the equation of the line of best fit.

5. Using MINITAB, create a fitted line plot of the data from the "Simple Linear" data set and identify the line of best fit.

6. Find a 98% confidence interval for the population slope parameter using the "Simple Linear" data set.

7. a. Using the statistics from Exercise 3, set up the appropriate null and alternative hypotheses test to determine if β_1 is significantly different from 0.

   b. Calculate by hand the value of the test statistic that would be used to determine if β_1 is significantly different from 0.

TABLE 5.5

Simple Linear Data Set

x	y
18.2	9.3364
29.6	12.1775
27.2	11.7275
21.8	9.5038
18.7	8.3875
16.8	8.43
25	9.8068
23.9	9.8372
15.6	8.3013
21.1	9.9421
13.2	7.0113
14.6	7.0859
20	8.6675
22.2	8.4263
30	10.5788
12.8	7.3939
23.4	11.5568
12.1	6.7498
19.2	8.61
15.6	7.635
18.2	8.315
17.4	8.385
16.5	7.83
13.1	7.415
29	11.075

c. Using ($\alpha = 0.05$), determine if the population parameter β_1 is significantly different from 0.

8. a. For the "Simple Linear" data set, calculate a 99% confidence interval to estimate the mean response for a predictor value of 20.

   b. Using the "Simple Linear" data set, calculate a 99% prediction interval to predict a single response for a predictor value of 20.

9. Does the unemployment rate have an impact on the crime rate?

   a. To investigate the answer to this question, run a simple regression analysis using the data set "Crime by County" and determine if a higher percentage of unemployed for the year 1991 is related to the crime rate for 1991 ($\alpha = 0.05$).

   b. Based on the analysis from part (a), do you believe that the percentage of unemployed people has a significant impact on the crime rate? Justify your answer.

   c. Do you think that a simple linear regression model is appropriate to answer this question? Why or why not?

10. a. Using the "Crime by County" data set, find a 95% confidence interval to estimate the mean crime rate if 10% of people are unemployed.

    b. Using the "Crime Rate by County" data set, find a 95% prediction interval to estimate the crime rate in a city that has 10% of people unemployed.

11. Do cars that have low annual fuel costs emit less greenhouse gases?

    a. Using the "Automobile Data" data set and MINITAB, find the equation of the line of best fit and determine if there is a relationship between the annual fuel cost and the amount of greenhouse gas emitted ($\alpha = 0.05$).

    b. Find a 95% confidence interval for the true population parameter β_1.

12. a. Using MINITAB, create a graph that consists of 99% confidence and prediction intervals for the relationship between the annual fuel cost and the amount of greenhouse gas emitted using the "Automobile" data set.

    b. Using this graph and the **Crosshairs** command in MINITAB, determine a 99% prediction interval for the amount of greenhouse gas emitted for an individual vehicle that has fuel costs of $1,000 per year.

    c. Find the exact confidence and prediction intervals for a fuel cost of $1,000 per year.

13. Do younger people incur more credit card debt?

   a. Using the data set "Credit Card Debt," run a regression analysis to determine if younger people tend to incur more credit card debt ($\alpha = 0.05$).

   b. Write the equation of the line of best fit.

   c. Draw a scatter plot with the line of best fit.

   d. Can you claim that younger people incur more debt? Explain your findings within the context of the problem.

14. Do people with more education incur less credit card debt?

   a. To answer this question, use the data set "Credit Card Debt" to run a regression analysis to determine if people with more education tend to incur less credit card debt ($\alpha = 0.05$).

   b. Write the equation of the line of best fit.

   c. Draw a scatter plot with the line of best fit.

   d. Can you claim that people with more education incur less debt? Explain your findings within the context of the problem.

15. Is the asking price of a home influenced by the age of the home?

   a. Using the "Expensive Homes" data set, run a regression analysis to determine if the asking price of a home is influenced by how old the home is ($\alpha = 0.01$).

   b. Write the equation of the line of best fit.

   c. Draw a scatter plot with the line of best fit.

   d. Can you claim that the asking price of a home is influenced by the age of the home? Explain your findings within the context of the problem.

6

More on Simple Linear Regression

6.1 Introduction

The focus of this chapter is to describe some different measures that can be used to assess model fit for a simple linear regression analysis. We will introduce measures to assess how much variability in the response variable is due to the given model by using the coefficient of determination, and the coefficient of correlation will also be discussed as a way to measure the linear association between two variables. Finally, exploratory graphs and formal tests will be introduced that can be used to verify the regression model assumptions that were presented in the last chapter, as well as how to assess the impact of outliers.

6.2 Coefficient of Determination

One simple strategy that can be used to predict the population mean value of y for a simple linear regression analysis is to use the sample mean $\bar{y}$. In using the "SAT–GPA" data set provided in Table 5.1, the value of $\bar{y}$ is the mean first-year grade point average for the given sample of ten students. The graph in Figure 6.1 illustrates the line that represents the sample mean of the y-values, $\bar{y} = 2.689$.

If we were to use only the sample mean $\bar{y}$ to predict y, then we could represent the sum of the squared differences between $\bar{y}$ and the observed y-value for each observation as follows:

$$S_{yy} = SST = \sum (y_i - \bar{y})^2$$

where $S_{yy} = SST$ represents the sum of the squares of the difference between each y-value and the mean of the y-values. However, by using only $\bar{y}$ to predict y, this clearly would not capture any impact that the variable x may have on y.

One way to determine whether the predictor variable x has an impact on the response variable y is to describe how well the regression model predicts y. This can be done by comparing how well the model that includes x predicts y to how well using $\bar{y}$ predicts y. The measure that does this is

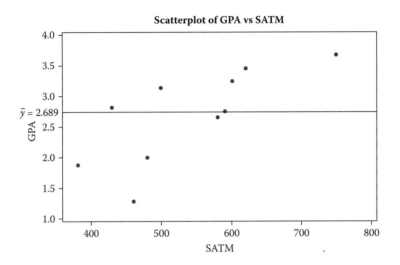

FIGURE 6.1
Scatter plot of GPA versus SATM with the line that represents the mean GPA of $\bar{y} = 2.689$.

called the *coefficient of determination*, which is denoted as R^2. The coefficient of determination is used to measure how well the regression model predicts y compared to using only the sample mean $\bar{y}$ to predict y. The value of R^2 is found by using the following formula:

$$R^2 = \frac{SST - SSE}{SST}$$

where $SST = \Sigma(y_i - \bar{y})^2$ is the sum of the squared differences between each observation and the mean $\bar{y}$, and $SSE = \Sigma\hat{e}_i^2 = \Sigma(y_i - \hat{y}_i)^2$ is the sum of the squared differences between each observation and the estimated (or fitted) y-value.

If the variable x does not contribute any information about y, then SSE (error sum of squares) will be equal to SST (total sum of squares) and the value of R^2 would be equal to 0. Otherwise, if x perfectly predicts y, then the observed y-value will equal the fitted y-value and SSE will equal 0, so the value of R^2 would be equal to 1. Therefore, the R^2 statistic can be used to measure the proportion of the variability in y that is explained by the given regression model. Notice that $0 \le R^2 \le 1$.

Example 6.1

For the data given in Table 5.1 that describe the score on the mathematics portion of the SAT examination and first-year grade point averages in college for a sample of ten students, to find the coefficient of determination R^2, since $SST = 5.228$ and $SSE = 2.357412$, then:

$$R^2 = \frac{SST - SSE}{SST} = \frac{5.228 - 2.357412}{5.228} \approx 0.549$$

Thus, the coefficient of determination $R^2 = 54.9\%$ suggests that 54.9% of the variability in first-year GPA (y) can be explained by using the score received on the mathematics portion of the SAT examination (x). The closer the value of R^2 is to 100%, the more useful the model is in predicting y.

6.3 Using MINITAB to Find the Coefficient of Determination

Every time you run a regression analysis, MINITAB will automatically provide the value of R^2. The highlighted portion of the MINITAB printout in Figure 6.2 shows the value of the coefficient of determination, R^2, for the regression analysis as well as $SSE = \Sigma(y_i - \hat{y}_i)^2$, $SST = \Sigma(y_i - \bar{y})^2$, and $SSR = SST - SSE$ using the data in Table 5.1.

Regression Analysis: GPA versus SATM

```
The regression equation is
GPA = - 0.096 + 0.00517 SATM

Predictor      Coef      SE Coef        T       P
Constant    -0.0965       0.9088    -0.11   0.918
SATM       0.005168      0.001656     3.12   0.014

S = 0.542841   R-Sq = 54.9%      R-Sq(adj) = 49.3%

Analysis of Variance

Source            DF      SS       MS       F       P
Regression         1    2.8707   2.8707    9.74   0.014
Residual Error     8    2.3574   0.2947
Total              9    5.2281

Unusual Observations

Obs   SATM     GPA     Fit  SE Fit   Residual   St Resid
2      460   1.280   2.281   0.216     -1.001      -2.01R

R denotes an observation with a large standardized residual.
```

FIGURE 6.2
Printout of the regression analysis for the data in Table 5.1 with the value of R^2 highlighted.

Because R^2 measures how well the regression model predicts y, it can also be calculated by taking the sum of squares due to the regression analysis, which is found by taking the difference between the total sum of squares and the error sum of squares ($SSR = SST - SSE$), and then dividing this difference by the total sum of squares (SST), as follows:

$$R^2 = \frac{SST - SSE}{SST} = \frac{SSR}{SST}$$

Example 6.2

For the regression analysis of the data used to model the relationship between the score received on the SAT mathematics examination and first-year grade point average, as given in Table 5.1, the "Analysis of Variance" table presented in Figure 6.2 shows that the total sum of squares is $SST = 5.2281$, and the error sum of squares is $SSE = 2.3574$. Using these values, calculate R^2 is as follows:

$$R^2 = \frac{SSR}{SST} = \frac{2.8707}{5.2281} = 0.549$$

6.4   Sample Coefficient of Correlation

The sample coefficient of correlation, denoted as r, can be used as a measure to describe a linear trend between two sample variables x and y. For a simple linear regression analysis, the value of r can be found by taking the square root of the coefficient of determination (R^2) and attaching to it the sign that indicates whether the slope of the regression line is positive or negative.

The sign of r, which can be obtained by looking at the slope of the line of best fit, describes the type of linear relationship between x and y. For instance, when r is negative, the x and y values are negatively correlated with each other. This suggests that an increase in x implies a decrease in y, as is illustrated in Figure 6.3.

If $r = 0$, this suggests that there is no linear association between the variables x and y, and thus a change in x does not bring about any change in y, as can be seen in Figure 6.4.

If r is positive, then there is a positive relationship between x and y. In other words, a positive increase in x brings about a positive increase in y, as is shown in Figure 6.5.

In addition to taking the square root of the coefficient of determination for a simple linear regression analysis, other formulas can be used to compute the value of r for two variables x and y, as follows:

$$r = \frac{S_{xy}}{\sqrt{S_{xx}} \cdot \sqrt{S_{yy}}}$$

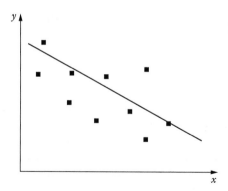

FIGURE 6.3
Negative relationship between x and y ($-1 \leq r < 0$).

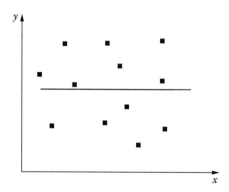

FIGURE 6.4
No relationship between x and y ($r = 0$).

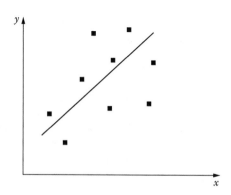

FIGURE 6.5
A positive relationship between x and y ($0 \leq r \leq 1$).

The coefficient of correlation, r, is related to the value of $\hat{\beta}_1$ and can be computed as follows:

$$r = \frac{s_x}{s_y} \hat{\beta}_1$$

where s_x is the sample standard deviation for the x-values, and s_y is the sample standard deviation for the y-values. The value of r does not have any units associated with it, and its value is restricted between -1 and $+1$ ($-1 \leq r \leq +1$).

Although two variables x and y can be highly correlated with each other, it is important to note that a strong correlation does not necessarily imply that a cause-and-effect relationship exists between x and y. Two variables could be highly correlated with each other because they are both associated with other unobserved variables. For instance, if the score on the mathematics portion of the SAT examination and first-year grade point average are both highly correlated with another variable, such as motivation, then a strong correlation between score received on the mathematics portion of the SAT exam and first-year GPA does not necessarily imply that a high SAT math score *causes* a high first-year grade point average. A strong correlation between the SAT math score and first-year GPA could also be due to the relationship that both the SAT math score and first-year grade point average have with other variables, such as motivation.

Also notice that r is not equal to the slope of the regression line. Although it is the case that positive values of r indicate that the slope of the simple linear regression line is positive (an increase in x implies an increase in y) and negative values of r indicate that the slope of the regression line is negative (an increase in x implies a decrease in y), the value of r is not the same as the estimate of the slope of the regression line ($\hat{\beta}_1$).

Example 6.3

We can use the data in Table 5.1 to calculate the value of the sample coefficient of correlation. To find the value of r we can take the square root of the value of the coefficient of determination (R^2) as follows:

$$r = \sqrt{0.549} = 0.741$$

Because the slope of the estimated line of best fit is positive, $r = +0.741$.

We can also find the value of the sample coefficient of correlation by using the formula that uses the values of S_{xx}, S_{yy}, and S_{xy}, as follows:

$$S_{xx} = 107490.00$$

$$S_{yy} = 5.228$$

$$S_{xy} = 555.290$$

Then:

$$r = \frac{S_{xy}}{\sqrt{S_{xx}} \cdot \sqrt{S_{yy}}} = \frac{555.290}{\sqrt{107490.000} \cdot \sqrt{5.228}} = \frac{555.290}{749.638} \approx 0.741$$

Recall from our regression analysis for the ten observations consisting of SAT math score and first-year grade point average that we found the line of best fit was

$$\hat{y} = -0.096 + 0.00517x$$

where the estimate of the slope parameter is $\hat{\beta}_1 = 0.00517$. Notice that the sample coefficient of correlation is not the same as the estimated slope parameter.

A strong correlation can only suggest whether a linear trend exists between two variables. Furthermore, a low value of r does not necessarily imply that there is no relationship between the two variables; it only suggests that no *linear* relationship exists. To illustrate what this means, consider the two scatter plots presented in Figures 6.6 and 6.7. Figure 6.6 illustrates a strong linear correlation between x and y. However, Figure 6.7 illustrates a weak correlation, but yet there appears to be a relationship between x and y, although not a linear relationship.

Notice in the Figure 6.7 that a weak correlation does not necessarily suggest that no relationship exists between x and y, but only that no linear relationship exists between x and y.

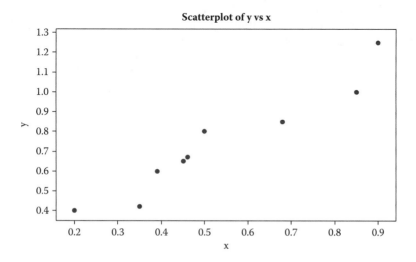

FIGURE 6.6

Scatter plot showing a strong linear trend of the relationship between x and y ($r \approx 0.96$).

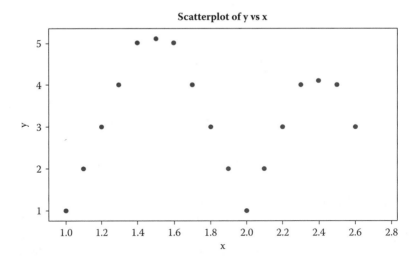

FIGURE 6.7
Scatter plot showing a weak coefficient of correlation but an apparent nonlinear relationship between x and y ($r \approx 0.09$).

6.5 Correlation Inference

Just like we used the estimated regression equation $\hat{y} = \hat{\beta}_0 + \hat{\beta}_1 x$ to estimate the true population equation $y = \beta_0 + \beta_1 x + \varepsilon$, and we used the sample mean $\bar{x}$ as an estimate of the true population mean μ, we can perform hypothesis tests for the true but unknown *population coefficient of correlation*. In other words, we can use the value r, the sample correlation coefficient, to estimate the true but unknown population coefficient of correlation ρ (pronounced "rho").

We can also test whether the true population coefficient of correlation is positive, negative, or different from 0 by running a hypothesis test using one of the three different null and alternative hypotheses, as follows:

$$H_0 : \rho = 0$$

$$H_A : \rho > 0$$

$$H_0 : \rho = 0$$

$$H_A : \rho < 0$$

$$H_0 : \rho = 0$$

$$H_A : \rho \neq 0$$

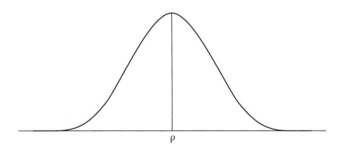

FIGURE 6.8
Sampling distribution of the sample coefficient of correlation *r*.

If we can show that $\rho > 0$ (for some specified level of significance), then we can say that the population variables x and y are *positively linearly correlated* with each other. Similarly, if $\rho < 0$, the population variables are *negatively linearly correlated* with each other. And if $\rho = 0$, the population variables are *uncorrelated* with each other, and thus no linear relationship exists between x and y.

Like many sampling distributions we have seen thus far, the sampling distribution of the sample coefficient of correlation has the *t*-distribution with $n - 2$ degrees of freedom, and it is centered about the true but unknown population coefficient of correlation ρ, as can be seen in Figure 6.8.

The following test statistic is used to test the true but unknown population coefficient of correlation (ρ):

$$T = \frac{r}{\sqrt{\dfrac{1 - r^2}{n - 2}}} \qquad df = n - 2$$

where r is the sample coefficient of correlation and n is the sample size.

Example 6.4

Suppose that we want to test whether the correlation between SATM and GPA from Table 5.1 is significantly different from 0 ($\alpha = 0.05$). We would set up the following null and alternative hypotheses:

$$H_0 : \rho = 0$$

$$H_A : \rho \neq 0$$

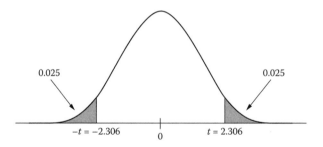

FIGURE 6.9
Rejection region for testing the population coefficient of correlation for $\alpha = 0.05$ with 8 degrees of freedom for a two-sided test.

Using the value of r obtained from the sample of size $n = 10$ gives the value of the test statistic, as follows:

$$T = \frac{r}{\sqrt{\dfrac{1-r^2}{n-2}}} = \frac{0.741}{\sqrt{\dfrac{1-(0.741)^2}{10-2}}} = 3.12$$

Comparing this value to the rejection region that is determined based on the direction of the alternative hypothesis and the level of significance, as illustrated in Figure 6.9, we see that we can reject the null hypothesis and accept the alternative hypothesis because the value of the test statistic T falls into the rejection region. Thus, the true population coefficient of correlation, ρ, is significantly different from 0, and we can infer that a linear trend exists between the score received on the SAT mathematics examination and first-year grade point average.

Since the two-sided test is significant, we could take this one step further and infer that the true but unknown population coefficient of correlation is significantly greater than 0 since the test statistic falls to the right of $t = 2.306$ (see Exercise 23 in Chapter 4).

However, notice that the population coefficient of correlation would not be significantly different from 0 if we used a level of significance of $\alpha = 0.01$. This is because the value of t that defines the rejection region would be defined by ± 3.355 and our test statistic would not fall into such a region.

Example 6.5

Suppose a sample of size $n = 18$ generates a sample coefficient of correlation of $r = .53$. We can test whether the true population coefficient of correlation, ρ, is significantly different from 0. The appropriate null and alternative hypotheses would be as follows:

$$H_0 : \rho = 0$$

$$H_A : \rho \neq 0$$

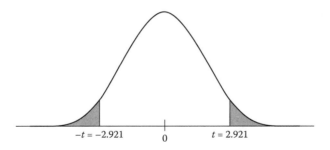

FIGURE 6.10
Rejection region for a two-tailed test of a population coefficient of correlation with a significance level of 0.01 and 16 degrees of freedom.

Given a level of significance of $\alpha = 0.01$, then:

$$T = \frac{0.53}{\sqrt{\dfrac{1-(0.53)^2}{18-2}}} \approx 2.50$$

for $18 - 2 = 16$ degrees of freedom.

We would then compare this value to $t = 2.921$, as illustrated in Figure 6.10.

Since the value of the test statistic T does not fall in the rejection region
defined by the alternative hypothesis, where $t = \pm 2.921$, we do not have
enough evidence to reject the null hypothesis and accept the alternative.
Therefore, we would infer that the true population coefficient of correlation
is not significantly different from 0.

Example 6.6

Suppose we want to test whether a population coefficient of correlation is
significantly greater than 0 ($\alpha = 0.05$). Given a sample of size $n = 8$ and a
sample coefficient of correlation of $r = .76$, the appropriate null and alternative hypotheses would be

$$H_0 : \rho = 0$$

$$H_A : \rho > 0$$

with a test statistic of

$$T = \frac{r}{\sqrt{\dfrac{1-r^2}{n-2}}} = \frac{0.76}{\sqrt{\dfrac{1-(0.76)^2}{8-2}}} = 2.86$$

and $n - 2 = 8 - 2 = 6$ degrees of freedom. We would then compare the value
of this test statistic to $t = 1.943$, as illustrated in Figure 6.11.

Since $T > t$, the test statistic falls in the rejection region, and thus we would
infer that the true population coefficient of correlation is significantly greater
than 0.

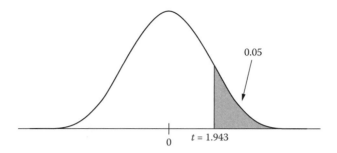

FIGURE 6.11
Rejection region for a one-tailed test of a population coefficient of correlation with a significance level of 0.05 and 6 degrees of freedom.

6.6   Using MINITAB for Correlation Analysis

To use MINITAB to find the sample coefficient of correlation for the data given in Table 5.1, simply enter the data in two separate columns and under the **Stat** tab select **Basic Statistics**, then **Correlation**, as illustrated in Figure 6.12. This gives the correlation dialog box that appears in Figure 6.13.

By selecting the variables for which you wish to find the correlation, simply highlight the variables and choose **Select**. Then, checking the box that says to **Display *p*-values** gives the MINITAB printout presented in Figure 6.14.

Notice that there are two values reported on the MINITAB printout, the value of the sample correlation $r = .741$ and a *p*-value. This *p*-value is the result of the hypothesis test of whether or not the true population coefficient of correlation is significantly different from 0; in other words,

$$H_0 : \rho = 0$$

$$H_A : \rho \neq 0$$

Since the *p*-value is .014, this suggests that the true population coefficient of correlation is significantly different from 0. Recall that the *p*-value gives the observed level of significance, and this value is less than $\alpha = 0.05$ but not less than $\alpha = 0.01$, as illustrated in Example 6.4.

6.7   Assessing Linear Regression Model Assumptions

Regression analysis involves so much more than simply entering data into a statistical program and performing a few hypothesis tests, constructing confidence intervals for parameters, and interpreting *p*-values. In order to make

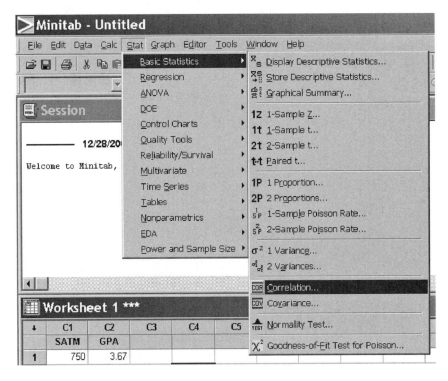

FIGURE 6.12
MINITAB commands for finding a sample coefficient of correlation.

meaningful inferences using the line of best fit, we need to be reasonably sure that the following four assumptions have been met:

1. The errors associated with any two observed values are independent of each other.
2. The error component is normally distributed.
3. The error component has constant variance.
4. The functional relationship between x and y is linear.

Exploratory data analysis and some formal tests can be used to assess whether or not these model assumptions have been reasonably met. One type of exploratory technique is to use *residual plots*, which are graphs of patterns of residual values. Using residual plots can give you some indication as to whether or not the model assumptions regarding the unknown error component may have been violated. For instance, an exploratory histogram of the residuals can be used to investigate whether the true but unknown error component is normally distributed. Formal tests of assumptions can also be

FIGURE 6.13
MINITAB dialog box for calculating the sample coefficient of correlation.

conducted if you are unsure of what you are seeing in the residual plots or if the sample size is too small to show any meaningful patterns.

Although the residual plots and formal tests of the model assumptions can be done by hand, it is much simpler to use MINITAB to generate the graphs of the residual plots and to do the formal test calculations.

6.8   Using MINITAB to Create Exploratory Plots of Residuals

We can use MINITAB to create a scatter plot, find the line of best fit, test the population parameters in our regression analysis, and check any regression assumptions.

Correlations: SATM, GPA

```
Pearson correlation of SATM and GPA = 0.741
P-Value = 0.014
```

FIGURE 6.14
MINITAB printout out of sample correlation coefficient and respective *p*-value.

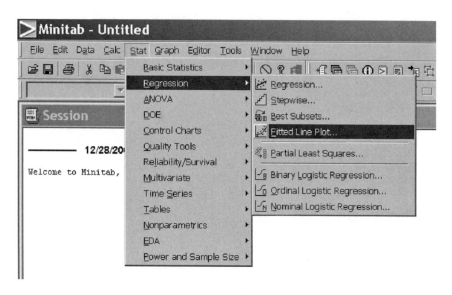

FIGURE 6.15
MINITAB commands to draw a fitted line plot.

Using the data given in Table 5.1 and MINITAB, the fitted line plot can be found by first going to the **Stat** tab; then select **Regression** and then **Fitted Line Plot**, as seen in Figure 6.15.

This generates the dialog box in Figure 6.16, which asks you to select the appropriate response and predictor variables along with the type of regression model that you are fitting.

Figure 6.17 gives the fitted line plot. Notice that the box on the right-hand side of the fitted line plot shows the root mean square error and the

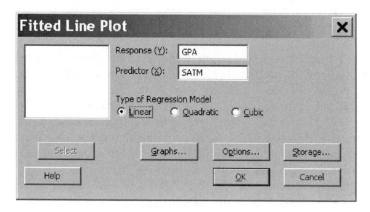

FIGURE 6.16
MINITAB dialog box to create a fitted line plot.

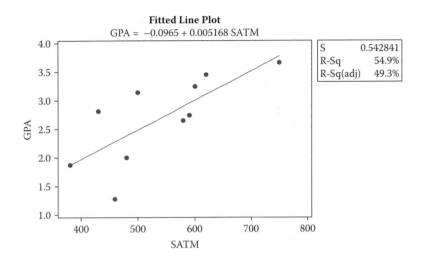

FIGURE 6.17
Scatter plot with fitted line superimposed for the data given in Table 5.1.

coefficient of determination (R^2). We will describe the adjusted R-squared statistic in more detail in Chapter 8.

Notice that all of the points on the scatter diagram in Figure 6.17 do not lie perfectly on the line of best fit, and the fitted line represents the line where the residual values above and below the line sum to 0, and the squared distance between the line and all of the data points is a minimum.

One of the regression assumptions that we are interested in verifying is whether the error component is normally distributed. We can graph a histogram of the residuals to assess whether this assumption about the true but unknown error component appears to hold true. Remember that the residuals give us the unexplained or residual variation, and the estimated model equation is based on the assumption that the error component is normally distributed. By plotting a histogram of the residuals (the difference between the observed and the predicted value), we can look to see if there are any pronounced outliers (those values that seem to be "far removed" from the rest of the data), and if the residuals appear roughly normally distributed. There are many different shapes of histograms of residuals, but again, we need to remember that we are looking for approximately normal distributions. We have to remember that since this is an exploratory process, we should not expect to see perfectly normal distributions.

Using the **Regression** tab from the **Stat** menu bar we get the regression dialog box as presented in Figure 6.18. We can then use MINITAB to generate the residual values. This is done by then selecting the **Storage** tab, as illustrated in Figure 6.19.

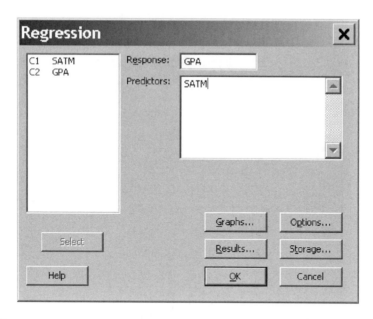

FIGURE 6.18
Regression dialog box.

By checking the **Residuals** box, as shown in Figure 6.19, MINITAB will display the residuals in the column labeled RESI1 within the worksheet illustrated in Figure 6.20.

Then by simply plotting a histogram with a normal curve superimposed on the residual values, we obtain the graph that is presented in Figure 6.21.

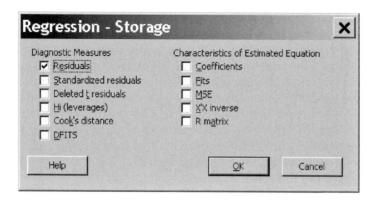

FIGURE 6.19
Regression storage options.

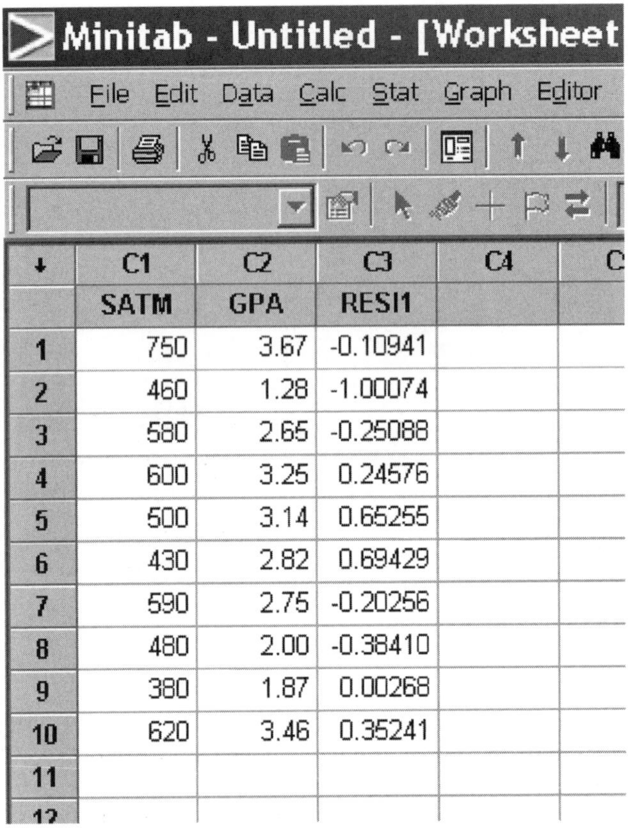

Minitab - Untitled - [Worksheet

File   Edit   Data   Calc   Stat   Graph   Editor

↓	C1	C2	C3	C4	C
	SATM	GPA	RESI1		
1	750	3.67	-0.10941		
2	460	1.28	-1.00074		
3	580	2.65	-0.25088		
4	600	3.25	0.24576		
5	500	3.14	0.65255		
6	430	2.82	0.69429		
7	590	2.75	-0.20256		
8	480	2.00	-0.38410		
9	380	1.87	0.00268		
10	620	3.46	0.35241		
11					
12					

FIGURE 6.20
MINITAB worksheet showing stored residual values.

We can also obtain a histogram of the residuals directly with MINITAB. To do this, we run a regression analysis as usual, then click on the **Graphs** tab in the regression dialog box to get the graphs dialog box for a regression analysis that is presented in Figure 6.22. Figure 6.23 gives the histogram of the residual values. If the assumption that the error component is normally distributed holds true, then the histogram of the residuals should resemble the shape of a normal (bell-shaped) curve. Obvious patterns or substantial clusters of data or other obvious departures from normality could be an indication that there are factors other than those we included in our model that may be influencing the response variable. Therefore, any inferences about the true but unknown population model that we may make could be suspect.

Another plot that can also be used to assess whether the error component is normally distributed is the *normal probability plot*. To create a normal

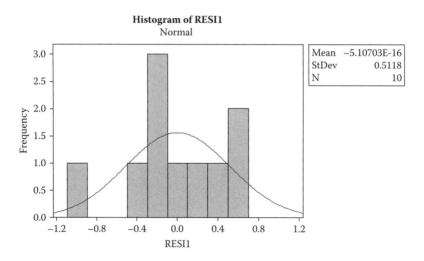

FIGURE 6.21
Histogram of the residual values for the data given in Table 5.1.

probability plot, the residuals are plotted in such a way that a straight-line pattern in the normal probability plot indicates normality. The normal probability plot can be created by plotting the residuals in numerical order versus what is called the *normal score* for each index in the sample. Normal scores are theoretical values that can be used to estimate the expected z-score for

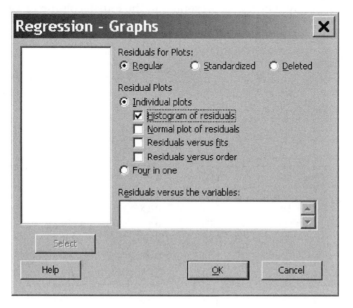

FIGURE 6.22
Regression graphs box to display a histogram of the residuals.

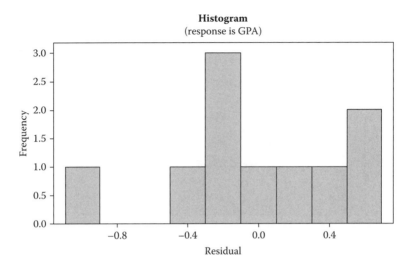

FIGURE 6.23
Histogram of the residual values as generated automatically by MINITAB.

a specific ordered data value on a standard normal curve. Normal scores are determined by the rank order of the data and are not determined using the actual values of the data themselves.

To find the normal scores for the data in Table 5.1, we first arrange the *residual data* in ascending order, as presented in the fourth column of Table 6.1.

TABLE 6.1

Residual Position Number Ranked in Ascending Order, Corresponding Predictor and Response Variables, Residual Values, Normal Scores, and Corresponding z-Scores for the Data from Table 5.1

Residual Position Number	Predictor Variable (x_i)	Response Variable (y_i)	$\hat{\varepsilon}_i = y_i - \hat{y}_i$	A_i	z_i
1	460	1.28	−1.00074146	0.0610	−1.545
2	480	2.00	−0.38409806	0.1585	−1.000
3	580	2.65	−0.25088101	0.2561	−0.655
4	590	2.75	−0.20255931	0.3537	−0.375
5	750	3.67	−0.10941204	0.4512	−0.120
6	380	1.87	0.00268490	0.5488	0.120
7	600	3.25	0.24576240	0.6463	0.375
8	620	3.46	0.35240581	0.7439	0.655
9	500	3.14	0.65254535	0.8415	1.000
10	430	2.82	0.69429342	0.9390	1.545

To create the column labeled A_i in Table 6.1, for each residual value we need to find the area in the lower tail of the standard normal distribution that is based on the rank of the residual observation. In other words, if the residual data come from a population that is normally distributed, then for the ith residual, the expected area to the left of the observation can be found by using the following formula:

$$A_i = \frac{i - 0.375}{n + 0.25}$$

where i is the position number of the ordered residual and n is the sample size.

Then, z_i, the normal score for the position of the ith residual, is the z-score that corresponds to this area under the standard normal curve.

For example, $A_4 = \dfrac{4 - 0.375}{10 + 0.25} = 0.3537$. This value corresponds to the area under the curve to the left of z_4, which corresponds to a z-score of approximately -0.375, as illustrated in Figure 6.24.

If the residual data are normally distributed, then by plotting the value of the ith residual versus the z-score for the rank ordering of each observation in the sample, all of these points should lie along a straight diagonal line, as illustrated in Figure 6.25.

We can also use MINITAB to generate the normal probability plot directly, by selecting the **Normal plot of residuals**, as illustrated in the regression graphs dialog box given in Figure 6.26.

The normal probability plot drawn directly from MINITAB is given in Figure 6.27. Notice that the normal probability plot given in Figure 6.27 does not plot the residual values versus the expected normal score, but instead plots the residual values versus the percent of A_i. Also notice that the normal probability plot generated directly from MINITAB provides a reference line that can be used to assess if the residuals versus the percentile lie along a

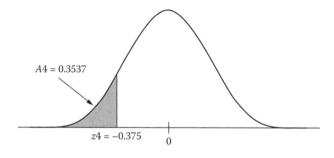

FIGURE 6.24
Area under the standard normal curve that corresponds to the fourth smallest residual value.

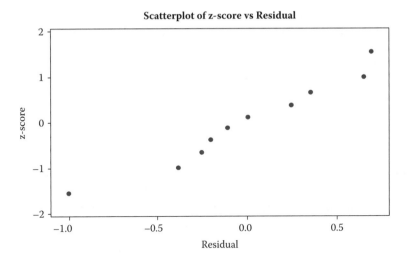

FIGURE 6.25
Scatter plot of z-score versus ordered residuals for the data in Table 6.1.

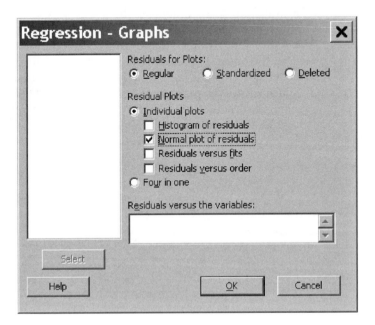

FIGURE 6.26
MINITAB graphs dialog box to select the normal plot of residuals.

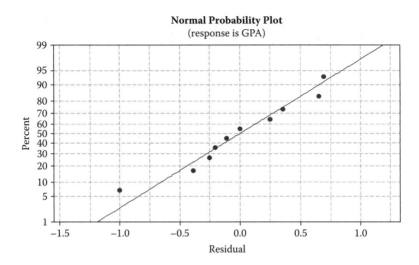

FIGURE 6.27
MINITAB-generated normal plot of the residuals versus the percentile of A_i.

diagonal straight line. This line can be obtained by finding the z-score for any given percentage and multiplying the z-score by the standard deviation of the residuals. For instance, if we consider 95% of the normal distribution, and if the standard deviation of the residuals is 0.512, as it is for this example, then the theoretical value of the residual that corresponds to 95% of the normal distribution is

$$1.645 \times 0.512 \approx 0.842$$

Therefore, one point on the diagonal reference line would be (0.842, 95).

The points in a normal probability plot should generally fall along a diagonal straight line if the residuals are normally distributed. If the points on the normal probability plot depart from the diagonal line, then the assumption that the errors are normally distributed may have been violated.

MINITAB also provides a scatter plot of the residual values versus the fitted or estimated values that can be used to assess the linearity and constant variance assumptions. This plot can be created by checking the **Residuals versus fits** box in the **Graphs** dialog box, as in Figure 6.28.

The graph of the residuals versus the fitted values is given in Figure 6.29. This plot should show a random pattern of residuals on both sides of the 0 line. There should not be any obvious patterns or clusters in the residual versus fitted plot. If the residual versus fitted plot shows any kind of pattern, this may suggest that the assumption of linearity has been violated. Furthermore, if there are differences in the amount of variation for certain

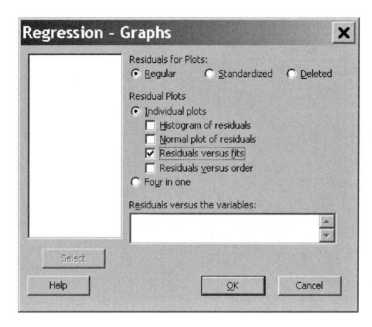

FIGURE 6.28
MINITAB graphs box for the residual versus the fitted values plot.

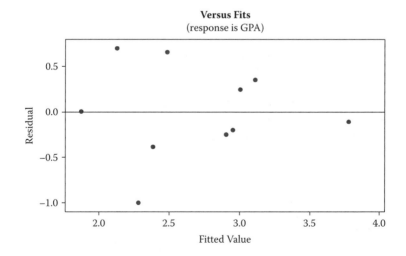

FIGURE 6.29
MINITAB-generated graph of the residual versus the fitted values.

patterns of fitted values, this could suggest that the assumption of constant variance has been violated.

6.9 A Formal Test of the Normality Assumption

Assessing residual plots by eye to determine if the normality assumption of the error component has been violated can be a very difficult task, especially when the sample size is small. Notice that in the last example, it was difficult to determine whether the normality assumption appears to have been violated by looking at the histogram of the residuals in Figure 6.23 and the normal probability plot in Figure 6.27.

The *Ryan–Joiner test* (Ryan and Joiner, 1976) can be used as a formal test of the normality assumption of the error component by comparing the distribution of the residuals to a normal distribution to see if they differ in shape. This can be done by first computing a measure of correlation between the residuals and their respective normal scores, and then using such a correlation as a test statistic. If we can infer that this correlation measure is close to 1.0, then the greater confidence we have that the unknown distribution of the error component is normally distributed. The Ryan–Joiner test statistic is calculated as follows:

$$RJ = \frac{\sum_{i=1}^{n} \hat{\varepsilon}_i \cdot z_i}{\sqrt{s^2(n-1)\sum_{i=1}^{n} z_i^2}}$$

where $\hat{\varepsilon}_i$ is the ith residual, z_i is the normal score for the ith residual (see Section 6.8 for a review of how to calculate normal scores), n is the sample size, and s^2 is the variance of the residuals.

The null and alternative hypotheses for the Ryan–Joiner test are given as follows:

H_0: The error component follows a normal distribution.

H_A: The error component does not follow a normal distribution.

Because the Ryan–Joiner test statistic represents a measure of the correlation between the residuals and data obtained from a normal distribution, the distribution of the test statistic has a different shape from what we have already seen. The distribution of the test statistic is skewed, where the rejection region is defined by the critical region rj, which is located in the left tail, as illustrated in Figure 6.30.

The value of the test statistic RJ is then compared to the value rj for a given level of significance using the table given in Table 3 of Appendix A. The value of rj defines the smallest correlation between the residuals and data obtained from a normal distribution. Notice that if the test statistic falls in the rejection region (in other words, if the test statistic RJ falls to the left of rj that defines

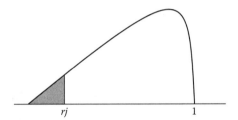

FIGURE 6.30
Distribution of the Ryan–Joiner test statistic.

the critical region), then it is less than the smallest correlation, and this leads to accepting the alternative hypothesis that the error component does not follow a normal distribution. If the test statistic falls to the right of the rejection region, then there is a strong correlation between the residuals and a normal distribution, which falls closer to the value of 1, and this leads to *not* accepting the alternative hypothesis, and thus we can infer that the assumption of normality of the error component does not appear to have been violated. Recall that two variables that are perfectly correlated have a correlation equal to 1.

Example 6.7

Using the data given in Table 6.1 we find that

$$\sum_{i=1}^{10} \hat{\varepsilon}_i \cdot z_i = 4.23220$$

$$\sum z_i^2 = 7.94215$$

$$s^2 = 0.262$$

So the test statistic would be calculated as follows:

$$RJ = \frac{\sum_{i=1}^{n} \hat{\varepsilon}_i \cdot z_i}{\sqrt{s^2 (n-1)\sum_{i=1}^{n} z_i^2}} = \frac{4.23220}{\sqrt{(0.262)(9)(7.94215)}} = 0.978$$

The value of this test statistic is then compared to the critical value *rj*, which defines the smallest correlation that can be used to infer normality for a given sample size. Thus, by comparing the value of the test statistic to the critical value of 0.9180 obtained from Table 3 of Appendix A* ($\alpha = 0.05$, $n = 10$), the test statistic *RJ* is greater than this value and thus does not fall in the rejection region. Notice that the test statistic suggests that there is a strong

---

* See https://www.minitab.com/resources/articles/normprob.pdf for a formal description of the Ryan–Joiner test and how the table in Appendix A was created.

correlation between the residuals and the data obtained from a normal distribution. Thus, the assumption that the true but unknown population error component is normally distributed does not appear to have been violated. In other words, because the test statistic is greater than the smallest correlation to infer normality defined by the corresponding value of rj for the given level of significance, we can infer that the assumption that the error component is normally distributed does not appear to have been violated.

The assumption of normality would likely be violated if the value of the Ryan–Joiner test statistic falls in the rejection region, which is less than the value rj that defines the rejection region for a given level of significance and sample size. This implies that the measure of correlation between the residual values and the values obtained from a normal distribution is less than the minimum value that is specified.

6.10 Using MINITAB for the Ryan–Joiner Test

To use MINITAB to run the Ryan–Joiner test for the data in Table 5.1, we first need to calculate the residual values. This can be done by selecting the **Residuals** using the **Storage** box of the **Regression** menu.

Then, once the residuals are found and stored in a column in the MINITAB worksheet, the Ryan–Joiner test can be performed by selecting the **Stat** command, then **Basic Statistics**, and then **Normality Test**, as illustrated in Figure 6.31.

Then we need to select to run the Ryan–Joiner test in the Normality test dialog box, as is illustrated in Figure 6.32.

Figure 6.33 gives the graph of the normality plot of the residuals along with the Ryan–Joiner test statistic and a corresponding p-value provided in the box on the right-hand side of the plot.

Notice in the box on the right-hand side of Figure 6.33 that we get the same value of the Ryan–Joiner test statistic as we did when we did the calculation by hand. Also notice that the p-value is greater than .100. This means that the test statistic falls outside of the rejection region, and thus we do not have reason to believe that the normality assumption has been violated.

6.11 Assessing Outliers

On the MINITAB printouts for a simple regression analysis, you may have noticed that the bottom part of the printout describes what are called "unusual observations." These unusual observations are also sometimes referred to as *outliers* or *extreme values*. Dealing with outliers or extreme values in a regression analysis can be very challenging, and often outliers or extreme values can be the cause of why a particular model does not fit well.

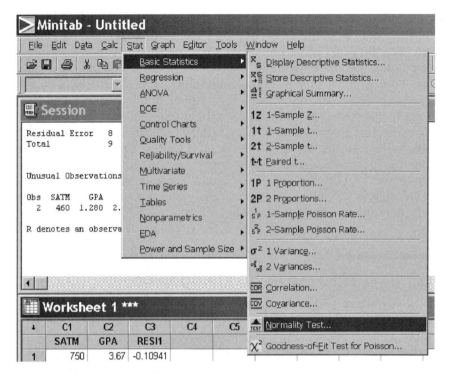

FIGURE 6.31
MINITAB commands to run a normality test.

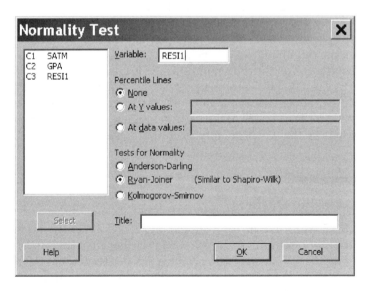

FIGURE 6.32
MINITAB dialog box for running the Ryan–Joiner test.

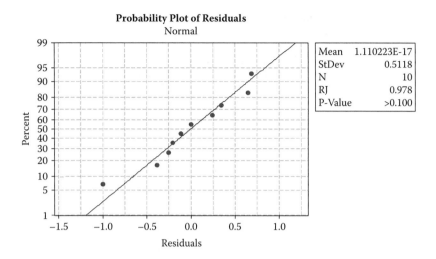

FIGURE 6.33
Normality plot and Ryan–Joiner test generated from MINITAB.

Outliers can be identified as observations that fall into either or both of the following two categories:

1. The leverage value for an observation is greater than $3 \cdot p/n$, where p is the number of beta parameters being estimated in the model and n is the sample size.
2. The standardized (or internally Studentized) residual for an observation is less than –2 or greater than +2.

6.12   Assessing Outliers: Leverage Values

One way to see if a particular observation is having a strong impact on the regression analysis is to determine the *leverage value* for the given observation. The leverage value for a given observation x_i for a simple linear regression model, $y = \beta_0 + \beta_1 x + \varepsilon$, is found using the following formula:

$$h_i = \frac{1}{n} + \frac{(x_i - \bar{x})^2}{S_{xx}}$$

where x_i is the observation of interest, $\bar{x}$ is the mean of all the x-values, n is the sample size, and S_{xx} is the sum of the squared differences of the individual x-values from the mean of the x-values.

TABLE 6.2

Table of Leverage Values for the Data
from Table 5.1

Observation Number	SATM	Leverage Value (h_i)
1	750	0.51419
2	460	0.15806
3	580	0.11564
4	600	0.13462
5	500	0.11415
6	430	0.21053
7	590	0.12420
8	480	0.13238
9	382	0.33519
10	620	0.16104

Leverage values describe whether or not a given observation, x_i, is far removed from the mean of the x-values. Values of h_i will fall between 0 and 1. When h_i is close to 1, this suggests that the observation x_i is far removed from the mean, and if h_i is close to 0, this indicates that the observation is not far removed from the mean.

Example 6.8

The data in Table 5.1 present a sample of the SAT mathematics examination scores and the first-year grade point averages for a sample of ten students. Suppose we want to find the leverage value for observation 2. This would be calculated as follows:

$$h_2 = \frac{1}{10} + \frac{(460 - 539)^2}{107490} = 0.15806$$

Table 6.2 presents the leverage values for all of the observations for Table 5.1.

6.13   Using MINITAB to Calculate Leverage Values

Using the storage tab in the regression dialog box, you can have MINITAB calculate the leverage values for you, as illustrated in Figure 6.34.

MINITAB will then calculate the leverage values and store them in the current MINITAB worksheet, as illustrated in Figure 6.35.

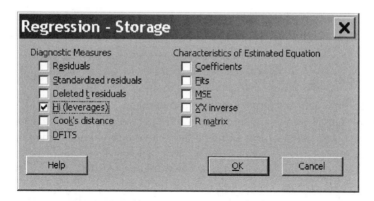

FIGURE 6.34
MINITAB storage dialog box to calculate leverage values.

When running a regression analysis, MINITAB will identify any observations that have a leverage value greater than $3 \cdot p/n$ where p is the total number of beta parameters being estimated in the model, and n is the sample size. Observations that have a leverage value greater than $3 \cdot p/n$ will be identified with an X under the "Unusual Observations" portion of the

↓	C1	C2	C3	C4	C5	C6
	SATM	GPA	HI1			
1	750	3.67	0.514187			
2	460	1.28	0.158061			
3	580	2.65	0.115639			
4	600	3.25	0.134617			
5	500	3.14	0.114150			
6	430	2.82	0.210531			
7	590	2.75	0.124198			
8	480	2.00	0.132384			
9	380	1.87	0.335194			
10	620	3.46	0.161038			
11						

FIGURE 6.35
MINITAB storage of leverage values in column C3.

MINITAB printout. It is important to note that the value of p is determined by counting all the beta parameters that are being estimated, and this includes β_0.

Example 6.9

Using the data set "Leverage" provided in Table 6.3 and the MINITAB printout of the regression analysis for this data set, as presented in Figure 6.36, we can see that observation 12 has a large leverage value, as highlighted in Figure 6.36.

In order for MINITAB to identify an influential observation with a large leverage value, the leverage value must be greater than $3 \cdot 2/13 = 0.46$. For this example, $p = 2$ because there are two parameters being estimated in the model: β_0 and β_1. MINITAB will identify such an influential observation with the value X on the regression printout. The leverage value for observation 12 is 0.852581; thus, it is flagged with an X in Figure 6.36 because its leverage value is greater than 0.46.

Example 6.10

For the example using the data in Table 5.1, a leverage value would have to be greater than $\dfrac{(3 \cdot p)}{n} = \dfrac{(3 \cdot 2)}{10} = (3 \cdot 2)/10$ in order for MINITAB to identify such an observation as having a large leverage value. As can be seen in Table 6.2, since there are no observations that have a leverage value greater than 0.60, there were no observations identified as having a large leverage value, as can be seen in the unusual observations portion of the MINITAB printout illustrated in Figure 6.37.

Recall that MINITAB identifies large leverage values with an X.

TABLE 6.3
"Leverage" Data Set

Observation Number	x	y
1	12	21
2	18	19
3	2	19
4	5	19
5	4	20
6	6	24
7	5	19
8	2	19
9	5	19
10	4	20
11	4	19
12	45	18
13	12	19

Regression Analysis: y versus x

```
The regression equation is
y = 20.0 - 0.0366 x
```

Predictor	Coef	SE Coef	T	P
Constant	19.9643	0.5489	36.37	0.000
x	-0.03658	0.03737	-0.98	0.349

S = 1.50476   R-Sq = 8.0%   R-Sq(adj) = 0.0%

Analysis of Variance

Source	DF	SS	MS	F	P
Regression	1	2.170	2.170	0.96	0.349
Residual Error	11	24.907	2.264		
Total	12	27.077			

Unusual Observations

Obs	x	y	Fit	SE Fit	Residual	St Resid
6	6.0	24.000	19.745	0.438	4.255	2.96R
12	45.0	18.000	18.318	1.389	-0.318	-0.55 X

```
R denotes an observation with a large standardized residual.
X denotes an observation whose X value gives it large leverage.
```

FIGURE 6.36
MINITAB printout for "Leverage" data given in Table 6.3.

6.14 Assessing Outliers: Internally Studentized Residuals

If a particular residual is more than two standard deviations away from the mean, then such an observation could have an impact on the regression analysis. We will be using what are called *internally Studentized residuals* as a way to quantify whether a particular observation could be affecting the regression analysis. The formula to calculate an internally Studentized residual for a given observation is as follows:

$$\varepsilon_i' = \frac{\hat{\varepsilon}_i}{S\sqrt{1-h_i}}$$

Unusual Observations

Obs	SATM	GPA	Fit	SE Fit	Residual	St Resid
2	460	1.280	2.281	0.216	-1.001	-2.01R

```
R denotes an observation with a large standardized residual.
```

FIGURE 6.37
Portion of the MINITAB printout that identifies an unusual observation for the data presented in Table 5.1.

TABLE 6.4

Leverage Values and Internally Studentized
Residuals for the Data Given in Table 5.1

Observation Number	SATM	Leverage Value (h_i)	Internally Studentized Residual (ε_i')
1	750	0.51419	–0.28917
2	460	0.15806	–2.00913
3	580	0.11564	–0.49145
4	600	0.13462	0.48667
5	500	0.11415	1.27720
6	430	0.21053	1.43947
7	590	0.12420	–0.39873
8	480	0.13238	–0.75964
9	382	0.33519	0.00607
10	620	0.16104	0.70876

where $\hat{\varepsilon}_i$ is the value of the ith residual, S is the root mean square error from the regression analysis, and h_i is the leverage value for the ith residual. The reason we choose to represent the residuals in this manner is because each internally Studentized residual follows a t-distribution with $n - 1 - p$ degrees of freedom, where p equals the total number of beta terms being estimated in the model, and n is the sample size.

Example 6.11

The internally Studentized residual for observation 2 given in Table 6.2 can be calculated as follows:

$$\varepsilon_i' = \frac{\hat{\varepsilon}_i}{S\sqrt{1-h_i}} = \frac{-1.0007}{(0.542841) \cdot \sqrt{1-0.15806}} \approx -2.0091$$

Table 6.4 presents the leverage values and the internally Studentized residuals for each observation for the data given in Table 5.1.

6.15 Using MINITAB to Calculate Internally Studentized Residuals

Figure 6.38 shows the regression storage dialog box where you can select to have MINITAB calculate the standardized residuals (the internally Studentized residuals) for each of the individual observations.

Using the data given in Table 5.1, Figure 6.39 shows that MINITAB calculates these residuals and stores them in the MINITAB worksheet.

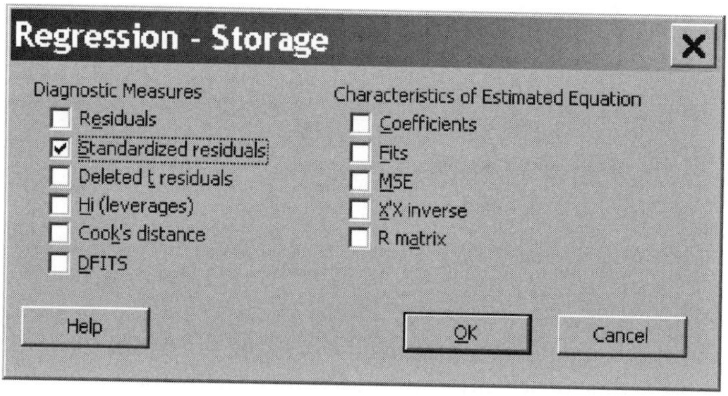

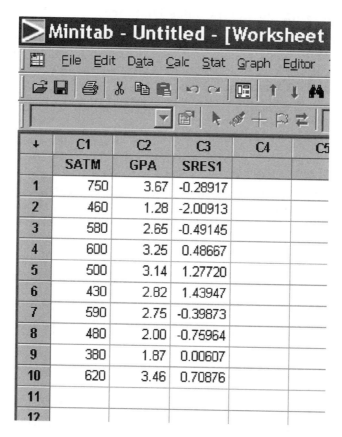

FIGURE 6.38
Regression dialog box to calculate the internally Studentized residuals.

↓	C1	C2	C3	C4	C5
	SATM	GPA	SRES1		
1	750	3.67	-0.28917		
2	460	1.28	-2.00913		
3	580	2.65	-0.49145		
4	600	3.25	0.48667		
5	500	3.14	1.27720		
6	430	2.82	1.43947		
7	590	2.75	-0.39873		
8	480	2.00	-0.75964		
9	380	1.87	0.00607		
10	620	3.46	0.70876		
11					
12					

FIGURE 6.39
MINITAB storage of internally Studentized residuals.

```
Unusual Observations

Obs    SATM     GPA     Fit    SE Fit   Residual   St Resid
 2     460    1.280   2.281    0.216    -1.001      -2.01R
```

R denotes an observation with a large standardized residual.

FIGURE 6.40
MINITAB printout indicating that observation 2 has a standardized residual that is less than –2.00.

Notice in the unusual observations portion of the MINITAB printout, as in Figure 6.40, that observation 2 is identified with the value of "R." The designation of R indicates that the observation has an internally Studentized residual that is either less than –2 or greater than +2.

6.16   Assessing Outliers: Cook's Distances

Another way to assess the impact that outliers have on a regression analysis is to look at a measure of influence for a given observation called *Cook's distance* (Cook, 1979). Cook's distance for a given observation is a measure of how much impact, including the observation in the regression analysis has on the estimates of the regression parameters. The Cook's distance for any given observation is given by the formula

$$D_i = \frac{\hat{\varepsilon}_i^2}{p \cdot S^2}\left[\frac{h_i}{(1-h_i)^2}\right]$$

where h_i is the leverage for the ith observation, S^2 is the mean square error from the regression analysis, p is the total number of beta parameters being estimated in the model, and $\hat{\varepsilon}_i$ is the residual for the ith observation.

Example 6.12

The value of Cook's distance for observation 2 from Table 5.1 can be calculated as follows:

$$D_i = \frac{\hat{\varepsilon}_i^2}{p \cdot S^2}\left[\frac{h_i}{(1-h_i)^2}\right] = \frac{(-1.0007)^2}{2 \cdot (0.29468)}\left[\frac{0.15806}{(1-0.15806)^2}\right] \approx 0.3789$$

6.17   Using MINITAB to Find Cook's Distances

Similar to leverage values and internally Studentized residuals, MINITAB can calculate Cook's distance for all of the observations by using the **Storage** command on the regression dialog box.

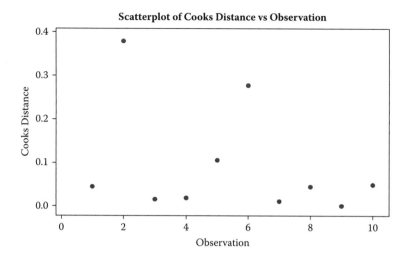

FIGURE 6.41
Scatter plot of Cook's distance versus the observation number.

One of the more interesting ways to use Cook's distances to assess the impact of outliers on a regression analysis is to draw a scatter plot of the observation number on the x-axis against the Cook's distance on the y-axis. Such a plot is illustrated in Figure 6.41.

This plot can be used to assess the impact that any outlier may have on the regression analysis by showing how far the individual Cook's distance falls away from 0. Notice in Figure 6.41 that observations 2 and 6 are the most influential because the Cook's distances for these plots fall the farthest away from 0.

6.18 How to Deal with Outliers

To determine the impact that an outlier has on a regression analysis, you can fit the model both with and without the outlier. Then you need to examine how the values of the estimated coefficients, along with the associated p-values and the R^2 statistic, have changed between the two models. Whether to delete an outlier permanently is a matter of personal conviction and tends to differ by field of study. The convention in some fields is to never delete an outlier, whereas others may do so if they believe the outlier was obtained either erroneously or through measurement error.

Example 6.13

The highlighted portion of Figures 6.42 and 6.43 illustrates how the value of the estimated coefficients, p-values, and R^2 statistics have changed by deleting the influential observation 2.

Regression Analysis: GPA versus SATM

```
The regression equation is
GPA = - 0.096 + 0.00517 SATM

Predictor        Coef   SE Coef       T        P
Constant      -0.0965    0.9088   -0.11    0.918
SATM         0.005168  0.001656    3.12    0.014


S = 0.542841   R-Sq = 54.9%   R-Sq(adj) = 49.3%

Analysis of Variance

Source          DF        SS       MS      F      P
Regression       1    2.8707   2.8707   9.74  0.014
Residual Error   8    2.3574   0.2947
Total            9    5.2281


Unusual Observations

Obs  SATM    GPA     Fit   SE Fit   Residual   St Resid
2     460  1.280   2.281    0.216     -1.001     -2.01R

R denotes an observation with a large standardized residual.
```

FIGURE 6.42
MINITAB printout of the regression analysis of first-year GPA versus score received on the mathematics portion of the SAT examination including observation 2.

Regression Analysis: GPA versus SATM

```
The regression equation is
GPA = 0.493 + 0.00429 SATM

Predictor        Coef   SE Coef       T        P
Constant       0.4933    0.7186    0.69    0.515
SATM         0.004294  0.001288    3.33    0.013


S = 0.408467   R-Sq = 61.4%   R-Sq(adj) = 55.8%

Analysis of Variance

Source          DF        SS       MS       F      P
Regression       1    1.8543   1.8543   11.11  0.013
Residual Error   7    1.1679   0.1668
Total            8    3.0222
```

FIGURE 6.43
MINITAB printout of the regression analysis of first-year GPA versus score received on the mathematics portion of the SAT examination deleting observation number 2.

By comparing the highlighted portions of Figures 6.42 and 6.43, notice that there is not a noticeable difference between the parameter estimates, p-values, and the value of the R^2 statistic of the two models. Since this unusual observation does not appear to be grossly affecting the analysis, it may be a good idea to keep this observation in the data set. Should there be a noticeable difference between the two models with respect to the R^2 statistic, parameter estimates, and p-values, you would want to investigate further why this particular observation is so influential.

Exercises

1. In Exercise 3 of Chapter 5, we were given a set of ordered pairs (x, y) that generated the following set of summary statistics:

$$n = 15, \quad \bar{x} = 50.33, \quad \bar{y} = 150.33$$

$$S_{xx} = 617.33, \quad S_{xy} = 2624.33, \quad S_{yy} = 16489.30$$

   Calculate the residual sum of squares $SSE = S_{yy} - \frac{S_{xy}^2}{S_{xx}}$, and an estimate of the error variance $S^2 = \frac{SSE}{n-2}$ and R^2, the coefficient of determination.

2. For the statistics given in Exercise 1,
   a. Find the value of the coefficient of correlation.
   b. Test to see if the population coefficient of correlation is significantly different from 0 ($\alpha = 0.01$).

3. For the data given in Exercise 1, find a 95% confidence interval for the mean value of y for a given x-value of 183.70.

4. Given the following MINITAB printout in Figure 6.44 for a regression analysis:
   a. What are the estimates of β_0 and β_1, the population intercept and slope?
   b. Write the population equation.
   c. Write the estimated equation.
   d. Determine whether or not β_1 is significantly different from 0 ($\alpha = 0.05$).
   e. Explain how you would determine whether or not you have a good model for predicting y based on x.

Regression Analysis: Y versus X

```
The regression equation is
Y = 2.94 - 0.00299 X

Predictor         Coef    SE Coef       T       P
Constant         2.941      1.636    1.80   0.086
X            -0.002986   0.003334   -0.90   0.380

S = 1.13141    R-Sq = 3.5%    R-Sq(adj) = 0.0%

Analysis of Variance

Source            DF       SS      MS      F       P
Regression         1    1.027   1.027   0.80   0.380
Residual Error    22   28.162   1.280
Total             23   29.189

Unusual Observations

Obs    X       Y     Fit   SE Fit   Residual   St Resid
18   510   3.640   1.418    0.245      2.222       2.01R

R denotes an observation with a large standardized residual.
```

FIGURE 6.44
MINITAB printout for Exercise 4.

5. The data set "Executive Bonus" gives the results of a questionnaire that was given to twenty executives in a specific field, asking them what factors influence their yearly bonuses. The data consist of survey responses that include yearly bonus amount (in thousands of dollars) and the number of employees supervised, as can be seen in Table 6.5.

   a. Write the population model describing the annual bonus (y) as a function of the number of employees supervised (x).

   b. Write the estimate of the model equation for describing annual earnings (y) as a function of the number of employees supervised (x).

   c. Fit the linear model using MINITAB. Justify whether or not you think the linear model reasonably fits the data. Highlight the value for the square root of the mean square error (MSE).

   d. Determine whether the number of employees supervised is significant in predicting executive salaries.

   e. Interpret the estimate of β_1 in your model.

   f. Describe if you believe that the model assumptions for the error component for a simple regression analysis hold true for this problem. Provide residual plots and MINITAB results from any formal tests and comment as necessary.

TABLE 6.5

"Executive Bonus" Data Set

Observation Number	Annual Bonus (thousands of dollars)	Number of Employees Supervised
1	25	152
2	30	216
3	72	358
4	15	95
5	20	126
6	52	350
7	10	65
8	20	78
9	25	145
10	43	291
11	50	300
12	30	162
13	15	59
14	20	163
15	45	221
16	50	264
17	45	351
18	25	180
19	48	291
20	23	176

6. Using the "Executive Bonus" data set and MINITAB, calculate the leverage values, the internally Studentized residuals, and Cook's distance for each of the observations. Create a Cook's distance plot and comment on the impact, if any, the outliers may have on the analysis.

7. Given the portion of a MINITAB printout in Figure 6.45, which provides details about unusual observations, what can you say about the leverage values and internally Studentized residual for this observation?

```
Unusual Observations
     Obs     x       y       Fit    SE Fit    Residual   St Resid
      20    2.2   108.65    47.76    12.07       60.89     4.22RX
```

FIGURE 6.45
Portion of MINITAB printout for Exercise 7.

8. Recall in the exercises in Chapter 3 that the kurtosis of a set of data was introduced. The *kurtosis* of a data set represents a measure of how the distribution of a data set is peaked (in other words, how "sharp" the data set appears when you consider its distribution). The kurtosis can also be used as a gauge to see how much the distribution of your data set differs from the normal distribution. For instance, the closer the kurtosis value is to 0, the more normally distributed the data are, and the closer the kurtosis value is to 1, the sharper the peak of the distribution is; a negative kurtosis value represents a flat distribution. We can also use the kurtosis as an exploratory technique to check on whether the residuals are normally distributed. The formula to calculate the kurtosis for the residuals is as follows:

$$Kurtosis = \left[\frac{n(n+1)}{(n-1)(n-2)(n-3)} \right] \cdot \sum_{i=1}^{n} \left(\frac{\hat{\varepsilon}_i - \overline{\hat{\varepsilon}}}{s} \right)^4 - \frac{3(n-1)^2}{(n-2)(n-3)}$$

where n is the sample size, $\hat{\varepsilon}_i$ is the ith residual, $\overline{\hat{\varepsilon}}$ is the mean of all of the residuals, and s is the standard deviation of the residuals. Using the kurtosis, decide whether the residual values for the "Executive Bonus" data set appear to follow a normal distribution.

9. Also recall in the exercises in Chapter 3 that the skewness of a set of data was introduced. The *skewness* of a data set can be used as a measure of how symmetrical a distribution is. If the skewness is positive, this indicates that more observations are below the mean than are above the mean, and if the skewness is negative, this indicates that there are more observations above the mean than are below the mean. A value of the skewness near 0 indicates that the distribution is symmetrical. The formula to calculate the skewness for the residuals is as follows:

$$Skewness = \frac{n}{(n-1)(n-2)} \cdot \sum_{i=1}^{n} \left(\frac{\hat{\varepsilon}_i - \overline{\hat{\varepsilon}}}{s} \right)^3$$

where n is the sample size, $\hat{\varepsilon}_i$ is the ith residual, $\overline{\hat{\varepsilon}}$ is the mean of all of the residuals, and s is the standard deviation of the residuals. Using the skewness, decide whether the residual values for the "Executive Bonus" data set appear to be symmetrical.

10. Table 6.6 provides a collection of correlations that describe the relationships between yearly salary and yearly bonus based on the number of hours worked each week, the number of employees supervised, corporate profits, and corporate assets for a sample of twenty-two executives.

TABLE 6.6

Coefficient of Correlation Describing the Relationship between Yearly Salary and Yearly Bonus Based on the Number of Hours Worked Each Week, Number of Employees Supervised, Corporate Profits, and Corporate Assets

	Coefficient of Correlation (r)	
	Salary	Bonus
Number of hours worked each week	.56	.61
Number of employees supervised	.63	−.46
Corporate profits	.21	−.12
Corporate assets	−.16	.08

a. Is it true that the yearly salary and the number of hours worked each week for the population are significantly correlated with each other ($\alpha = 0.05$)? Justify your answer.

b. Determine if any of the other population coefficients of correlation are significant. Show all relevant work.

11. The correlation inference that was described in Section 6.5 can only test whether a population coefficient of correlation is different from 0. We may also be interested in testing hypotheses for population correlations other than 0. In order to do this we need to use *Fisher's Z-transformation* so that we can obtain a test statistic that is approximately normally distributed. However, this transformation only gives an approximate normal distribution when the sample size is greater than 20.

The test statistic for testing correlations other than 0 is as follows:

$$Z = \frac{\frac{1}{2}\ln\left[(1+r)\Big/(1-r)\right] - \frac{1}{2}\ln\left[(1+\rho_0)\Big/(1-\rho_0)\right]}{\frac{1}{\sqrt{n-3}}}$$

where ln is the natural logarithm and ρ_0 is the population coefficient of correlation being tested under the null hypothesis.

This test statistic is used in a manner similar to that of other test statistics that we have calculated throughout this text. However, because this test statistic follows a standard normal distribution, we will be using the standard normal tables so we do not have to be concerned with the degrees of freedom.

Suppose we have a sample of size $n = 160$, and we want to test whether the true population correlation between two variables, *x* and *y*, is significantly different from 0.80. Suppose, based on sample data, we found that $r = .75$ ($\alpha = 0.05$).

a. Set up the appropriate null and alternative hypotheses.

b. Calculate the test statistic Z.

c. Using a standard normal distribution, shade in the appropriate rejection region(s) and label the corresponding values of *z*.

d. Find the *p*-value and interpret your results.

12. For two samples of size n_1 and n_2 from two different populations, we can also test whether the two population coefficients of correlation are different from each other. To do this, we first need to find the appropriate Z-transformations for each of the different populations we are sampling from:

$$Z_1 = \frac{1}{2}\ln\frac{1+r_1}{1-r_1}$$

$$Z_2 = \frac{1}{2}\ln\frac{1+r_2}{1-r_2}$$

where r_1 is the sample coefficient of correlation from the first population, r_2 is the sample coefficient of correlation from the second population, and ln is the natural logarithm.

Given each of these transformations, can use the following test statistic, which follows a standard normal distribution:

$$Z = \frac{Z_1 - Z_2}{\sqrt{\dfrac{1}{(n_1 - 3)} + \dfrac{1}{(n_2 - 3)}}}$$

a. Suppose that from two different samples of size $n_1 = 89$, $n_2 = 58$ we have two sample correlations, $r_1 = 0.450$ and $r_2 = 0.350$, and we want to determine if ρ_1 is significantly greater than ρ_2 ($\alpha = 0.05$). State the appropriate null and alternative hypotheses.

b. Calculate the appropriate Z-transformation for each correlation.

c. Calculate the test statistic Z.

d. Using a standard normal curve, shade in the appropriate rejection region(s) and label the corresponding values of *z* that define the rejection region(s).

e. Find the *p*-value and interpret your results.

13. What a difference a single outlier can make! The "Single Outlier" data set in Table 6.7 presents a collection of ordered pairs.

   a. Using MINITAB, run a simple linear regression analysis to develop a model that describes the relationship between x and y.

   b. Check graphically to see if the regression model assumptions have been reasonably met.

   c. Run the Ryan–Joiner test of normality, and comment on the results.

   d. Determine the leverage values, studentized residuals, and Cook's distances; and draw a Cook's distance plot.

TABLE 6.7

"Single Outlier" Data Set

x	y
80	83
54	59
78	86
71	63
68	68
73	77
82	83
74	79
78	78
59	65
97	68
73	82
80	78
54	59
75	80
66	61
72	78
55	63
70	75
73	80
68	69
78	82
64	69
72	74
62	65
81	84
79	92
57	63
68	68
77	84

e. Notice the outlier that has the largest Cook's distance. Remove this outlier from the data set, and rerun steps (a) through (c). Comment on the impact that keeping this point in the data set has on your analysis.

14. Exercise 9 in Chapter 5 asked you to determine if the unemployment rate has an impact on the crime rate. Using the data set "Crime by County," run a regression analysis and comment on whether you believe that the model assumptions for a regression analysis have been reasonably met. Generate residual plots and do any necessary formal tests of the assumptions.

15. Exercise 11 in Chapter 5 asked you to determine if cars that get better fuel economy emit less greenhouse gases. Using the "Automobile" data set, run a regression analysis and comment on whether you believe that the model assumptions for a regression analysis have been reasonably met. Generate residual plots and do any necessary formal tests of the assumptions.

16. Exercise 13 in Chapter 5 asked you to determine whether younger people incur more credit card debt. Using the data set "Credit Card Debt," run a regression analysis and comment on whether you believe that the model assumptions for a regression analysis have been reasonably met. Generate residual plots and do any necessary formal tests of the assumptions.

17. Exercise 14 in Chapter 5 asked you to determine if people with more education incur less credit card debt. Using the data set "Credit Card Debt," run a regression analysis and comment on whether you believe that the model assumptions for a regression analysis have been reasonably met. Generate residual plots and do any necessary formal tests of the assumptions.

18. Exercise 15 in Chapter 5 asked you to determine if the asking price of a home was influenced by the size of the home. Using the "Expensive Homes" data set, run a regression analysis and comment on whether you believe that the model assumptions for a regression analysis have been reasonably met. Generate residual plots and do any necessary formal tests of the assumptions.

References

Cook, R. 1979. Influential observations in linear regression. *Journal of the American Statistical Association* 74:169–74.

Ryan, T., and Joiner, B. 1976. *Normal probability plots and tests for normality*. Technical report. University Park, PA: The Pennsylvania State University.

7

Multiple Regression Analysis

7.1 Introduction

Often it is the case when more than one predictor, or x-variable, has an impact on a given response, or y-variable. For example, suppose we are interested in developing a model that can be used to predict executive salaries. An executive's salary is determined not only by his or her level of education and the number of years of experience, but also by other factors, such as whether the business is profit or nonprofit, and the number of employees the executive manages. Furthermore, you can probably even think of more factors that could also be used in determining an executive's salary besides those just mentioned. For instance, the gender of the executive and the region of the country where he or she is employed could also be factors that impact salaries. In order to account for multiple factors having an impact on some outcome or response variable of interest, multiple regression analysis can be used.

This chapter begins by describing the basics of multiple regression analysis. We will work through some examples, and provide an explanation of how to use MINITAB to run a multiple regression analysis. We will also describe some measures that can be used to assess how well a multiple regression model fits the underlying set of data, how multicollinearity can impact the inferences made from a multiple regression analysis, and elaborate on how to use MINITAB to assess the model assumptions. Finally, the chapter concludes with a discussion of how to include higher-order powers of a predictor variable in a multiple regression analysis.

7.2 Basics of Multiple Regression Analysis

In order to obtain a useful regression model that includes more than one independent or x-variable, it is important to consider all of the variables that may impact the outcome measure of interest. Thus, in modeling executive salaries, besides collecting data on the level of education and the number of years of experience, we also may want to collect data on any other factors that could impact executive salaries, such as the type of business (profit or nonprofit), the numbers of employees managed, employee gender, the region where employed, etc.

The set of predictor variables, or x-variables, that may impact a given response variable, or y-variable, are also called *control variables* or *covariates*. A control variable (or covariate) is a variable that must be included in a regression analysis because it is likely to have a significant impact on the response variable of interest. *Multiple regression analysis* is a statistical technique that can be used to create a regression model that includes more than one predictor variable, or x-variable, where such variables can be used to model a single continuous response, or y-variable.

The basic form of a linear multiple regression model is similar to that of a simple linear regression model, except that more than one predictor variable is included in the population model equation, as described below:

$$y = \beta_0 + \beta_1 x_1 + \beta_2 x_2 + \cdots + \beta_j x_j + \varepsilon,$$

where y is the continuous response (or dependent) variable, $x_1, x_2, \cdots x_j$ are the predictor (or independent) variables, β_0 is the constant term, $\beta_1, \beta_2 \cdots \beta_j$ are the unknown population coefficients that correspond to the effect that each of the different independent variables has on the response variable, and ε is the random error component. The assumptions underlying the random error component in a multiple regression analysis are similar to the assumptions underlying the random error component in a simple linear regression analysis. Thus, it is assumed that the random error component is independent, normally distributed, and has constant variance.

The population linear multiple regression equation as just described stipulates that the response variable is linearly related to each of the independent variables when the other independent variables are held fixed. In other words, the effect of the ith independent variable, x_i, is found by estimating the population parameter β_i, and the estimate of β_i, denoted as $\hat{\beta}_i$, describes the estimated effect that the variable x_i has on the response variable when all of the other predictor variables are held fixed.

Just like with simple linear regression analysis, the method of least squares will be used to determine the equation that best fits the data, and confidence intervals and hypothesis tests can be performed on the unknown population regression parameters by using the statistics obtained from an estimated model equation. Different strategies and techniques, such as exploratory analyses and formal tests, can be used to assess the model fit. However, by including more than one predictor variable in the model, the calculations used to find the equation of best fit become much more tedious and very complicated. This is where MINITAB or any other statistical program can be very useful.

One way to become more familiar with multiple regression analysis is to work through an example. We will focus on describing the basic multiple regression equation, how to determine the estimates of the parameters, how to assess the model fit, and how to interpret the estimates of the parameters in order to make inferences regarding the relationships that we are interested in.

Example 7.1

Suppose we are interested in modeling the effect that the number of years of education and the number of years of experience has on executive salaries. Table 7.1 consists of the number of years of education, the number of years of experience, and the yearly salary for a random sample of fifteen executives from a specific type of business.

We want to develop a model that predicts the effect that the number of years of education and the number of years of experience has on executive salaries.

TABLE 7.1

Education, Experience, and Yearly Salary for a Sample of Fifteen Executives

Observation Number	Education in Years (x_1)	Experience in Years (x_2)	Yearly Salary in Thousands of Dollars (y)
1	12	12	95
2	13	19	145
3	16	20	164
4	14	24	186
5	15	30	197
6	12	16	139
7	13	19	163
8	15	15	125
9	16	25	122
10	13	26	173
11	14	25	152
12	12	5	75
13	13	19	165
14	13	22	167
15	14	28	187

Because we have two independent variables (education and experience), the population linear model equation that we are interested in estimating would be written as follows:

$$y = \beta_0 + \beta_1 x_1 + \beta_2 x_2 + \varepsilon$$

Similar to a simple linear regression analysis, we will use MINITAB to find the estimated equation that best fits the data. In other words, we will use MINITAB to find the estimated linear model equation that is described as

$$\hat{y} = \hat{\beta}_0 + \hat{\beta}_1 x_1 + \hat{\beta}_2 x_2$$

where $\hat{\beta}_1$ and $\hat{\beta}_2$ are the estimated effects of education and experience on executive salaries, respectively.

However, before we use MINITAB to find the estimated model equation, we may first want to take a look at how each of the predictor variables is related to the response variable. We will do this by creating what is called a matrix plot.

7.3   Using MINITAB to Create a Matrix Plot

In simple linear regression where there is only one predictor variable, we used a scatter plot to assess the relationship between the predictor and response variables. However, with multiple regression analysis there is more than one independent variable. Although a scatter plot can only assess the relationship between two variables, we can use a collection of scatter plots to display the relationship between each of the individual predictor variables and the response variable. To create a matrix plot using MINITAB, select **Graph** and then **Matrix Plot**, as illustrated in Figure 7.1.

In order to graph each predictor variable versus the response variable, we need to select a matrix plot for y versus each x (and with a smoother), as presented in Figure 7.2. To draw a matrix plot for all of the x- and y-variables, highlight and select the appropriate variables, as presented in Figure 7.3.

Figure 7.4 presents the graph of the matrix plot, which consists of individual scatter plots for each of the predictor variables versus the response variable along with a smoothed curve that approximates the functional relationship between each x and y. The smoothed curve (the type of curve that MINITAB draws is called a lowess smoother) provides an empirical approximation of the functional form of the relationship between each predictor variable and the response variable without fitting any specific model.

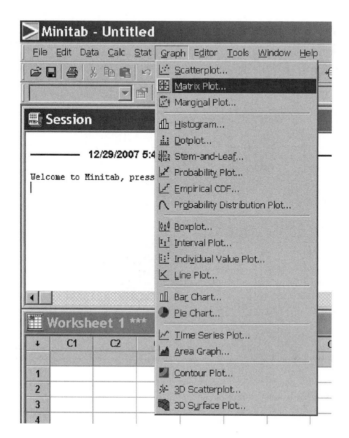

FIGURE 7.1
MINITAB commands to draw a matrix plot.

Notice that the left panel of the matrix plot in Figure 7.4 describes the relationship between the years of education and yearly salary. The shape of the scatter plot and smoother suggest that the relationship between education and yearly salary may not be linear, but instead may be represented by a quadratic or some other function. However, notice that the right panel of Figure 7.4 suggests more of a linear trend between the number of years of experience and yearly salary.

Although a matrix plot cannot tell you the best model that should be used in a regression analysis, it can be used in an exploratory manner to indicate what the form of the relationship may be between each of the individual predictor variables and the response variable. More details on how to include nonlinear predictor variables in a regression model will be discussed in Section 7.14.

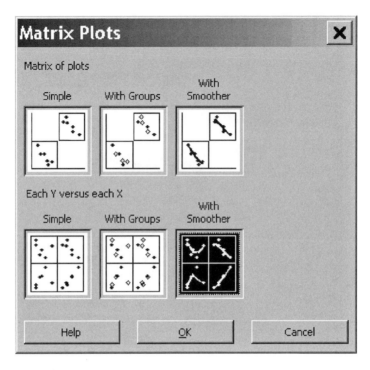

FIGURE 7.2
MINITAB dialog box to select a plot of each *y* versus each *x* with a smoother.

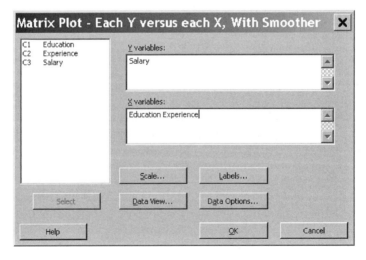

FIGURE 7.3
Matrix plot dialog box to select appropriate *y*- and *x*-variables.

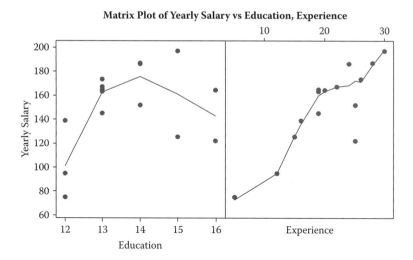

FIGURE 7.4

Matrix plot and smoother describing the relationship between salary and education and salary and experience.

7.4 Using MINITAB for Multiple Regression

The same steps are required to use MINITAB to run a multiple regression analysis as with a simple regression analysis. Under the **Stat** menu, select **Regression**, as shown in Figure 7.5. This gives the regression dialog box presented in Figure 7.6.

Then selecting the appropriate response and predictor variables gives the MINITAB output, as presented in Figure 7.7. The highlighted section on the top portion of the MINITAB printout in Figure 7.7 gives the estimated model equation:

$$\hat{y} = \hat{\beta}_0 + \hat{\beta}_1 x_1 + \hat{\beta}_2 x_2 \approx 108 - 4.18x_1 + 4.88x_2$$

Also highlighted in Figure 7.7 are the parameter estimates, standard errors for each of the coefficients, values of the test statistic for each coefficient, and the respective p-values. Table 7.2 summarizes the estimated regression parameters, the corresponding standard errors for each of the individual coefficients, the values of a test statistic T, and p-values for each of the individual predictors.

Before we describe in more detail how to make inferences using the information contained in Table 7.2, we will first describe a way to assess the initial fit of a multiple regression model by using the coefficient of determination and conducting an overall test of the model.

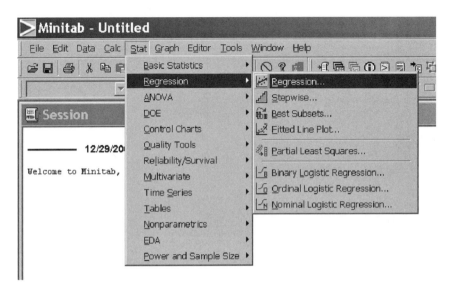

FIGURE 7.5
MINITAB commands for running a regression analysis.

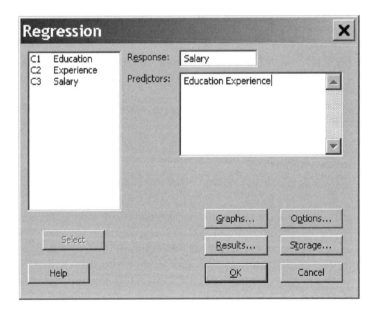

FIGURE 7.6
MINITAB regression dialog box.

TABLE 7.2

Table Summarizing the Output of the Multiple Regression Analysis in Figure 7.7

Parameter Estimate	Estimated Standard Error	Test Statistic, T	p-value
$\hat{\beta}_0 = 108.21$	54.01	2.00	.068
$\hat{\beta}_1 = -4.182$	4.490	−0.93	.370
$\hat{\beta}_2 = 4.8824$	0.9215	5.30	.000

Regression Analysis: Salary versus Education, Experience

```
The regression equation is
Salary = 108 - 4.18 Education + 4.88 Experience

Predictor      Coef   SE Coef      T      P
Constant     108.21     54.01   2.00  0.068
Education     -4.182     4.490  -0.93  0.370
Experience    4.8824    0.9215   5.30  0.000

S = 19.1341   R-Sq = 73.4%   R-Sq(adj) = 68.9%

Analysis of Variance

Source           DF        SS      MS      F      P
Regression        2   12096.0  6048.0  16.52  0.000
Residual Error   12    4393.4   366.1
Total            14   16489.3

Source       DF   Seq SS
Education     1    1819.0
Experience    1   10276.9

Unusual Observations

Obs  Education  Salary     Fit  SE Fit  Residual  St Resid
9         16.0  122.00  163.36   10.23    -41.36    -2.56R

R denotes an observation with a large standardized residual.
```

FIGURE 7.7
MINITAB printout for multiple regression analysis.

7.5 Coefficient of Determination for Multiple Regression

As with simple linear regression, the coefficient of determination, R^2, can be used to determine the proportion of variability in y that is explained by the estimated regression model, and it is calculated by using the following formula:

$$R^2 = \frac{SSR}{SST} = \frac{SST - SSE}{SST}$$

where SST is the total sum of squares, SSR is the regression sum of squares, and SSE is the error sum of squares.

Example 7.2

To calculate the coefficient of determination for the results of the multiple regression analysis presented in Figure 7.7, we can take the ratio of the regression sum of squares to the total sum of squares as follows:

$$R^2 = \frac{SSR}{SST} = \frac{SST - SSE}{SST} = \frac{16489.3 - 4393.4}{16489.3} = 0.734$$

Thus, 73.4% of the variability in executive salaries is attributed to the model that includes education and experience as predictor variables. Recall that an R^2 value near 100% suggests that a large amount of variability in the response variable is due to the given model, whereas an R^2 value near 0 suggests that the model does not account for a large amount of variability in the response variable.

7.6 Analysis of Variance Table

When using MINITAB to run a regression analysis, you may have noticed that an analysis of variance table also appears in the printout (see the bottom of the printout in Figure 7.7). We can use the information in the analysis of variance table to conduct an *F-test* to determine whether the multiple regression model that we have specified thus far is useful in predicting the response variable. In other words, we want to know how well the overall model (which includes all of the given predictor variables described in our model) contributes to predicting the response variable.

Because we are interested in determining whether the overall model contributes to predicting the response, y, first we need to test whether at least one of the predictor variables is significantly different from 0. To do this we will consider the following null and alternative hypotheses:

$$H_0 : \beta_1 = \beta_2 = \cdots = \beta_j = 0$$

$$H_A : \text{At least one } \beta_i \text{ is different from } 0$$

The test statistic that we will use consists of the ratio of the mean square error due to the regression (MSR) and the mean square error due to the residual error (MSE), as follows:

$$F = \frac{MSR}{MSE}$$

In performing this F-test we will be using the *F-distribution*. An F-distribution has the shape of a skewed curve, as presented in Figure 7.8.

The value of f_α in Figure 7.8 defines the rejection region for a significance level of α. Notice that the rejection region for an F-distribution is always in the right tail.

As with the t-distribution, there are many F-distributions, each determined by their degrees of freedom. There are two parameters for F-distributions, which represent the degrees of freedom, instead of only one, as with the t-distribution. The first number of degrees of freedom for the F-distribution is called the *degrees of freedom for the numerator*, and the second is called the *degrees of freedom for the denominator*. Because our test statistic is the ratio of the mean square error due to the regression and the mean square error due to the residual error, the number of degrees of freedom for the numerator would be $p - 1$, and the number of degrees of freedom for the denominator would be $n - p$, where p is the total number of parameters being estimated in the model (the total number of parameters being estimated includes β_0), and n is the sample size.

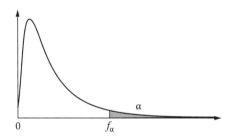

FIGURE 7.8
F-distribution.

We can use the following notation to describe the value of f for a specified level of significance and number of degrees of freedom for the numerator and the denominator as follows:

$$f_\alpha (p - 1, n - p)$$

where α is the given level of significance.

Example 7.3

Using a significance level of .05 for the executive salary data and the MINITAB printout given in Figure 7.7, suppose we want to determine if the overall model contributes to predicting y. The null and alternative hypotheses we would be testing are

$$H_0 : \beta_1 = \beta_2 = 0$$

$$H_A : \text{At least one } \beta_i \text{ is different from 0.}$$

In order to do this we need to find the ratio of the mean square error due to the regression and the mean square error due to the residual error as follows:

$$F = \frac{MSR}{MSE}$$

The mean square error due to the regression (MSR) can be found by taking the regression sum of squares (SSR) and dividing it by $p - 1$, as follows:

$$MSR = \frac{SSR}{p - 1}$$

where $SSR = \sum_{i=1}^{n} (\hat{y}_i - \bar{y})^2$, and p is the total number of beta parameters being estimated.

The mean square error (MSE) can be found by diving the sum of the squares of the errors (SSE) by $n - p$, as follows:

$$MSE = \frac{SSE}{n - p}$$

where $SSE = \sum_{i=1}^{n} (y_i - \hat{y}_i)^2$, p is the total number of beta parameters being estimated, and n is the sample size.

Table 7.3 illustrates these calculations done by hand (rounded to four decimal places).

TABLE 7.3

Estimated Value ($\hat{y}_1$), Difference between Estimated and Observed Values ($y_i - \hat{y}_i$), and Difference between Estimated Values and the Mean ($\hat{y}_i - \bar{y}$)

Education in Years (x_1)	Experience in Years (x_2)	Yearly Salary in Thousands of Dollars (y)	Estimated Value ($\hat{y}$)	Difference between Observed and Estimated Values ($y_i - \hat{y}_i$)	Difference between Estimated Values and Mean ($\hat{y}_i - \bar{y}$)
12	12	95	116.6167	−21.6167	−33.7167
13	19	145	146.6115	−1.6115	−3.7219
16	20	164	138.9478	25.0522	−11.3855
14	24	186	166.8415	19.1585	16.5081
15	30	197	191.9539	5.0461	41.6206
12	16	139	136.1463	2.8537	−14.1870
13	19	163	146.6115	16.3885	−3.7219
15	15	125	118.7178	6.2822	−31.6155
16	25	122	163.3598	−41.3598	13.0265
13	26	173	180.7883	−7.7883	30.4550
14	25	152	171.7239	−19.7239	21.3905
12	5	75	82.4398	−7.4398	−67.8935
13	19	165	146.6115	18.3885	−3.7219
13	22	167	161.2587	5.74131	10.9254
14	28	187	186.3711	0.6289	36.0378

Now use the data given in Table 7.3 to perform the following calculations:

$$SSR = \sum_{i=1}^{15} (\hat{y}_i - \bar{y})^2 \approx 12095.97$$

$$MSR = \frac{SSR}{p-1} = \frac{12095.97}{3-1} \approx 6047.99$$

Note that $p = 3$ because there are three parameters that are being estimated, β_0, β_1, and β_2.

Similarly,

$$SSE = \sum_{i=1}^{15} (y_i - \hat{y}_i)^2 \approx 4393.37$$

$$MSE = \frac{SSE}{n-p} = \frac{4393.37}{15-3} = 366.11$$

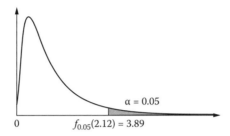

FIGURE 7.9
F-distribution for a significance level of .05 with 2 degrees of freedom for the numerator and 12 degrees of freedom for the denominator.

Therefore,

$$F = \frac{MSR}{MSE} = \frac{6047.99}{366.11} = 16.52$$

We now compare the value of this test statistic to the F-distribution with $p - 1 = 3 - 1 = 2$ degrees of freedom for the numerator and $n - p = 15 - 3 = 12$ degrees of freedom for the denominator using the F-distribution in Table 4 of Appendix A, as shown in Figure 7.9.

Since the value of the test statistic $F = 16.52$ is greater than the value of $f = 3.89$, which defines the rejection region for $\alpha = 0.05$ with 2 degrees of freedom for the numerator and 12 degrees of freedom for the denominator, we can accept the alternative hypothesis and reject the null hypothesis. Thus, we can infer that at least one of the population predictor parameters is significantly different from 0. In essence, this test suggests that the overall model that includes the independent variables of years of education and years of experience is useful in predicting yearly salary.

The highlighted portion of the analysis of variance table obtained from the MINITAB printout, given in Figure 7.10, provides approximately the same values for SSR, SSE, MSR, and MSE along with the value of the F-statistic (the slight difference is due to round-off error).

Notice that the p-value for the F-test is .000. Since the p-value is less than our predetermined level of significance of .05, we can accept the alternative

Analysis of Variance

Source	DF	SS	MS	F	P
Regression	2	12096.0	6048.0	16.52	0.000
Residual Error	12	4393.4	366.1		
Total	14	16489.3			

FIGURE 7.10
Analysis of variance table for the regression analysis predicting executive salaries based on experience and education.

hypothesis that at least one β_i is different from 0, and reject the null hypothesis that all of the β_i's are equal to 0. Thus, at least one of our predictor variables is useful in predicting y.

Just because the model appears to be useful for predicting executive salaries, the F-test does not suggest that the specified model is the best, nor does it suggest that the predictor variables of years of education and years of experience are the only variables that are useful in predicting executive salaries. Using the F-test for a multiple regression analysis basically serves as a first step in determining if at least one of the population predictor variables in the model is statistically significant in predicting the response variable. If the p-value for the F-test is less than the specified level of significance, this means that at least one of the predictor variables is significant in predicting the response, and the analysis should be pursued further. If the p-value for the F-test is greater than the specified level of significance, this implies that none of the predictor variables are significant in predicting the response, and you may want to consider formulating a different model.

7.7 Testing Individual Population Regression Parameters

After conducting an F-test and if the overall model is useful in predicting y, we may be interested in determining the effect that each of the individual predictor variables has on the response variable. For the last example, we might be interested in estimating the effect that the individual predictor variable of years of education or years of experience has on an executive's yearly salary.

If we can infer that the population parameter for a single independent variable is not significantly different from 0, then we might expect that this variable does not have an impact on the response variable. If we can infer that the population parameter for a single independent variable is significantly different from 0, then we might expect that this variable does have an impact on the response variable. We can assess the effect of the ith predictor variable on the response variable by conducting the following hypothesis test:

$$H_0 : \beta_i = 0$$
$$H_A : \beta_i \neq 0$$

In order to determine which of the individual population predictor variables has an impact on the response variable, we can test each of them individually by using the following test statistic:

$$T = \frac{\text{Parameter estimate for predictor } i}{\text{Estimated standard error for predictor } i}$$

Similar to many hypothesis tests we have done thus far, this test statistic is centered about the true but unknown population parameter β_i and has the t-distribution with $n - p$ degrees of freedom, where p is the total number of beta parameters that are being estimated in the model and n is the number of observations.

Example 7.4

To test whether the level of education has an impact on executive salaries (holding experience fixed), we would conduct a hypothesis test to see if the population parameter that represents the effect of the number of years of education on yearly salaries is significantly different from 0. Thus, our null and alternative hypotheses would be stated as follows:

$$H_0 : \beta_1 = 0$$
$$H_A : \beta_1 \neq 0$$

Using the parameter estimate $\hat{\beta}_1$ for the number of years of education and the standard error for this coefficient that were obtained from the MINITAB printout given in Figure 7.7, the test statistic would be

$$T = \frac{-4.182}{4.490} \approx -0.931$$

$$d.f = n - p = 15 - 3 = 12$$

Then by comparing the value of the test statistic $T = -0.931$ to the value $t = 2.179$ that defines the rejection region ($\alpha = 0.05$), as illustrated in Figure 7.11, we would claim that we do not have enough evidence to suggest that the true but unknown population parameter β_1 is significantly different from 0. This is because the test statistic does not fall into the rejection region.

Thus, by using the estimated multiple regression equation, we can infer that the years of education parameter does not have a significant effect in predicting yearly executive salaries, holding years of experience fixed.

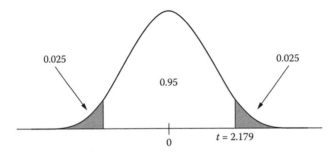

FIGURE 7.11
The t-distribution for a significance level of .05 with 12 degrees of freedom.

Similarly, to test whether years of experience is significant in predicting executive salaries (holding education fixed), we would compare the following test statistic to the t-distribution with 12 degrees of freedom, illustrated in Figure 7.11:

$$T = \frac{4.8824}{0.9215} \approx 5.30$$

Since the value of $T = 5.30$ falls in the rejection region, we would conclude that years of experience is significant in predicting yearly executive salaries. To interpret this within the context of our problem, we could infer that for every additional year of experience, the yearly salary would increase by an average of approximately $4,882 dollars per year, holding the variable of education fixed.

In addition to testing the effect of each of the individual predictor variables, we can calculate a confidence interval for any of the individual population regression parameters by using the estimated values as follows:

$$\hat{\beta}_i \pm t_{\frac{\alpha}{2}} \cdot (\text{standard error of the estimate for } \hat{\beta}_i)$$

where $\hat{\beta}_i$ is the estimated coefficient for the ith predictor variable and t_α is described by the desired level of significance with $n - p$ degrees of freedom, where p is the total number of beta parameters being estimated in the model and n is the sample size.

Example 7.5

Suppose that we want to calculate a 95% confidence interval for the true but unknown population parameter β_2, which represents the effect of the years of experience on executive salaries. To do this, we would add and subtract the standard error of the estimate for the appropriate variable from the value of the estimated parameter, as follows:

$$4.8824 \pm 2.179 \times 0.9215 = (2.874, 6.890)$$

Interpreting this confidence interval would suggest that we are 95% confident that for every additional year of experience, this results in a salary increase between approximately $3,000 and $7,000 per year, holding the variable of education fixed.

If we calculated a 95% confidence interval for β_1, the effect of the years of education on executive salaries, we would find that our confidence interval covers both positive and negative values, as follows:

$$-4.182 \pm 2.179 \times 4.490 = (-13.966, 5.602)$$

And just as you might expect, because this confidence interval contains both positive and negative values, and hence the value of 0, there is no significant effect in using the years of education to predict the yearly salary of executives.

7.8 Using MINITAB to Test Individual Regression Parameters

Whenever a regression analysis is run using MINITAB, it automatically tests whether each of the individual regression parameters is significantly different from 0. This is highlighted in the MINITAB printout given in Figure 7.12.

Notice that the values of the test statistics for each of the regression parameters are the same as when we calculated them by hand, and the *p*-values correspond to the same conclusions that the number of years of experience has a significant effect on yearly salaries, but the number of years of education does not.

Regression Analysis: Salary versus Education, Experience

```
The regression equation is
Salary = 108 - 4.18 Education + 4.88 Experience

Predictor        Coef   SE Coef        T       P
Constant       108.21     54.01     2.00   0.068
Education      -4.182     4.490    -0.93   0.370
Experience     4.8824    0.9215     5.30   0.000

S = 19.1341     R-Sq = 73.4%     R-Sq(adj) = 68.9%

Analysis of Variance

Source             DF        SS       MS       F       P
Regression          2   12096.0   6048.0   16.52   0.000
Residual Error     12    4393.4    366.1
Total              14   16489.3

Source          DF    Seq SS
Education        1    1819.0
Experience       1   10276.9

Unusual Observations

Obs   Education   Salary      Fit   SE Fit   Residual   St Resid
  9        16.0   122.00   163.36    10.23     -41.36     -2.56R

R denotes an observation with a large standardized residual.
```

FIGURE 7.12
MINITAB printout highlighting the *t*-test for each of the individual beta parameters for the executive salary data in Table 7.1.

7.9 Multicollinearity

The reason for including numerous predictors in a multiple regression model is because these predictor variables are believed to be related to the response, or dependent, variable. And in order to develop a good model, we need to be sure that we have included all possible predictor variables that could have a significant impact on the response variable. However, we need to be careful that the individual predictor variables themselves are not highly correlated with each other.

Some very serious consequences may occur if a regression model consists of predictor variables that are highly correlated with each other. In particular, the estimates of the population regression parameters (the betas) can be very unstable. What this means is that if highly correlated variables are included as predictor variables, the regression parameter estimates may fluctuate dramatically when such terms are added to or dropped from the model. This can happen when the standard errors become very large.

Using the executive salary data, we are now going to look at the impact that including highly correlated predictor variables can have on a regression analysis.

Example 7.6

Consider the data set provided in Table 7.4. This data set consists of some of the same data given in Table 7.1, but with the addition of the variable of age in years (x_3).

Now notice what happens when we use MINITAB to run two different regression analyses: one analysis (Figure 7.13) is the MINITAB printout from the regression analysis that includes the predictor variables for education and experience but does not include the variable of age, and the other analysis (Figure 7.14) is the MINITAB printout from the regression analysis that includes the predictor variables for education and experience along with the variable of age.

Table 7.5 summarizes the results of the regression analyses from both Figures 7.13 and 7.14, which compare the parameter estimates, standard errors, and p-values for these two separate models.

As can be seen in Table 7.5, the effect of years of experience on yearly salary was significant in the model that did not include the variable of age (Figure 7.13), but this same predictor was not significant in the model that included the variable of age (Figure 7.14). The reason for this difference in the parameter estimates is because by adding the variable of age, which is highly correlated with years of experience, the standard errors became much larger and the estimates of the effects of the independent variables became unstable. Therefore, by adding variables that are highly correlated with each other, we can generate opposite interpretations of the observed effect. The

TABLE 7.4

Executive Salary Data Consisting of the Variables of Years of Education, Years of Experience, Age in Years, and Yearly Salary in Thousands of Dollars

Observation Number	Education in Years (x_1)	Experience in Years (x_2)	Age in Years (x_3)	Yearly Salary in Thousands of Dollars (y)
1	12	12	42	95
2	13	19	48	145
3	16	20	51	164
4	14	24	53	186
5	15	30	61	197
6	12	16	46	139
7	13	19	50	163
8	15	15	45	125
9	16	25	56	122
10	13	26	56	173
11	14	25	55	152
12	12	5	35	75
13	13	19	48	165
14	13	22	52	167
15	14	28	57	187

standard error columns in Table 7.5 illustrate how large the standard errors became when the predictor age was added to the regression model.

7.10   Variance Inflation Factors

There is a measure, called the *variance inflation factor* (VIF), that can be used to determine whether or not a high degree of multicollinearity between the predictor variables can be affecting the estimates of the standard error for the estimated parameters. A high VIF suggests that the parameter estimates may become unstable because the variables that are highly correlated with each other can generate very large standard errors for the individual regression parameters. A low VIF would suggest that the correlation between the predictor variables is not large enough to significantly impact the estimates of the standard error of the estimated parameters.

Regression Analysis: Salary versus Education, Experience

```
The regression equation is
Salary = 108 - 4.18 Education + 4.88 Experience
```

Predictor	Coef	SE Coef	T	P
Constant	108.21	54.01	2.00	0.068
Education	-4.182	4.490	-0.93	0.370
Experience	4.8824	0.9215	5.30	0.000

```
S = 19.1341   R-Sq = 73.4%   R-Sq(adj) = 68.9%
```

Analysis of Variance

Source	DF	SS	MS	F	P
Regression	2	2096.0	6048.0	16.52	0.000
Residual Error	12	4393.4	366.1		
Total	14	16489.3			

Source	DF	Seq SS
Education	1	1819.0
Experience	1	10276.9

Unusual Observations

Obs	Education	Salary	Fit	SE Fit	Residual	St Resid
9	16.0	122.00	163.36	10.23	-41.36	-2.56R

R denotes an observation with a large standardized residual.

FIGURE 7.13
Regression analysis of years of education and years of experience on yearly salary.

Variance inflation factors for each individual predictor variable (excluding β_0) can be calculated as follows:

$$VIF_j = \frac{1}{1 - R_j^2}$$

where $j = 1, 2, \cdots, p{-}1$, p is the total number of beta parameters being estimated in the model, and R_j^2 is the coefficient of determination for the model in which the variable x_j is used as the response variable and all the other variables in the model are used as predictor variables.

To calculate the variance inflation factors, first we need to calculate the value of R^2 for each of the regression analyses in which each predictor variable is used as a response variable with all the other variables as predictor variables. For example, to find the variance inflation for each of the predictor variables of years of education (x_1), years of experience (x_2), and age (x_3)

Regression Analysis: Salary versus Education, Experience, Age

```
The regression equation is
Salary = 240 - 2.65 Education + 9.76 Experience - 5.01 Age

Predictor      Coef   SE Coef      T      P
Constant      240.4     213.9   1.12   0.285
Education    -2.646     5.193  -0.51   0.620
Experience    9.762     7.685   1.27   0.230
Age          -5.014     7.837  -0.64   0.535

S = 19.6231    R-Sq = 74.3%    R-Sq(adj) = 67.3%

Analysis of Variance

Source            DF        SS       MS      F      P
Regression         3   12253.6   4084.5  10.61  0.001
Residual Error    11    4235.7    385.1
Total             14   16489.3

Source         DF    Seq SS
Education       1    1819.0
Experience      1   10276.9
Age             1     157.6

Unusual Observations

Obs  Education  Salary     Fit   SE Fit   Residual   St Resid
  9       16.0  122.00  161.30    10.98     -39.30     -2.42R

R denotes an observation with a large standardized residual.
```

FIGURE 7.14
Regression analysis of years of education, years of experience, and age in years on yearly salary.

using the executive salary data given in Table 7.4, we would have to run three separate regression analyses to find the value of the R^2 statistic for each of the three models, as described in Table 7.6.

Example 7.7

To find the VIF for the variable of years of education (x_1), we will use MINITAB to run a regression analysis (Figure 7.15) in which education is treated as the response variable and experience and age are the predictor variables. Then using the value of $R^2 = 0.436$ from such an analysis, as described in the MINITAB

TABLE 7.5

Parameter Estimates, Standard Errors, and *p*-Values for the Regression Model That Does Not Include the Variable of Age (Figure 7.13) versus the Regression Model That Does Include the Variable of Age (Figure 7.14)

	Parameter Estimate		Standard Error		*p*-value	
	Without Age Figure 7.13	With Age Figure 7.14	Without Age Figure 7.13	With Age Figure 7.14	Without Age Figure 7.13	With Age Figure 7.14
Constant	108.21	240.4	54.0	213.9	.068	.285
Education	−4.182	−2.646	4.490	5.193	.370	.620
Experience	4.8824	9.762	0.9215	7.685	.000	.230
Age	—	−5.014	—	7.837	—	.535

printout in Figure 7.15, we can calculate the variance inflation for the variable of education as follows:

$$VIF_{\text{Education}} = \frac{1}{1 - R^2_{\text{Education}}} = \frac{1}{1 - 0.436} \approx 1.773$$

A similar analysis can be used to find the VIFs for the variable that represents years of experience.

But how do you know when the value of a VIF is too high? As a general rule of thumb, a VIF above 5 indicates that there may be some strong correlations between the predictor variables and that such correlations could be affecting the estimated standard errors. Even modest correlations among the predictor variables can attribute to the estimated standard errors becoming very large. For many situations, variance inflation factors above 10 usually require further investigation.

7.11 Using MINITAB to Calculate Variance Inflation Factors

Under the **Options** box in the **Regression** dialog box, check the **Variance inflation factors** box to provide VIFs for a regression analysis, as can be seen in Figure 7.16.

TABLE 7.6

Regression Models to Calculate the Variance Inflation Factors

Response	Predictor	
x_1	x_2	x_3
x_2	x_1	x_3
x_3	x_1	x_2

Regression Analysis: Education versus Experience, Age

```
The regression equation is
Education = - 9.4 - 0.593 Experience + 0.698 Age

Predictor        Coef   SE Coef        T       P
Constant        -9.39     11.58    -0.81   0.433
Experience    -0.5932    0.3914    -1.52   0.156
Age            0.6977    0.3863     1.81   0.096


S = 1.09090    R-Sq = 43.6%     R-Sq(adj) = 34.2%


Analysis of Variance

Source            DF        SS      MS       F       P
Regression         2    11.052   5.526    4.64   0.032
Residual Error    12    14.281   1.190
Total             14    25.333

Source        DF   Seq SS
Experience     1    7.171
Age            1    3.882
```

FIGURE 7.15
MINITAB printout of the regression analysis of years of experience and age in years on years of education.

Checking this box generates the MINITAB printout presented in Figure 7.17, where the highlighted values indicate the variance inflation factors for each of the three predictor variables.

Recall that the R^2 statistic is the ratio of the sum of squares of the regression analysis to the total sum of squares, and thus provides a measure of the proportion of the variability in y that is explained by the given regression model. If R^2 is close to 0, then the given regression model does not contribute to predicting y, and thus the VIF will be a small number. However, if R^2 is close to 1 (or 100%), then the given regression model does contribute to predicting y, and thus the value of the VIF will be very large. Notice in Figure 7.17 that the variance inflation factors for experience and age are very large. Large values of VIFs indicate that these variables are highly correlated with each other, and such a strong correlation could be having an impact on the parameter estimates. Thus, including both of these predictor variables in a regression model can generate very large standard errors, as was seen by comparing the parameter estimates, standard errors, and p-values, as displayed in Figures 7.13 and 7.14. Also, note that the hand calculations for the variance inflation factors may not always exactly match the value of the variance inflation factors generated by MINITAB because of round-off error.

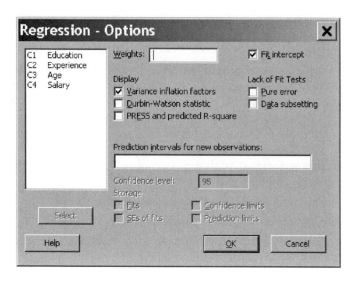

FIGURE 7.16
MINITAB regression options box to display variance inflation factors.

Regression Analysis: Salary versus Education, Experience, Age

```
The regression equation is
Salary = 240 - 2.65 Education + 9.76 Experience - 5.01 Age

Predictor      Coef   SE Coef      T      P     VIF
Constant      240.4     213.9   1.12  0.285
Education    -2.646     5.193  -0.51  0.620   1.774
Experience    9.762     7.685   1.27  0.230  92.239
Age          -5.014     7.837  -0.64  0.535  98.468

S = 19.6231   R-Sq = 74.3%   R-Sq(adj) = 67.3%

Analysis of Variance

Source          DF        SS      MS      F      P
Regression       3   12253.6  4084.5  10.61  0.001
Residual Error  11    4235.7   385.1
Total           14   16489.3

Source      DF    Seq SS
Education    1    1819.0
Experience   1   10276.9
Age          1     157.6

Unusual Observations
Obs  Education  Salary     Fit   SE Fit  Residual  St Resid
  9       16.0  122.00  161.30    10.98    -39.30     -2.42R
R denotes an observation with a large standardized residual.
```

FIGURE 7.17
MINITAB printout for a regression analysis highlighting the variance inflation factors for each of the three predictor variables.

7.12 Multiple Regression Model Assumptions

Similar to simple linear regression analysis, before we can use an estimated model to make inferences regarding the unknown population model, we need to make sure that the following model assumptions are reasonably met. These assumptions are based on the distribution of the random error component and are identical to those model assumptions for a simple regression analysis:

1. The errors associated with any two observed values are independent of each other.
2. The error component is normally distributed.
3. The error component has constant variance.
4. The functional relationship between the predictor variables (the x's) and the response (y) can be established.

Just as with simple linear regression analysis, we can use MINITAB to create residual plots and perform some formal tests of these assumptions.

7.13 Using MINITAB to Check Multiple Regression Model Assumptions

In the same fashion as with simple regression analysis, the **Graphs** tab on the **Regression** dialog box provides three plots that can be used to assess the assumptions of normality and constant variation of the error component. The residual graphs for the model that predicts executive salaries based on the two predictor variables of years of education and years of experience are given in Figures 7.18 to 7.20.

We can also use the Ryan–Joiner test as a formal check on the assumption of normality of the error component. To do this, we first need to select the **Storage** tab from the **Regression** dialog box, and then check the **Residuals** box to store them, as shown in Figure 7.21.

Then by running the Ryan–Joiner test from the **Normality Test** selection under **Stat** and **Basic Statistics**, we get the Ryan–Joiner dialog box, as illustrated in Figure 7.22.

This gives the normal probability plot, Ryan-Jointer test statistics, and p-value, as in Figure 7.23.

Recall that the null and alternative hypotheses for the Ryan–Joiner test are as follows:

H_0 : The error component follows a normal distribution.

H_A : The error component does not follow a normal distribution.

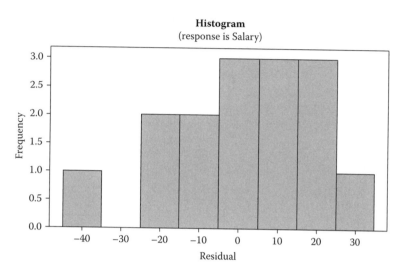

FIGURE 7.18
Histogram of the residual values.

Since the *p*-value is greater than our predetermined level of significance of $\alpha = 0.05$, this indicates that there is not enough evidence to reject the null hypothesis, and thus the assumption of normality does not appear to have been violated. In other words, because the value of the test statistic is greater

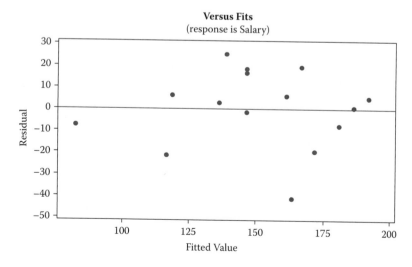

FIGURE 7.19
Residual versus fitted values plot.

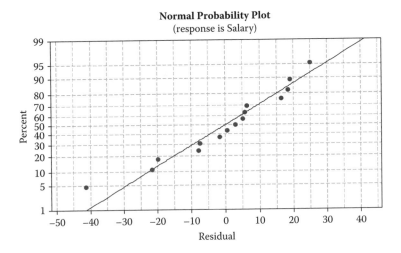

FIGURE 7.20
Normal probability plot of residuals.

than the smallest correlation defined by the corresponding critical value, we can infer that the assumption that the error component is normally distributed does not appear to have been violated.

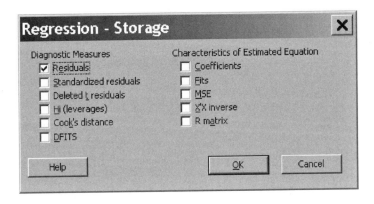

FIGURE 7.21
Regression dialog box to store residuals.

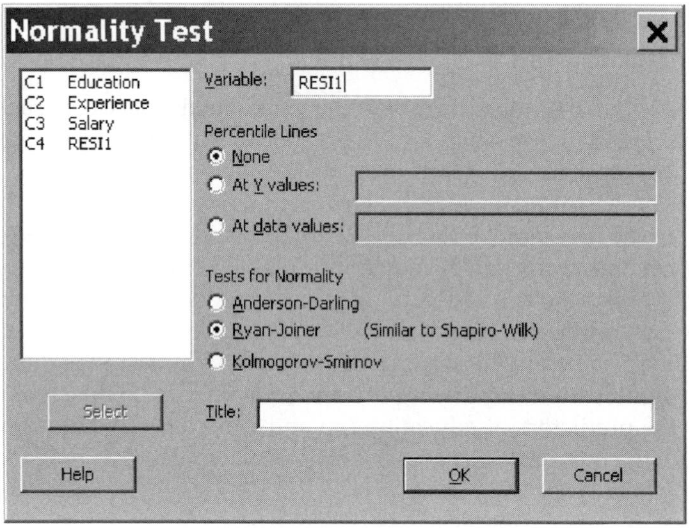

FIGURE 7.22
MINITAB dialog box to run the Ryan–Joiner test for normality of the residuals.

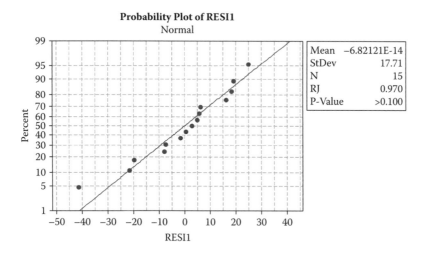

FIGURE 7.23
Normal probability plot, Ryan–Joiner test statistic, and p-value.

7.14   Quadratic and Higher-Order Predictor Variables

Recall that a matrix plot illustrates the relationship between each of the individual predictor variables and the response variable. The matrix plot in Figure 7.24 illustrates the years of education versus yearly salary and the years of experience versus yearly salary from the data provided in Table 7.1.

Notice in the first panel of Figure 7.24 that the relationship between years of education and yearly salary may not resemble a linear one, but appears to resemble a quadratic relationship. One way to model such a nonlinear relationship using multiple regression analysis could be to include the quadratic term of the variable of education in the model and determine whether such a term is significant in predicting the response variable. Thus, a model that includes the predictors of education (x_1) and experience (x_2) along with the addition of the quadratic term for education (x_1^2) would be as follows:

$$y = \beta_0 + \beta_1 x_1 + \beta_2 x_1^2 + \beta_3 x_2 + \varepsilon$$

7.15   Using MINITAB to Create a Quadratic Variable

We can use MINITAB to create a second-order term of a variable by using the calculator function under **Calc**, as illustrated in Figure 7.25.

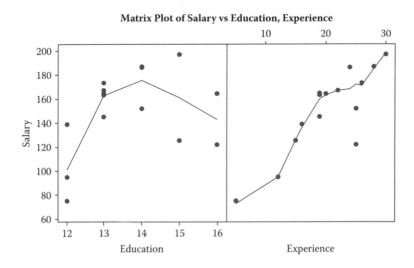

FIGURE 7.24
Matrix plot of salary versus education (x_1) and salary versus experience (x_2) from the data in Table 7.1.

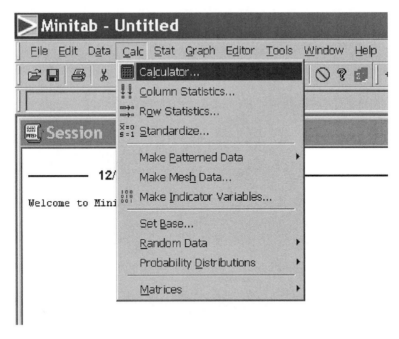

FIGURE 7.25
MINITAB commands for the calculator function.

This brings up the calculator dialog box in Figure 7.26.

We can use the calculator function to create the square of the education variable by describing the calculation to be done in the expression box and by specifying where to store the results of the calculations. By storing the result in column 4 (C4), we can create the second power of the variable of education by squaring the variable using the ** button to raise it to the second power, as illustrated in Figure 7.27.

This creates a new column of data in column 4 of the MINITAB worksheet, which consists of the square of the variable of education, as illustrated in Figure 7.28.

Now in order to test whether the quadratic predictor of education is significant in predicting executive salaries, we can run a regression analysis that includes the quadratic predictor, as in Figure 7.29. This gives the MINITAB printout in Figure 7.30.

Notice that based on the *F*-test, the overall model is useful in predicting yearly salary. However, both the years of education and the years of education-squared predictor variables are not significant in predicting the

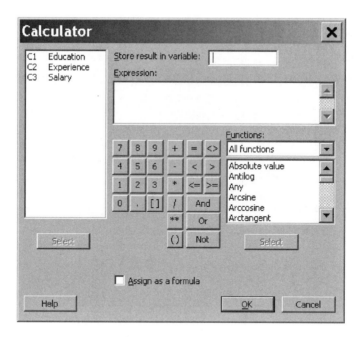

FIGURE 7.26
MINITAB dialog box for the calculator function.

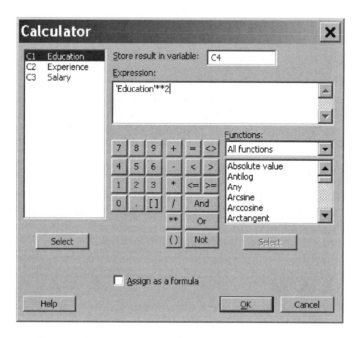

FIGURE 7.27
MINITAB calculator function to create the squared value of the variable "education."

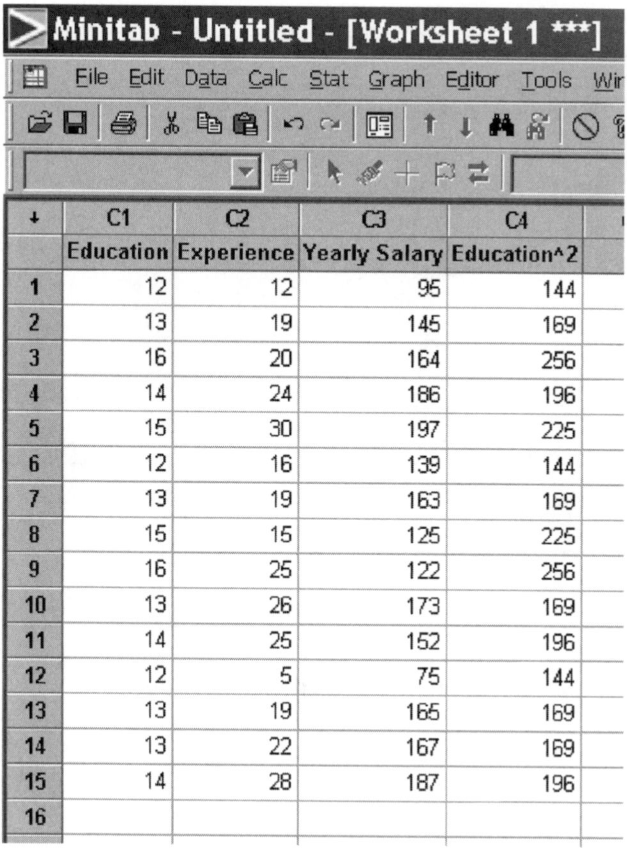

	C1	C2	C3	C4	
	Education	Experience	Yearly Salary	Education^2	
1	12	12	95	144	
2	13	19	145	169	
3	16	20	164	256	
4	14	24	186	196	
5	15	30	197	225	
6	12	16	139	144	
7	13	19	163	169	
8	15	15	125	225	
9	16	25	122	256	
10	13	26	173	169	
11	14	25	152	196	
12	12	5	75	144	
13	13	19	165	169	
14	13	22	167	169	
15	14	28	187	196	
16					

FIGURE 7.28
MINITAB worksheet showing the square of the variable "education."

yearly salary (the p-values for both of these parameters are larger than any reasonable significance level).

When modeling a nonlinear relationship, the interpretation of the parameters is no longer as simple as if we were modeling a strict linear relationship. For instance, only the sign of the estimated coefficient for a quadratic term has any interpretive value, and this only suggests whether the quadratic function opens upwards or downwards. Thus, when we include a nonlinear power of a predictor variable in a multiple regression model, we can no longer interpret the estimate of the effect of the quadratic term in the same manner as if it were a linear term. Similar considerations would hold true for any other higher-order powers of the independent variables.

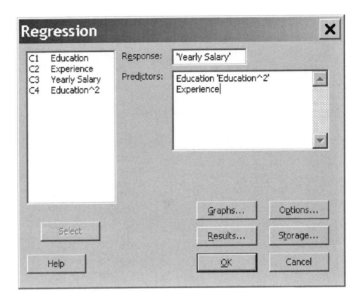

FIGURE 7.29
Regression dialog box including the square of education.

Regression Analysis: Salary versus Education, Education^2, Experience

```
The regression equation is
Salary = - 993 + 155 Education - 5.62 Education^2 + 3.82 Experience

Predictor      Coef   SE Coef       T       P
Constant     -992.7     753.6   -1.32   0.215
Education     155.4     109.1    1.42   0.182
Education^2   -5.617     3.836   -1.46   0.171
Experience    3.816     1.142    3.34   0.007

S = 18.2823    R-Sq = 77.7%    R-Sq(adj) = 71.6%

Analysis of Variance

Source            DF         SS        MS       F       P
Regression         3    12812.7    4270.9   12.78   0.001
Residual Error    11     3676.7     334.2
Total             14    16489.3

Source          DF   Seq SS
Education        1   1819.0
Education^2      1   7263.8
Experience       1   3729.8

Unusual Observations

Obs   Education   Salary     Fit   SE Fit   Residual   St Resid
  3        16.0   164.00  132.36    12.08      31.64       2.31R
  9        16.0   122.00  151.44    12.72     -29.44      -2.24R

R denotes an observation with a large standardized residual.
```

FIGURE 7.30
MINITAB printout including the quadratic of the variable "education."

Exercises

1. The data set "Expensive Homes" gives the asking prices for a random sample of 115 single-family homes. In addition to the asking price of the home, the data set has measures on the number of bedrooms, number of bathrooms, square footage, year the home was built, age of the home, lot size, number of garages, and number of stories.

   a. Write a population model equation that describes the relationship between the asking price of homes (y) as a linear function of the number of bedrooms (x_1), number of bathrooms (x_2), square footage (x_3), age of the home (x_4), lot size (x_5), number of garages (x_6), and the number of stories (x_7).

   b. Using MINITAB, draw a matrix plot that shows the relationship between the response variable for each of the predictor variables.

   c. Using MINITAB, find the estimated model of the function that relates the asking price to the other seven independent variables.

   d. Using a significance level of .05, determine whether the given model is useful in predicting the asking price of homes.

   e. Using both exploratory plots and formal tests, check to see if your model assumptions have been reasonably met.

   f. Determine which, if any, predictor variables are significant in predicting the asking price. Interpret your findings.

   g. Determine if any of the independent variables are highly correlated with each other, and decide if this has had an impact on your findings.

   h. Can you identify any other factors that you think would influence the asking price of homes? If so, what are they, and how do you think they could impact your analysis?

2. What factors influence how much residents of Hawaii are willing to spend on a mortgage or rent payment? The data set "Hawaii Mortgage-Rent" consists of 210 individuals living in Hawaii and was obtained by taking a sample from the 2003 U.S. Census (http://www.census.gov/acs/www/Products/PUMS/C2SS/minmaxval3.htm). This data set contains measures of the number of individuals in the household, the total commute time to work, the rent/mortgage payment, the number of vehicles in the household, the monthly cost of utilities, the household income, and the number of workers in the family.

   a. Draw a matrix plot that shows the relationship between the response and each of the predictor variables.

b. Use this data set to develop a linear model that can be used to predict which factors, if any, impact the amount spent on mortgage or rent payments ($\alpha = 0.10$).

c. Using exploratory plots and formal tests, check the regression model assumptions and comment on whether you believe these model assumptions hold true.

d. Do you think any higher-order terms need to be included in the model? Justify your response.

3. You will notice in the "Hawaii Mortgage-Rent" data set that many of the variables have missing values. Data sets with missing values are not uncommon and tend to be due to individuals not providing responses to all the questions in a survey. One strategy that can be used to deal with missing values is *imputation*. Imputation entails imputing a value in place of each missing value. One typical imputation strategy would be to replace all of the missing values of a variable with the mean value of that variable.

a. For the "Hawaii Mortgage-Rent" data set, use the strategy of imputing the mean value for the missing values for each of the variables that has missing data, and then run a regression analysis ($\alpha = 0.10$).

b. Comment on what you think has changed by imputing the mean value for the missing values of each variable ($\alpha = 0.10$).

c. Did imputing the mean change the conclusions you found in Exercise 2 regarding what factors contribute to how much Hawaii residents are willing to pay for mortgage or rent?

4. The data set "Credit Card Debt" in Table 7.7 gives the amount of credit card debt for a sample of twenty-four individuals along with measures of their age, education level, and yearly salary.

a. Using this data set, develop a model that can be used to predict credit card debt based on age, education, and salary ($\alpha = 0.05$).

b. Draw a matrix plot that shows the relationship between the response and each of the predictor variables.

c. Using exploratory techniques and formal tests, check relevant model assumptions and check for multicollinearity.

d. What factors are significant in predicting credit card debt? Interpret your findings in the context of the problem, and comment on whether you think your conclusions make sense.

e. Using your estimated regression equation, estimate what the average credit card debt would be for an individual who is 25 years old, with 20 years of education, and whose salary is $50,000 per year.

f. Can you think of any other factors that could influence the amount of credit card debt?

TABLE 7.7

"Credit Card Debt" Data Set

Credit Card Debt (in dollars)	Age (in years)	Education Level (in years)	Yearly Salary (in thousands)
5,020	45	16	55
8,550	50	19	75
6,750	48	18	68
1,255	29	18	42
0	19	12	25
0	20	13	38
4,685	35	16	50
5,790	33	14	57
2,150	30	16	42
550	19	12	35
6,850	22	16	42
1,250	45	16	58
460	50	16	62
1,280	35	12	39
500	28	14	52
430	32	14	58
250	44	12	35
650	49	12	41
1,285	32	14	57
690	50	12	35
500	18	13	41
6,000	30	17	90
7,500	39	16	54
12,850	32	20	85

5. The data set "Ferry" consists of the revenue, time of day the ferry left the dock, number of passengers on the ferry, number of large objects on the ferry, weather conditions, and number of crew for a random sample of twenty-seven ferry runs over a 1-month period.

    a. Write the population model equation that relates the revenue generated from the ferry (y) as a linear function of the number of passengers transported (x_1), the number of large objects transported on the ferry (x_2), and the number of crew (x_3).

b. Using MINITAB, draw a matrix plot that shows the relationship between the response variable and each of the predictor variables.

c. Using MINITAB, find the estimated model of the linear function that relates the revenue generated to all of the independent variables.

d. Using a significance level of .05, determine if the model is useful in predicting the revenue generated by the ferry.

e. Using exploratory plots and formal tests, check to see if your model assumptions have been reasonably met.

f. Determine which, if any, of the independent variables are highly correlated with each other, and decide if this has had an impact on your findings.

6. The data set "Standardized Test" consists of a random sample of twenty-four high school students' scores on a state standardized test (score) along with descriptive measures of gender (female), family income in thousands of dollars (income), race (minority = W if identified as white, minority = NW if identified as nonwhite), and the score received on a previous aptitude test (aptitude).

a. Write the population model equation that relates the score received on the standardized test (y) as a linear function of income (x_1), and the score received on a previous aptitude test (x_2).

b. Using MINITAB, draw a matrix plot that shows the relationship between the response variable and each of the predictor variables.

c. Using MINITAB, find the estimated model of the linear function that relates the score received on the standardized test to income and aptitude scores.

d. Using a significance level of .05, determine if the model is useful in predicting standardized test scores.

e. Using exploratory graphs and formal tests, check to see if your model assumptions have been reasonably met.

f. Determine which, if any, of the independent variables are highly correlated with each other, and decide if this has had an impact on your findings.

8

More on Multiple Regression

8.1 Introduction

As we have seen from the different multiple regression analyses we have performed thus far, in some ways it may seem like we are on a fishing expedition trying to discover a useful relationship between the response and predictor variables. Not only do we need to find the appropriate model that describes the relationship between the set of predictor variables and the response variable, but we also need to be sure that the regression assumptions are reasonably met, and that we have not included predictor variables in the regression model that are highly correlated with each other.

You may have also noticed that all of the predictor and response variables that we have considered thus far have been continuous. However, there may be occasions when including categorical predictors in a multiple regression analysis may be of interest.

In this chapter we will describe ways to include categorical predictor variables in a regression model and will also describe a more systematic approach to finding the model with the best fit. This chapter will also describe how confidence and prediction intervals can be found with a multiple regression analysis.

8.2 Using Categorical Predictor Variables

There may be occasions when you want to include a predictor variable that describes some characteristic of interest, but such a characteristic can only be represented as a categorical variable. Recall from Chapter 1 that a categorical variable can be used to represent different categories. For instance, the variable "gender" is a categorical variable because it represents two different categories: male and female. We can code this variable in such a way that the number 1 indicates the presence of the characteristic of interest (being female) and the number 0 indicates the absence of the characteristic of interest (being male). Even though gender can be coded numerically, you may recall from Chapter 1 that the numbers used to code a categorical variable are only used for reference, and they cannot be manipulated by using basic arithmetic. In other words, a coding scheme for a categorical variable only represents a way to characterize different categories, but such codes have no mathematical properties.

Categorical predictor variables can be included in a multiple regression analysis provided that they are described in a specific manner, as will be illustrated in the next two examples.

Example 8.1

Consider the portion of the "Executive Salary" data set presented in Table 8.1. This data set is the same one we used in the last chapter (Table 7.1), with the addition of the categorical variable "female." The variable "female" in Table 8.1 is used to represent the gender of the executive. This variable is coded in such a way that if *Female* = 1, then the subject is identified as female, and if *Female* = 0, then the subject is identified as male.

Notice that for any observation in the data set, the assignment of the value *Female* = 1 represents the presence of the characteristic of interest (being a female), and the assignment of the value *Female* = 0 represents the absence of the characteristic of interest (not being a female). Using the numbers 1 and 0 to represent the presence or absence of some characteristic of interest will allow us to include such variables in a multiple regression model because using the values of 1 and 0 will provide the opportunity to make meaningful inferences using the estimated parameters. Such variables that only consist

TABLE 8.1

Executive Salary Data Including a Variable to Represent Gender Where *Female* = 1 if the Subject Is Identified Female, and *Female* = 0 if the Subject Is Identified as Male

Observation Number	Education in Years (x_1)	Experience in Years (x_2)	Female (x_3)	Yearly Salary in Thousands of Dollars (y)
1	12	12	1	95
2	13	19	0	145
3	16	20	0	164
4	14	24	0	186
5	15	30	0	197
6	12	16	1	139
7	13	19	0	163
8	15	15	1	125
9	16	25	1	122
10	13	26	0	173
11	14	25	1	152
12	12	5	1	75
13	13	19	0	165
14	13	22	0	167
15	14	28	1	187

of the numbers 1 or 0 to represent the presence or absence of some character-istic are called *binary* or *indicator variables*. A binary or indicator variable can take on only one of two possible values, either a 1 or a 0.

The linear population regression model that includes the variables of years of education (x_1) and years of experience (x_2), along with the gender of the executive (x_3) would be written as follows:

$$y = \beta_0 + \beta_1 x_1 + \beta_2 x_2 + \beta_3 x_3 + \varepsilon$$

where the variable x_3 now represents the categorical variable "female," which describes whether or not the subject is identified as female.

We will use MINITAB to find the parameter estimates for the regression model, which includes a categorical predictor variable, and we will also describe how to interpret the effect of a categorical predictor variable in a multiple regression analysis.

8.3   Using MINITAB for Categorical Predictor Variables

Figure 8.1 shows the data from Table 8.1 entered into MINITAB. Notice that the variable "female" is represented as a numeric variable because we are using the numeric values of 1 to represent female and 0 to represent male; there is no text or date associated with this variable.

To include a categorical predictor variable in a multiple regression model, all we need to do is run a regression analysis in the exact same way as if we had only numeric predictors. Thus, all we need to do is to include the predic-tor "female" along with all of the other predictor variables, as illustrated in Figure 8.2. Then, running the regression analysis yields the MINITAB print-out presented in Figure 8.3.

Notice that the highlighted portion of Figure 8.3 represents the effect of gender on executive salaries. The analysis suggests that "female" is a sig-nificant predictor of executive salaries ($p < .05$). Interpreting the parameter estimates for a categorical predictor variable is very similar to what we have done in interpreting the parameter estimates for numeric predictor variables. However, to interpret the parameter estimate for a categorical predictor that consists of only two possible values, the category that has the value of 1 is interpreted with respect to the category that has the value of 0. Thus, for this example, being a female executive (*Female* = 1) would suggest that the yearly salary is $24,954 less than that of male executives (holding all other variables fixed). Because gender is represented by a binary or indicator variable, we can only interpret the finding for the characteristic that has been assigned the value of 1, which in this case represents being identified as female. The interpretation of a parameter estimate for a categorical predictor variable is always given with respect to the baseline characteristic, and the baseline characteristic is that which has been designated with the value of 0.

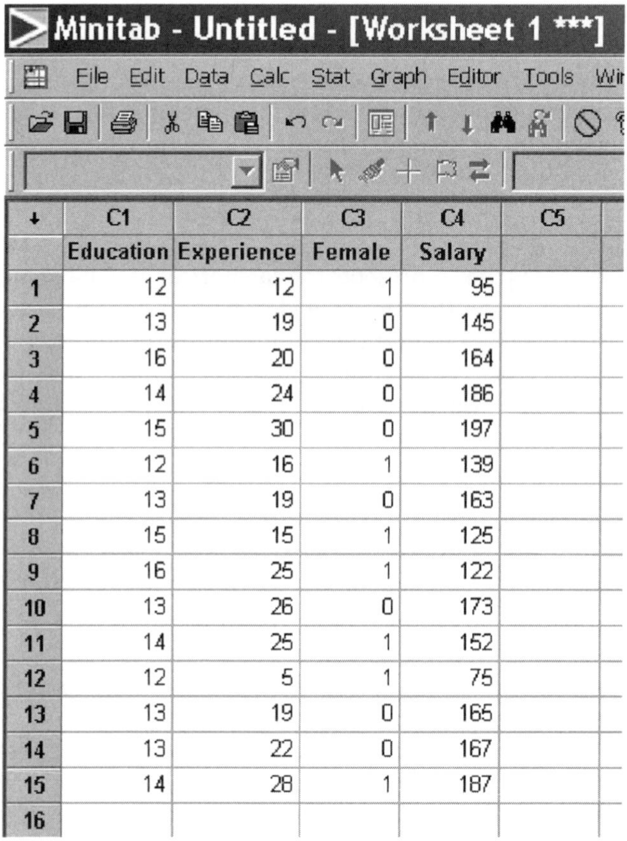

↓	C1	C2	C3	C4	C5
	Education	Experience	Female	Salary	
1	12	12	1	95	
2	13	19	0	145	
3	16	20	0	164	
4	14	24	0	186	
5	15	30	0	197	
6	12	16	1	139	
7	13	19	0	163	
8	15	15	1	125	
9	16	25	1	122	
10	13	26	0	173	
11	14	25	1	152	
12	12	5	1	75	
13	13	19	0	165	
14	13	22	0	167	
15	14	28	1	187	
16					

FIGURE 8.1
Executive salary data including qualitative variable "female" in column C3 of a MINITAB worksheet.

Although it may seem that you can only compare two categories because the category that is described by the value 1 is compared to the category that is described by the value 0, this does not have to be the case. There may be occasions when you want to include a categorical predictor variable that has more than two distinct categories. We can extend using indicator variables to include categorical predictors that have more than two distinct categories. However, when a variable has more than two categories, it is important to note that only the values of 0 and 1 can be used to describe the characteristics of a categorical predictor variable. The following example illustrates how to include a categorical predictor variable in a multiple regression analysis when there are three distinct categories, along with a description of how to interpret the parameter estimate for such a categorical predictor variable.

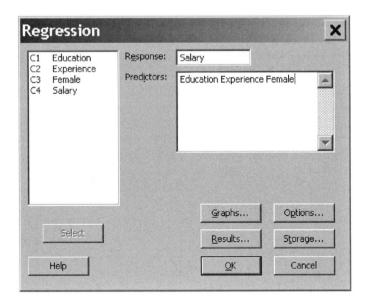

FIGURE 8.2
MINITAB dialog box including the qualitative predictor of female.

Example 8.2

Suppose that in addition to education, experience, and gender, we want to include a variable that describes an executive's political party affiliation, as described in Table 8.2. For this example, the categorical predictor variable has three distinct categories: *D* represents that the executive identifies as a Democrat, *I* represents that the executive identifies as an Independent, and *R* represents that the executive identifies as a Republican.

Notice that since there are three categories of political party instead of two, when the data are entered into MINITAB, the variable that describes an executive's political affiliation will be represented as a text column, as can be seen in Figure 8.4.

If you were to try and run a multiple regression analysis, the text data in column 4 would not show up in the regression dialog box, as can be seen in Figure 8.5. This is because a regression analysis can only be run with data that are numeric in form.

In order to include a categorical predictor variable that has more than two distinct categories in a multiple regression model, a set of indicator variables needs to be created that describes the different categories. In order to do this, we will use the **Make Indicator Variables** command from the **Calc** menu in MINITAB, as shown in Figure 8.6. This gives us the indicator dialog box illustrated in Figure 8.7.

We first need to specify the variable that is to be converted into a set of indicator variables, which in our case is the variable political party (C4 Political Party).

Regression Analysis: Salary versus Education, Experience, Female

```
The regression equation is
Salary = 116 - 2.65 Education + 4.04 Experience - 25.0 Female

Predictor       Coef   SE Coef       T      P
Constant       116.07    41.67    2.79  0.018
Education      -2.647     3.494   -0.76  0.465
Experience     4.0370    0.7621   5.30  0.000
Female        -24.954     8.209  -3.04  0.011

S = 14.7331   R-Sq = 85.5%   R-Sq(adj) = 81.6%

Analysis of Variance

Source            DF        SS       MS      F      P
Regression         3   14101.6   4700.5  21.65  0.000
Residual Error    11    2387.7    217.1
Total             14   16489.3

Source        DF    Seq SS
Education      1    1819.0
Experience     1   10276.9
Female         1    2005.7

Unusual Observations

Obs  Education  Salary     Fit  SE Fit  Residual  St Resid
  9       16.0  122.00  149.69    9.07    -27.69     -2.39R
```

FIGURE 8.3
MINITAB output from a regression analysis modeling executive salaries as a function of education, experience, and the categorical predictor for gender.

Upon specifying this variable in the top panel of the dialog box, MINITAB will then create the appropriate number of indicator variables depending on the number of categories of this variable, as seen in Figure 8.8.

MINITAB will generate a set of three different indicator variables for each of the three categories, as can be seen in Figures 8.8 and 8.9. In column C6 (Political Party_D) in Figure 8.9, a 1 represents the category "Democrat" and a 0 is assigned otherwise. Similarly, column C7 in Figure 8.9 (Political Party_I) is where a 1 represents the category "Independent" and a 0 otherwise, and for column C8 in Figure 8.9 (Political Party_R), the value of 1 is assigned to "Republican" and 0 otherwise.

To include these categorical variables in our regression analysis, we only need to include two of these three columns of indictor variables. Notice in Figure 8.10 that the two categories of Democrat and Independent were included as predictors in the regression model. This will give the MINITAB

TABLE 8.2

Executive Salary Data That Includes Years of Education, Years of Experience, Gender, and Political Party Affiliation

Observation Number	Education in Years (x_1)	Experience in Years (x_2)	Female (x_3)	Political Party (x_4)	Yearly Salary in Thousands of Dollars (y)
1	12	12	1	D	95
2	13	19	0	I	145
3	16	20	0	I	164
4	14	24	0	R	186
5	15	30	0	R	197
6	12	16	1	I	139
7	13	19	0	I	163
8	15	15	1	D	125
9	16	25	1	D	122
10	13	26	0	R	173
11	14	25	1	I	152
12	12	5	1	D	75
13	13	19	0	R	165
14	13	22	0	R	167
15	14	28	1	R	187

printout given in Figure 8.11. If you include all three columns of indicator variables, then MINITAB will automatically discard the last variable that was entered because this variable will always be perfectly negatively correlated with the sum of the other two indicator variables.

Notice in Figure 8.10 that the variable that represents the category "Republican" was not included in the regression model, and because this category was the one not included, it becomes the baseline or reference category. Although we did not include the column that represents the category of Republican, notice that when the value of 0 is entered for both Democrat and Independent, as seen in columns 6 and 7 of Figure 8.9, this would represent the baseline reference category of Republican.

We can interpret the findings for each of the two categories entered in the model with respect to the reference category, which in this case is Republican. Thus, the effect of being a Democrat on executive salaries, as seen in Figure 8.11, is estimated to be −60.54 ($\alpha = 0.05$), compared to Republican executives. This suggests that executives who identify as Democrats tend to make $60,540 less than executives who identify as Republicans (holding all other factors fixed). Similarly, executives who identify as Independent tend to make $19,618 less than Republicans.

When there are more than two categories for a given categorical variable, you can make any category you wish the reference category. This entails creating the appropriate number of indicator variables to match the number of

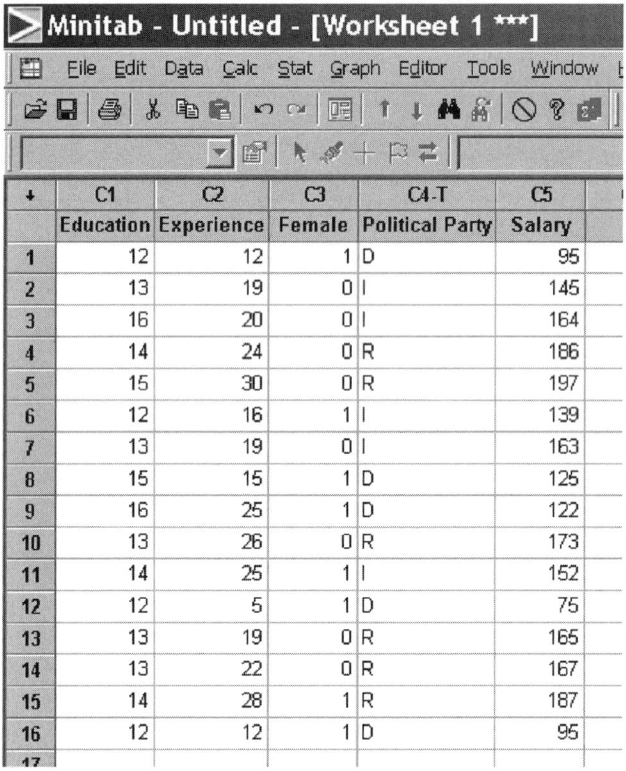

↓	C1	C2	C3	C4-T	C5	
	Education	**Experience**	**Female**	**Political Party**	**Salary**	
1	12	12	1	D	95	
2	13	19	0	I	145	
3	16	20	0	I	164	
4	14	24	0	R	186	
5	15	30	0	R	197	
6	12	16	1	I	139	
7	13	19	0	I	163	
8	15	15	1	D	125	
9	16	25	1	D	122	
10	13	26	0	R	173	
11	14	25	1	I	152	
12	12	5	1	D	75	
13	13	19	0	R	165	
14	13	22	0	R	167	
15	14	28	1	R	187	
16	12	12	1	D	95	

FIGURE 8.4
Executive salary data including the categorical variable that describes political affiliation as text in column 4.

categories and excluding the particular category from the regression model, and then all the other categories would then be compared to the omitted category. For instance, if we wanted "Independent" to be the reference category, we would include the indicator variables for both "Democrat" and "Republican" and exclude the indicator variable for "Independent." Then any inferences made regarding the other categories would be with respect to the omitted category, which would be "Independent."

8.4 Adjusted R^2

One key point to remember when creating a regression model is not to include variables in the regression model that are highly correlated with each other, and not to include so many variables in the regression model such that there are more variables than there are observations in the data set. Adding more variables to a regression model will always cause the R^2 statistic to increase

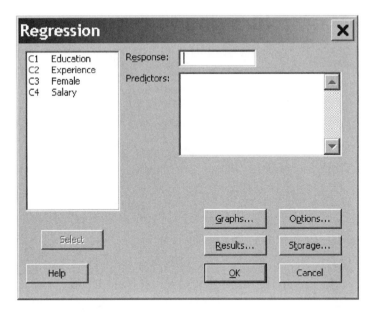

FIGURE 8.5
Regression dialog box for the data in Table 8.2, which does not show text variables.

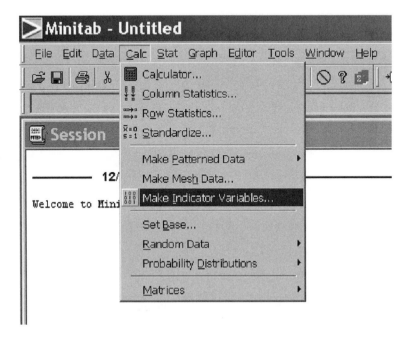

FIGURE 8.6
MINITAB commands to make indicator variables.

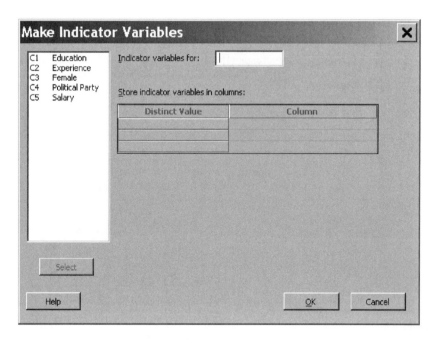

FIGURE 8.7
MINITAB dialog box to make indicator variables.

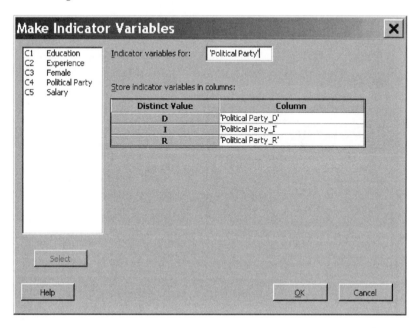

FIGURE 8.8
MINITAB dialog box specifying how to create three individual indicator variables for the categorical variable "political party."

↓	C1	C2	C3	C4-T	C5	C6	C7	C8	
	Education	Experience	Female	Political Party	Salary	Political Party_D	Political Party_I	Political Party_R	
1	12	12	1	D	95	1	0	0	
2	13	19	0	I	145	0	1	0	
3	16	20	0	I	164	0	1	0	
4	14	24	0	R	186	0	0	1	
5	15	30	0	R	197	0	0	1	
6	12	16	1	I	139	0	1	0	
7	13	19	0	I	163	0	1	0	
8	15	15	1	D	125	1	0	0	
9	16	25	1	D	122	1	0	0	
10	13	26	0	R	173	0	0	1	
11	14	25	1	I	152	0	1	0	
12	12	5	1	D	75	1	0	0	
13	13	19	0	R	165	0	0	1	
14	13	22	0	R	167	0	0	1	
15	14	28	1	R	187	0	0	1	
16	12	12	1	D	95	1	0	0	
17									

FIGURE 8.9

MINITAB worksheet that includes indicator variables for each of the three distinct categories of political party.

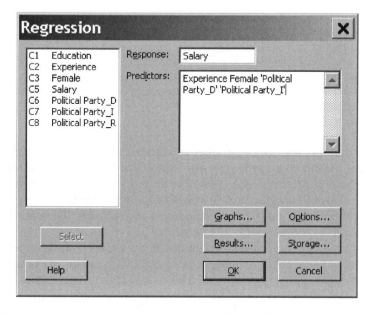

FIGURE 8.10

MINITAB regression dialog box including the two categorical variables for Democrat and Independent.

Regression Analysis: Salary versus Education, Experience, ...

```
The regression equation is
Salary = 69.9 + 5.94 Education + 1.15 Experience - 3.19 Female
           - 60.5 Political Party_D - 19.6 Political Party_I

Predictor                 Coef   SE Coef       T       P
Constant                 69.90     26.94    2.59   0.029
Education                5.937     2.821    2.10   0.065
Experience              1.1540    0.7868    1.47   0.177
Female                  -3.192     6.810   -0.47   0.650
Political Party_D       -60.54     13.26   -4.57   0.001
Political Party_I      -19.618     7.111   -2.76   0.022


S = 8.84827 R-Sq = 95.7% R-Sq(adj) = 93.4%


Analysis of Variance

Source            DF        SS      MS       F       P
Regression         5   15784.7  3156.9   40.32   0.000
Residual Error     9     704.6    78.3
Total             14   16489.3


Source             DF    Seq SS
Education           1    1819.0
Experience          1   10276.9
Female              1    2005.7
Political Party_D   1    1087.2
Political Party_I   1     595.9
```

FIGURE 8.11
MINITAB printout highlighting indicator variables for Democrat and Independent political affiliations.

or stay the same, regardless of whether or not the additional variables are significant in influencing the response variable y. Furthermore, when there are more variables than observations, the R^2 statistic could be forced toward 100%. A value of $R^2 = 1.00$, or 100%, would suggest that the underlying regression model perfectly predicts the response variable, and this is not likely a situation that would be found in practice.

If there are numerous variables included in a regression model, then using another statistic, called the adjusted R^2, denoted as R^2_{adj}, can be a more appropriate measure for assessing which variables to include in a regression model.

The R^2_{adj} statistic adjusts the value of R^2 to account for both the total number of parameters being estimated in the model and the sample size. This is done by dividing SST and SSE by their respective degrees of freedom, as seen in the following equation:

$$R^2_{adj} = \frac{\dfrac{SST}{(n-1)} - \dfrac{SSE}{(n-p)}}{\dfrac{SST}{(n-1)}}$$

where SST is the total sum of squares, SSE is the error sum of squares, n is the sample size, and p is the total number of beta parameters being estimated in the model.

Example 8.3

For the regression analysis in Figure 8.11, the approximate values of R^2 and R^2_{adj} can be calculated as follows:

$$R^2 = \frac{SST - SSE}{SST} = \frac{16489.3 - 704.6}{16489.3} \approx 0.957$$

$$R^2_{adj} = \frac{\dfrac{SST}{(n-1)} - \dfrac{SSE}{(n-p)}}{\dfrac{SST}{(n-1)}} = \frac{\dfrac{16489.3}{14} - \dfrac{704.6}{15-6}}{\dfrac{16489.3}{14}} \approx \frac{1099.52}{1177.81} \approx 0.934$$

For this example, $n = 15$ and $p = 6$ because there are six beta values being estimated in the model (including the constant term β_o).

You may have noticed that the value of R^2 will increase or remain relatively the same when additional variables are added to the model. This will hold true even if adding other variables only contributes a minimal amount in predicting y. However, R^2_{adj} will only increase if the addition of the new variable improves the fit of the model by more than just a small amount. In fact, the adjusted R-squared statistic can actually decrease if a variable is added to a model that does not improve the fit by more than a small amount. Thus, it is possible that negative values of R^2_{adj} could be obtained. If this were the case (although this usually does not happen), then MINITAB would report the value of $R^2_{adj} = 0.0$.

8.5 Best Subsets Regression

There are many different techniques that can be used to take a more systematic approach to selecting which variables to include in a multiple regression model other than by simply using trial and error. One such technique that we will be considering is called *best subsets regression*. Best subsets regression is simply a collection of the best regression analyses that is based on different subsets of the predictor variables. The best-fitting regression model can then be selected based on three different model selection criteria.

The three different model selection criteria that can be used to distinguish between competing models in a best subsets regression analysis are as follows:

1. R^2 values
2. R^2_{adj} values
3. C_p statistic (also called Mallow's C_p statistic)

The values of R^2 and R^2_{adj} are the familiar R-squared and adjusted R-squared statistics that have been previously described. Recall that the R-squared statistic tells us the proportion of variability in the response that is attributed to the given model. Recall that a higher value of R^2 implies that more variability in y is attributed to the model and less variability is due to error. Furthermore, the value of the R^2_{adj} statistic can also be used to see if including additional predictor variables influences the response variable by more than just a random amount.

The C_p statistic can be used as a measure to help decide the best predictor variables to include in a regression model. The C_p statistic compares the full model, which includes all of the predictor variables, to those models that include the various subsets of the predictor variables, and thus it provides a way to balance including too many predictor variables in the model (overfitting the model) versus including too few predictor variables in the model (underfitting the model). Overfitting a model can cause the parameter estimates to be unstable by introducing large standard errors, and underfitting the model can introduce bias in the parameter estimates. Note that the full model is defined to be the model that includes the entire set of predictor variables of interest, or it can also represent a subset of all of the available predictor variables.

In a best subsets regression, the C_p statistic can be used such that the model that both minimizes the value of C_p and is close to the total number of beta parameters estimated in the model is the best with respect to balancing overfitting and underfitting the model. In other words, the smallest value of C_p that is also close to the number of beta parameters being estimated in the model is the model that minimizes the standard errors that can be introduced by including too many predictor variables, and minimizes the bias that can be introduced if there are too few predictor variables.

The C_p statistic can be calculated as follows:

$$C_p = \frac{SSE_p}{MSE_{FULL}} - n + 2p$$

where p is the total number of beta parameters being estimated in the subset model, n is the sample size, SSE_p is the sum of squares of the error component for the regression that has p beta variables in the subset model, and MSE_{FULL} is the mean square error for the full model, which includes all of the predictor variables we are interested in.

We will use best subsets regression as a statistical modeling technique that can be used in providing a systematic way to select a model with a given number of predictor variables that maximizes R^2 and R^2_{adj}, and minimizes C_p, where C_p is close to the total number of beta parameters being estimated in the model.

Example 8.4

Using the data from Table 8.1, suppose we are interested in finding the best-fitting model for one, two, and three predictors. Recall that x_1 represents years of education, x_2 represents years of experience, and x_3 represents gender. To find the best-fitting model consisting of only a single predictor, we could run three separate regression analyses and estimate each of the following subset population regression equations:

1. $y = \beta_0 + \beta_1 x_1 + \varepsilon$
2. $y = \beta_0 + \beta_2 x_2 + \varepsilon$
3. $y = \beta_0 + \beta_3 x_3 + \varepsilon$

Figures 8.12 to 8.14 give the MINITAB printouts for each of these regression analyses using the executive salary data from Table 8.1.

From each of these three estimated subset models, we could collect the values for R^2 and R^2_{adj}. We can also calculate the value of C_p for each of these models by running the full model to find MSE_{FULL} (see Figure 8.15 for the MINITAB printout for the full model), and then calculating C_p for each model as follows:

For Model 1:

$$C_p = \frac{SSE_p}{MSE_{FULL}} - n + 2p = \frac{14670.0}{217.10} - 15 + 2(2) = 56.57$$

For Model 2:

$$C_p = \frac{SSE_p}{MSE_{FULL}} - n + 2p = \frac{4711.0}{217.10} - 15 + 2(2) = 10.70$$

For Model 3:

$$C_p = \frac{SSE_p}{MSE_{FULL}} - n + 2p = \frac{9858.9}{217.10} - 15 + 2(2) = 34.41$$

Regression Analysis: Salary versus Education

```
The regression equation is
Salary = 34.5 + 8.47 Education

Predictor      Coef   SE Coef      T       P
Constant      34.53     91.63   0.38   0.712
Education     8.474     6.674   1.27   0.226

S = 33.5929    R-Sq = 11.0%    R-Sq(adj) = 4.2%

Analysis of Variance

Source            DF      SS     MS      F      P
Regression         1    1819   1819   1.61  0.226
Residual Error    13   14670   1128
Total             14   16489

Unusual Observations

Obs  Education  Salary      Fit   SE Fit   Residual   St Resid
 12       12.0   75.00   136.21    14.11     -61.21      -2.01R

R denotes an observation with a large standardized residual.
```

FIGURE 8.12
Regression analysis only including the variable for years of education (x_1).

The values of R^2, R^2_{adj}, and C_p are summarized in Table 8.3.

Notice that for these last three calculations $p = 2$ because there are two variables being estimated in our subset models, namely, one predictor variable and the constant term, β_0.

Using the values of R^2, R^2_{adj}, and C_p provided in Table 8.3, the best one-variable model would be Model 2 because this model is the one that maximizes the values of R^2 and R^2_{adj}, and the value of C_p for this model is the smallest, and it is also the model such that the value of the C_p statistic is closest to the

TABLE 8.3

R^2, R^2_{adj}, and C_p Statistics for the One-Variable Models for the Executive Salary Data

Model	Variable	R_2	R^2_{adj}	C_p
(1)	x_1	11.0	4.2	56.6
(2)	x_2	71.4	69.2	10.7
(3)	x_3	40.2	35.6	34.4

Regression Analysis: Salary versus Experience

```
The regression equation is
Salary = 60.3 + 4.43 Experience

Predictor      Coef   SE Coef     T       P
Constant      60.34     16.53   3.65   0.003
Experience   4.4257    0.7763   5.70   0.000

S = 19.0364     R-Sq = 71.4%    R-Sq(adj) = 69.2%

Analysis of Variance

Source             DF     SS      MS       F       P
Regression          1   11778   11778   32.50   0.000
Residual Error     13    4711     362
Total              14   16489

Unusual Observations

Obs  Experience    Salary    Fit    SE Fit  Residual  St Resid
  9        25.0   122.00 170.99    6.11    -48.99    -2.72R
 12         5.0    75.00  82.47   12.88     -7.47    -0.53 X

R denotes an observation with a large standardized residual.
X denotes an observation whose X value gives it large leverage.
```

FIGURE 8.13
Regression analysis including only the variable for years of experience (x_2).

number of parameters that are being estimated in the subset models (namely, $p = 2$), compared to the other models.

We could then repeat this process for the following collection of all possible two-variable models:

4. $y = \beta_0 + \beta_1 x_1 + \beta_2 x_2 + \varepsilon$
5. $y = \beta_0 + \beta_1 x_1 + \beta_3 x_3 + \varepsilon$
6. $y = \beta_0 + \beta_2 x_2 + \beta_3 x_3 + \varepsilon$

For this scenario, $p = 3$ because there are three predictors in each subset model (two variables and the constant term β_0).

The values of R^2, R^2_{adj}, and C_p for Models 4, 5, and 6 are summarized in Table 8.4.

Using R^2, R^2_{adj}, and C_p from Table 8.4, the best two-variable model is Model 6, which includes x_2 and x_3. This is because Model 6 has the maximum values of R^2 and R^2_{adj} and the value of the C_p statistic is the minimum. Furthermore, this

Regression Analysis: Salary versus Female

```
The regression equation is
Salary = 170 - 42.1 Female

Predictor      Coef   SE Coef       T       P
Constant    170.000     9.736   17.46   0.000
Female       -42.14     14.25   -2.96   0.011

S = 27.5386    R-Sq = 40.2%    R-Sq(adj) = 35.6%

Analysis of Variance

Source             DF        SS       MS      F       P
Regression          1    6630.5   6630.5   8.74   0.011
Residual Error     13    9858.9    758.4
Total              14   16489.3

Unusual Observations

Obs   Female   Salary      Fit   SE Fit   Residual   St Resid
 12     1.00    75.00   127.86    10.41     -52.86     -2.07R
 15     1.00   187.00   127.86    10.41      59.14      2.32R
R denotes an observation with a large standardized residual.
```

FIGURE 8.14
Regression analysis including only the variable for female (x_3).

model would be better than the best one-variable model because the value of C_p is much closer to the number of beta parameters being estimated (which in this case, there are three parameters being estimated: β_0, β_2, and β_3). Thus, we can use the value of the C_p statistic to determine the model that balances too many versus too few predictor variables.

Since there is only one three-variable model (represented by Model 7), the values of R^2, R^2_{adj}, and C_p are in Table 8.5.

$$7.\ y = \beta_0 + \beta_1 x_1 + \beta_2 x_2 + \beta_3 x_3 + \varepsilon$$

TABLE 8.4

R^2, R^2_{adj}, and C_p Statistics for All Two-Variable Models for the Executive Salary Data

Model	Variables	R^2	R^2_{adj}	C_p
(4)	x_1, x_2	73.4	68.9	11.2
(5)	x_1, x_3	48.6	40.0	30.1
(6)	x_2, x_3	84.8	82.2	2.6

Regression Analysis: Salary versus Education, Experience, Female

```
The regression equation is
Salary = 116 - 2.65 Education + 4.04 Experience - 25.0 Female

Predictor          Coef   SE Coef      T      P
Constant         116.07     41.67   2.79  0.018
Education        -2.647      3.494  -0.76  0.465
Experience       4.0370     0.7621   5.30  0.000
Female          -24.954      8.209  -3.04  0.011

S = 14.7331   R-Sq = 85.5%    R-Sq(adj) = 81.6%

Analysis of Variance

Source             DF        SS      MS      F      P
Regression          3   14101.6  4700.5  21.65  0.000
Residual Error     11    2387.7   217.1
Total              14   16489.3

Source          DF    Seq SS
Education        1    1819.0
Experience       1   10276.9
Female           1    2005.7

Unusual Observations

Obs  Education  Salary     Fit   SE Fit  Residual  St Resid
  9       16.0  122.00  149.69     9.07    -27.69    -2.39R
R denotes an observation with a large standardized residual.
```

FIGURE 8.15
Full regression model including all the predictor variables, x_1, x_2, and x_3.

By combining all three of these subset regressions, which look at all the one-, two-, and three-variable models, we can select which of these different subsets of models best represents the relationship we are looking to describe by using the R^2, R^2_{adj}, and C_p statistics.

Based on the above model fits for one, two, and three variables, we could conclude that there are two models that seem to have the best fit: Model 6, which has two-variables, and Model 7, the three-variable model. However, distinguishing

TABLE 8.5

R^2, R^2_{adj}, and C_p Statistics for the Three-Variable Model (7) Using the Executive Salary Data

Model	Variables	R^2	R^2_{adj}	C_p
(7)	x_1, x_2, x_3	85.5	81.6	4.0

between which of these two models to use is more than a simple matter of finding the model that best meets the three-model criteria; it is also a matter of professional judgment and known theory. For instance, the best two-variable model includes the predictors of experience and female, whereas the best three-variable model consists of education, experience, and female. Choosing between whether to use the two- or three-variable model depends on whether education is a variable that needs to be included in the model. The choice of whether to include the variable for education would be guided by what previous studies have done and what theory suggests, and also by analyzing what the impact would be from including this extra predictor variable in the model. Thus, good model fitting requires so much more than simply running a best subsets regression analysis and using the three-model criteria to find the best model fit; it relies on the researcher investigating the significance and practical impact of including all the relevant variables in a model.

8.6   Using MINITAB for Best Subsets Regression

MINITAB can be used to perform all of the above calculations for the one-, two-, and three-variable models, which include all of the different subsets of the predictor variables. Under the **Stat** menu, select **Regression** and then **Best Subsets**, as presented in Figure 8.16.

This then gives the dialog box in Figure 8.17, where the response and predictor variables need to be specified. The **Free predictors** box is used to specify those predictors that are free to be added or dropped when running

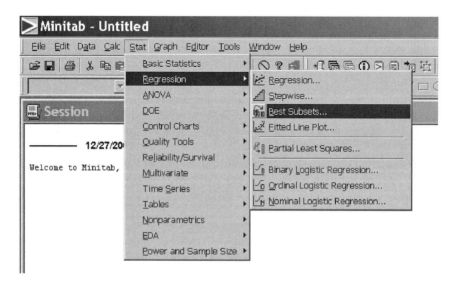

FIGURE 8.16
MINITAB commands to run a best subsets regression analysis.

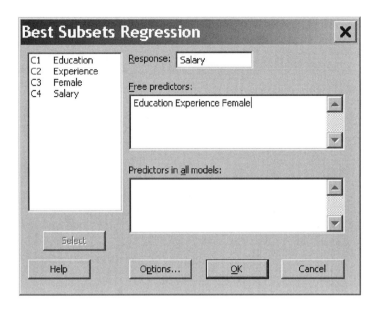

FIGURE 8.17
MINITAB best subsets regression dialog box.

the best subsets regression analyses with different numbers of predictors. The **Predictors in all models** box is used to specify those predictors that will be included in every subset model.

Notice in Figure 8.17 that the best subsets regression can be run by including the three variables of education, experience, and female in the **Free predictors** box to give the MINITAB printout in Figure 8.18.

The rows of the MINITAB printout in Figure 8.18 give the model selection criteria for the different subset models. Also notice that the left-most column gives the number of variables (Vars) that are included in each subset model (the MINITAB default is to display the two best models with respect to each of the number of variables in the model based on the models that give the maximum value of the R^2 statistic). The MINITAB printout in Figure 8.18 also gives the values of R^2, R^2_{adj}, and C_p along with the root mean square error, S, for each of the models. The last three columns represent the different variables that are included in each of the models. For instance, the first two-variable model includes the variables of experience and female, and the second two-variable model includes the variables of education and experience. Lining up the X values under the given variable with the number of variables in the model tells you which variables are included in each of the different subset models.

Although it is very simple to run a best subsets regression analysis using a statistics package to find the best-fitting model, such a mechanical process does not account for any of the other aspects of regression analysis, such as

Best Subsets Regression: Salary versus Education, Experience, Female

```
Response is Salary
```

			Mallows		E d u c a t i o n e	E x p e r i e n c e	F e m a l e
Vars	R-Sq	R-Sq(adj)	Cp	S			
1	71.4	69.2	10.7	19.036	X		
1	40.2	35.6	34.4	27.539		X	
2	84.8	82.2	2.6	14.469		X	X
2	73.4	68.9	11.2	19.134	X	X	
3	85.5	81.6	4.0	14.733	X	X	X

FIGURE 8.18
MINITAB output of best subsets regression analysis for the executive salary data in Table 8.1.

checking the model assumptions and verifying that there is not a problem with multicollinearity. These other aspects of model fitting always need to be considered, and best subsets regression does not address any of these issues.

8.7 Confidence and Prediction Intervals for Multiple Regression

Recall when we were studying simple linear regression analysis that we described two types of intervals: confidence intervals that can be used for estimating the mean population response for a specified value of the predictor variable, and prediction intervals that can be used for predicting a specific response value for a single observation at a specified value.

We can calculate such confidence and prediction intervals in a multiple regression analysis. In calculating confidence and prediction intervals for multiple regression analysis, the specified value has to consist of the specified values for all of the predictor variables that are included in the model. In other words, the "specified value" is really a vector of values in a multiple regression analysis that represents a specific value for each of the individual predictor variables included in the model.

For instance, suppose we want to find confidence and prediction intervals using the executive salary data given in Table 8.1. If we are interested in calculating confidence and prediction intervals for executives with 16 years of education (this would be equivalent to having a bachelor's degree), 10 years of experience, and who are males, our specified value would correspond to the vector $(x_1, x_2, x_3) = (16, 10, 0)$.

Because calculations for confidence and prediction intervals are very cumbersome to do by hand for a multiple regression analysis, we will use MINITAB to calculate these intervals for us.

8.8 Using MINITAB to Calculate Confidence and Prediction Intervals for a Multiple Regression Analysis

From the **Regression** dialog box in Figure 8.19, select the **Options** tab, as illustrated in Figure 8.20.

To have MINITAB calculate confidence and prediction intervals, the specific values for the set of predictor variables need to be entered, where each value is separated by a space. The variables must also be entered in the order in which the variable was entered in the regression model. So for our example, to find confidence and prediction intervals for executives with 16 years of education, 10 years of experience, and who are males, we would enter the values as $(x_1, x_2, x_3) = (16, 10, 0)$, which can be seen in Figure 8.20.

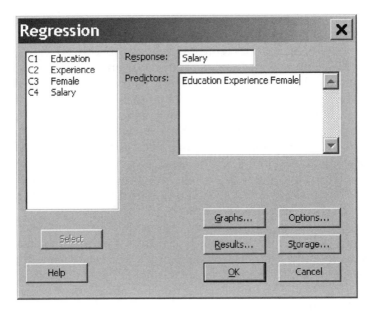

FIGURE 8.19
MINITAB dialog box for a regression analysis.

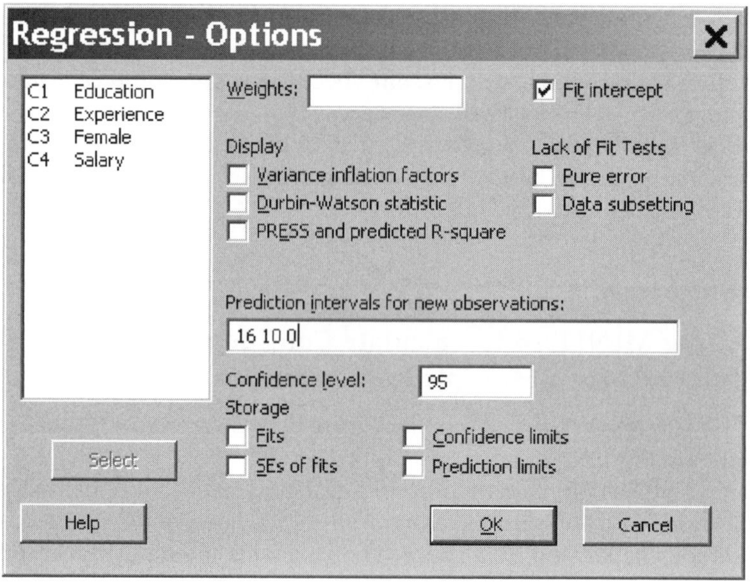

FIGURE 8.20
MINITAB options dialog box for specifying the specific values for the set of predictor variables.

Figure 8.21 gives the MINITAB printout with the confidence and prediction intervals highlighted.

The confidence interval in Figure 8.21 suggests that we are 95% confident that for all executives with 16 years of education and 10 years of experience who are males, the mean population yearly salary will be between $78,710 and $149,460. For the prediction interval, we can predict with 95% confidence that an executive who has 16 years of education and 10 years of experience who is male will earn a yearly salary between $66,090 and $162,080. Similar to simple linear regression analysis, the prediction interval is much wider than the confidence interval for the same specific values of the predictor variables. This is because there is more uncertainty in estimating a single response from a single observation versus estimating the average response by using the entire set of observations.

Also notice in Figure 8.21 that there is a symbol X on the outside of the prediction interval. This symbol means that there is a point in the data set that is an extreme outlier with respect to the predictors. When using confidence and prediction intervals, the X does not mean that the leverage value is high, as is indicated by the X symbol in the unusual observations portion of the MINITAB printout.

Regression Analysis: Salary versus Education, Experience, Female

```
The regression equation is
Salary = 116 - 2.65 Education + 4.04 Experience - 25.0 Female

Predictor         Coef   SE Coef       T       P
Constant        116.07     41.67    2.79   0.018
Education       -2.647      3.494   -0.76   0.465
Experience      4.0370     0.7621    5.30   0.000
Female         -24.954      8.209   -3.04   0.011

S = 14.7331    R-Sq = 85.5%    R-Sq(adj) = 81.6%

Analysis of Variance

Source              DF         SS       MS       F       P
Regression           3    14101.6   4700.5   21.65   0.000
Residual Error      11     2387.7    217.1
Total               14    16489.3

Source       DF    Seq SS
Education     1    1819.0
Experience    1   10276.9
Female        1    2005.7

Unusual Observations
Obs   Education   Salary      Fit   SE Fit   Residual   St Resid
  9        16.0   122.00   149.69     9.07     -27.69      -2.39R

R denotes an observation with a large standardized residual.

Predicted Values for New Observations

New
Obs      Fit   SE Fit        95% CI              95% PI
  1   114.08    16.07   (78.71, 149.46)   (66.09, 162.08)X

X denotes a point that is an outlier in the predictors.

Values of Predictors for New Observations

New
Obs   Education   Experience     Female
  1        16.0         10.0   0.000000
```

FIGURE 8.21
MINITAB regression printout with 95% prediction and confidence intervals for executives
with 16 years of education, 10 years of experience, and who are males.

Similar to simple linear regression analysis, one thing to keep in mind when making inferences with confidence and prediction intervals is that it is not usually a good idea to make inferences where the specified vector of the predictor values is too far removed from the range of the data used when fitting the regression model. Using a set of predictor values where even one of the predictor values is out of range can cause the confidence and prediction intervals to become very wide and unstable.

8.9 Assessing Outliers

We can use MINITAB in the exact same manner as with simple linear regression to calculate outlier statistics. For instance, we can create a Cook's distance plot by plotting the observation number against the value of the Cook's distance statistic for the regression model specified in Figure 8.21, as can be seen in Figure 8.22.

As expected, observation 9 is an extreme value and may impact the estimates of the regression parameters. Also, the value of the standardized residual is either less than –2 or more than +2, as can be seen in Figure 8.21.

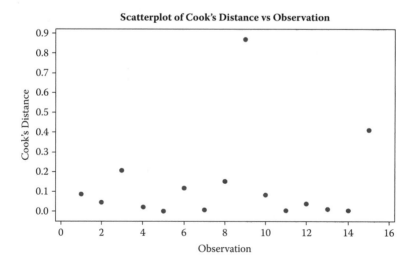

FIGURE 8.22
Cook's distance plot for the regression analysis in Figure 8.21, which models executive salaries based on education, experience, and gender.

Exercises

1. What factors predict how well a high school student does on a standardized test? With the introduction of the No Child Left Behind legislation, researchers have been working on examining whether there is a difference in standardized test scores based on various factors, such as gender, ability, socioeconomic status, and race. The data set "Standardized Test" in Table 8.6 represents a random sample of twenty-four high school students' scores on a state standardized test (score) along with descriptive measures of gender (female), family income in thousands of dollars (income), race (minority = W

TABLE 8.6

"Standardized Test" Data Set

Observation Number	Female (x_1)	Income (x_2)	Minority (x_3)	Aptitude (x_4)	Score (y)
1	M	38	W	85	88
2	M	47	W	57	72
3	M	52	NW	91	94
4	F	68	NW	57	70
5	F	52	W	68	78
6	M	49	W	94	88
7	F	78	NW	75	83
8	M	65	W	84	64
9	F	120	NW	80	77
10	F	141	W	65	74
11	F	35	W	90	87
12	M	200	W	75	76
13	M	157	NW	68	79
14	F	63	NW	71	76
15	F	87	W	84	87
16	M	79	W	80	65
17	F	81	W	55	69
18	M	95	NW	64	50
19	M	36	W	37	48
20	F	45	W	81	85
21	F	79	NW	32	67
22	M	67	NW	45	70
23	M	81	W	67	76
24	F	128	NW	43	59

if identified as white, minority = NW if identified as nonwhite), and the score received on a previous aptitude test (aptitude).

a. Using MINITAB, run a regression analysis on the score received on the standardized test including all four of the predictor variables of gender, family income, race, and the score received on a previous aptitude test. This entails creating indicator variables for the categorical predictor variables of female and minority.

b. For $\alpha = 0.05$, determine which, if any, predictor variables are significant in predicting y.

c. Check the regression model assumptions, and comment on whether you think they are reasonably met.

d. Find the variance inflation factors and comment on whether you think there is an issue with multicollinearity.

e. Interpret which variables significantly impact students' scores on the standardized test, holding all other variables fixed.

2. Using the "Standardized Test" data set, create a binary variable to look at whether the nonwhite students tend to perform better on the standardized test, as compared to white students. Determine whether you think this variable is significant, and comment on the magnitude of the effect, if there is one.

3. Using the "Standardized Test" data set, run a best subsets regression analysis and identify the models that you believe best meet the three criteria used in a best subsets regression analysis.

4. Using the "Standardized Test" data set, create a Cook's distance plot using the model that you believe best meets the three criteria from a best subsets regression (from Exercise 3), and comment on what you see. Does this plot make sense given the results of your regression analysis?

5. An island ferry company wants to determine what factors impact the revenue it makes from running its ferry boats to a local island in the ocean. The ferry charges for passengers to ride the ferry and also charges for delivering large items to the island, such as automobiles and heavy construction equipment. The data set "Ferry," presented in Table 8.7, consists of the revenue (in dollars), time of day the ferry left the dock, number of passengers on the ferry, number of large objects on the ferry, weather conditions, and number of crew for a random sample of twenty-seven ferry runs over a 1-month period.

a. Using MINITAB, run a multiple regression analysis on the total revenue, including the numeric and categorical predictor variables of time, number of passengers, number of large objects, weather conditions, and number of crew members that are given in the "Ferry" data set ($\alpha = 0.05$).

TABLE 8.7

"Ferry" Data Set

Revenue	Time of Day	Number of Passengers	Number of Large Objects	Weather Conditions	Number of Crew
7,812	Morning	380	6	Calm	4
6,856	Noon	284	4	Calm	4
9,568	Evening	348	10	Calm	3
10,856	Morning	257	16	Rough	5
8,565	Morning	212	12	Rough	4
8,734	Noon	387	6	Calm	3
9,106	Evening	269	13	Calm	5
8,269	Evening	407	8	Rough	3
6,373	Noon	385	7	Calm	4
5,126	Noon	347	4	Calm	3
6,967	Noon	319	9	Rough	4
8,518	Morning	297	7	Rough	3
7,229	Evening	345	6	Rough	4
6,564	Morning	287	8	Rough	4
8,168	Morning	348	7	Calm	4
9,879	Evening	189	20	Calm	6
10,288	Evening	215	22	Calm	5
11,509	Morning	247	21	Rough	5
5,254	Noon	345	9	Rough	3
9,895	Morning	348	4	Calm	3
8,434	Evening	451	6	Calm	3
7,528	Noon	378	8	Rough	4
7,667	Morning	346	10	Calm	3
9,591	Evening	287	18	Calm	6
8,543	Evening	245	12	Rough	5
9,862	Morning	274	10	Calm	5
10,579	Morning	245	20	Rough	7

b. Check any relevant model assumptions and comment on whether you believe such assumptions have been violated.

c. Identify any outliers or extreme values.

d. Using the results from the regression analysis, determine if the number of large items shipped affect the amount of revenue, and if the condition of the sea has an effect on the amount of revenue.

e. Do you think that multicollinearity is a problem?

f. Using best subsets regression analysis, determine the models that best fit the data according to the three criteria for model selection. Justify why you chose these particular models.

g. Would the ferry operators make more money if they only ran the ferry boats in either the morning or the evening, compared to the afternoon?

6. Using the "Expensive Homes" data set and MINITAB, run a multiple regression analysis to develop a model that predicts the asking price based on the number of bedrooms, number of bathrooms, square footage, age of the home, lot size, number of garages, and number of stories ($\alpha = 0.05$).

    a. Check any relevant model assumptions and comment on whether you believe such assumptions have been violated.

    b. Identify any outliers or extreme values.

    c. Using the results from the regression, determine if newer homes have higher asking prices, and if larger homes have higher asking prices.

    d. Do you think that multicollinearity is a problem?

    e. Using best subsets regression analysis, determine the models that best fit the data according to the three criteria for model selection. Justify why you chose these particular models.

7. Using the "Credit Card" data set and MINITAB, run a multiple regression analysis to develop a model that predicts the amount of credit card debt based on age, education, and yearly salary ($\alpha = 0.05$).

    a. Check any relevant model assumptions and comment on whether you believe such assumptions have been violated.

    b. Identify any outliers or extreme values.

    c. Using the results from the regression, determine if people who make more money incur more credit card debt.

    d. Do you think that multicollinearity is a problem?

    e. Using best subsets regression analysis, determine the models that best fit the data according to the three criteria for model selection. Justify why you chose these particular models.

9

Analysis of Variance (ANOVA)

9.1 Introduction

Up to now, we have been concerned with making inferences about population parameters based on sample data by calculating confidence intervals and conducting hypothesis tests. We have also used regression analysis to develop models to predict one variable (y) based on another variable (x), or set of variables ($x_1, x_2, \cdots, x_k$). In this chapter we will be discussing ways to use sample data to make inferences about more than two different population means. But before we begin discussing how to compare more than two population means, we first need to describe some basic concepts of designing an experiment.

9.2 Basic Experimental Design

Often, one needs to design an *experiment* to address some question of interest. For example, suppose that we are interested in comparing the effect of several different brands of fertilizer on the growth rate of grass. To conduct an experiment to do such a comparison, we could start by partitioning a plot of grass into several different sections and then apply the different brands of fertilizer to each section. Then after a period of time has elapsed, we could measure the amount of growth of the grass for each of the different brands of fertilizer. What underlies conducting such an experiment is that there are no other factors that could have contributed to the growth rate of the grass other than the application of the fertilizer. Thus, the goal in designing an experiment is to be able to isolate the effect of a given factor or factors of interest.

There are different types of experimental designs to consider that deal with isolating the factor or factors of interest. We will be describing two basic types of experimental designs: randomized designs and randomized block designs.

A *randomized design* is when all factors except for a single factor of interest are held constant. For instance, a randomized design that could be used to compare the effect of different brands of fertilizer on the growth rate of grass could be to take a large plot of grass, divide it into sections that represent the number of fertilizer brands that are being compared, and then randomly apply the brands of fertilizer to each of the different sections of the plot of land.

Then if all other factors, such as sunlight, water, pests, etc., were the same for each of these sections, this would constitute a randomized design.

Figure 9.1 illustrates what a randomized design would look like if we were to partition a plot of grass into six equal sections and then randomly assign the six different brands of lawn fertilizer to each of the sections.

However, one problem with using a randomized design occurs if there is more than a single factor that may be different among the different sections. When this is the case we would need to consider using a *randomized block design*.

A randomized block design can be used when there may be factors other than a single factor of interest that may not be constant. If this were the case, then we would need to control for these other factors so that they do not affect the outcome of interest.

For example, suppose that our section of grass was half in the sun and half in the shade. If we were to use the randomized design that is illustrated in Figure 9.1, then we could not be sure that we are teasing out the effect of the fertilizer because we cannot be certain whether it was the sun or the lack of sun that had an impact on the growth rate of the grass. In order to control for the difference in sunlight among the different sections of grass, we would use a randomized block design that would entail partitioning the section of grass into both sunny and shady portions, and then randomly assigning all of the different brands of fertilizer to both of these sections, as illustrated in Figure 9.2.

Once we have designed an experiment where we are able to control for any extraneous factors that could affect our outcome of interest (which for our example is the growth rate of grass), we can then begin to compare whether there is a difference in the true but unknown population means based on the given factor. Thus, for our example, we can test to see if there is a difference in the mean growth rate of grass based on the different brands of fertilizer. The statistical method that we will use to compare whether three or more population means are significantly different from each other with respect to one factor is called a *one-way analysis of variance* (ANOVA).

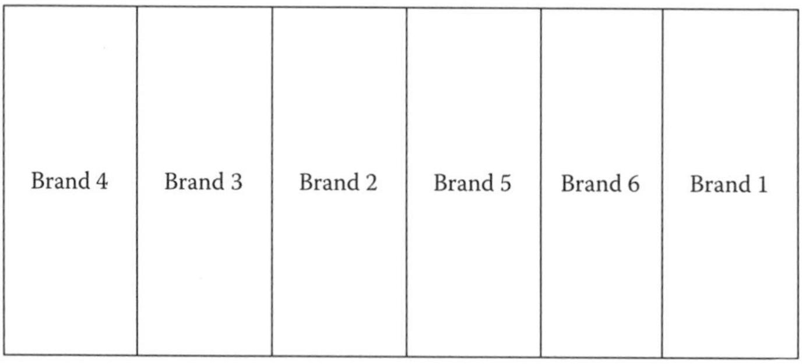

Brand 4	Brand 3	Brand 2	Brand 5	Brand 6	Brand 1

FIGURE 9.1
Random assignment of six different brands of fertilizer to six different sections of land.

Sun	Shade
Brand 3	Brand 2
Brand 4	Brand 6
Brand 1	Brand 3
Brand 5	Brand 5
Brand 2	Brand 1
Brand 6	Brand 4

FIGURE 9.2
Randomized block design blocking out the effect of sun and shade.

9.3 One-Way ANOVA

You may recall that in regression analysis, we used an ANOVA table to describe how the total sum of squares was partitioned into the regression sum of squares and the error sum of squares. Similarly, we can also perform an ANOVA to determine whether there is a difference between the true but unknown population means from several different populations by partitioning the variability based on some factor of interest and the error.

A one-way ANOVA is very similar to a regression analysis where the only independent variables in the model are indicator (or binary) variables. A one-way ANOVA compares the means of a variable that is classified by only one other variable, which is called the *factor*. The possible values of the factor are called the *levels* of the factor. We will be dealing with what is referred to as a *fixed-effects ANOVA model*, which means that the factors are fixed.

For instance, we could use a one-way ANOVA to compare the effect of several brands of lawn fertilizer on the mean growth rate of grass. We would consider the variable "fertilizer" the factor or treatment, and the different brands of fertilizer would correspond to the different levels of the factor.

Suppose we are interested in comparing the population means based on k different levels of some factor or treatment. We could draw a random sample from each of these k different levels of the population, as illustrated in Table 9.1.

In Table 9.1, $y_{i,j}$ represents the ith observation from the jth level. Notice that we do not necessarily need to take equal sample sizes from each level. A *balanced* one-way ANOVA is when each of the different levels have the same sample size.

We want to know whether there is a difference in the population means among the k factor levels. To do this, we will conduct a one-way ANOVA. A one-way ANOVA can be described as a special case of a regression analysis, as presented in the following model:

$$y_{i,j} = \mu + \tau_j + \varepsilon_{i,j}$$

where j ranges from $1,\dots,k$, where k is the number of factor levels; $y_{i,j}$ is the ith response observation from level j; μ is the mean common to all levels of the factor; τ_j is the effect of the jth factor level; and $\varepsilon_{i,j}$ is the error component for the ith observation from level j. Thus, the treatment means for each of the k different levels can be described as follows:

$$\mu_1 = \mu + \tau_1$$

$$\mu_2 = \mu + \tau_2$$

$$\vdots$$

$$\mu_k = \mu + \tau_k$$

TABLE 9.1

Samples Drawn from k Different Populations

Sample 1	Sample 2	Sample 3	...	Sample k
$y_{1,1}$	$y_{1,2}$	$y_{1,3}$	...	$y_{1,K}$
$y_{2,1}$	$y_{2,2}$	$y_{2,3}$	...	$y_{2,K}$
...	...	...	...	...
...	...	...	...	...
...	$y_{n_2,2}$	...	...	...
...		$y_{n_3,3}$	...	...
$y_{n_1,1}$			...	$y_{n_k,k}$

For example, in considering the effect of k different brands of lawn fertilizer on the growth rate of grass, μ would be the average growth rate common to all the brands of fertilizer, and τ_j would be the effect of the jth brand of fertilizer on the growth rate of grass.

For instance, suppose that the mean growth rate common to all brands of fertilizer is 25 mm/week. If brand j contributes, on average, an additional 5 mm/week to the growth rate, then $\mu_j = \mu + \tau_j = 25 + 5 = 30$ mm would be the mean growth rate for the grass treated with fertilizer brand j.

The main idea behind a one-way ANOVA is to determine whether there is a difference in the true but unknown population treatment effects for the k different levels of the treatment as described by the following null and alternative hypotheses:

$$H_0 : \tau_1 = \tau_2 = \tau_3 = \ldots = \tau_k$$

$$H_A : \text{At least one pair of population effects differ.}$$

The null hypothesis states that all of the true but unknown population effects are equal to each other, whereas the alternative hypothesis states that at least two population effects are different from each other.

Similarly, the difference in treatment effects can also be stated in terms of the population means for the k different levels of the treatment. In other words, we can test whether there is at least one pair of population means that are significantly different from each other. The null and alternative hypotheses would then be stated as follows:

$$H_0 : \mu_1 = \mu_2 = \mu_3 = \ldots = \mu_k$$

$$H_A : \text{At least one pair of population means differ.}$$

In performing a one-way ANOVA, we will be using the *F-distribution*. Recall that a random variable has the *F*-distribution if its distribution has the shape of a skewed curve, as presented in Figure 9.3.

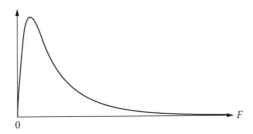

FIGURE 9.3
F-distribution.

Just as with t-distributions, there are alternative F-distributions that are identified by their degrees of freedom. Recall that F-distributions have two parameters to represent the degrees of freedom instead of only one, as with the t-distribution. The first degree of freedom for the F-distribution is called the *degrees of freedom for the numerator*, and the second is called the *degrees of freedom for the denominator*.

For a one-way ANOVA, we will use the following notation to describe the rejection region of an F-distribution:

$$f_\alpha\left(k-1, n-k\right)$$

where $k-1$ is the number of degrees of freedom for the numerator, $n-k$ is the number of degrees of freedom for the denominator, and α is the level of significance. The variable k is the number of populations we are sampling from, which represents the different levels of the factor, and n is the total sample size, which represents the total number of observations collected across all levels of the factor.

Just as with any other probability distribution, the total area under the F-distribution equals 1. But unlike the z- and t-distributions, the F-distribution starts at 0 on the horizontal axis and goes infinitely to the right approaching the horizontal axis, and thus is right skewed.

The following test statistic, which has the F-distribution, is used for a one-way ANOVA:

$$F = \frac{MSTR}{MSE}$$

The test statistic, F, describes the ratio of two mean square terms, where the term $MSTR$ corresponds to the mean square due to the treatment or factor, and MSE corresponds to the mean square due to the error component. If some of the means (or effects) from the populations we are sampling from are significantly different from each other, then the MSTR would not likely be due to sampling error and would tend to be larger than the MSE, and this would lead to F being large, and thus to rejecting the null hypothesis.

The test statistic F is then compared to the value of f_α with $(k-1, n-k)$ degrees of freedom, where k is the number of populations being sampled from, and n is the total number of observations collected.

Example 9.1

Suppose that we are interested in whether there is a difference in the mean driving distance between three different brands of golf balls. Consider the data set "Golf Balls" that is presented in Table 9.2. This data set consists of the driving distances (in yards) for three different brands of golf balls, which were collected under identical conditions using an automatic professional driving machine.

TABLE 9.2

Random Sample of the Driving Distances for
Three Different Brands of Golf Balls

Brand 1	Brand 2	Brand 3
285	292	291
290	272	269
271	279	295
288	283	273

Because our response variable is the driving distance, for each of these three different brands of golf balls the mean driving distance would be as follows:

$$\bar{y}_1 = 283.50$$

$$\bar{y}_2 = 281.50$$

$$\bar{y}_3 = 282.00$$

The sample size for each of the three different brands is

$$n_1 = 4$$

$$n_2 = 4$$

$$n_3 = 4$$

To use a one-way ANOVA to see if there is a significant difference in the driving distances between the three different brands of golf balls, our null and alternative hypotheses would be stated as follows:

$$H_0 : \mu_1 = \mu_2 = \mu_3$$

H_A : At least one pair of population means differ.

The alternative hypothesis could also be expressed as showing that at least one pair of population means is different, as follows:

$$H_A : \mu_1 \neq \mu_2, \text{ or } \mu_1 \neq \mu_3, \text{ or } \mu_2 \neq \mu_3$$

To calculate the test statistic, we need to find the value of MSTR (the mean square for the treatment) by first calculating the overall or grand mean for all of the observations, $\bar{y}_G$, as follows:

$$\bar{y}_G = \frac{sum\ of\ all\ observations}{n_1 + n_2 + \cdots + n_k}$$

where n_j is the sample size from level *j*.

To find the mean square for the treatment (MSTR), we would need to first find the sum of squares for the treatment (SSTR). The sum of squares for the treatment can be found by taking the sum of the squared differences between the mean from each factor level and the grand mean squared multiplied by the sample size for each of the given factor levels, as follows:

$$SSTR = \sum_{j=1}^{k} n_j (\bar{y}_j - \bar{y}_G)^2 = n_1(\bar{y}_1 - \bar{y}_G)^2 + n_2(\bar{y}_2 - \bar{y}_G)^2 + \cdots + n_k(\bar{y}_k - \bar{y}_G)^2$$

where n_j is the sample size from factor level j, $\bar{y}_j$ is the mean of the observations from factor level j, and $\bar{y}_G$ is the grand mean for all the observations.

Because SSTR represents the sum of squares for the treatment that measures the difference between the mean of the jth group and the grand mean, SSTR can be used to estimate the variability across all of the different populations.

The mean square for the treatment is then found by dividing the sum of squares for the treatment by $k - 1$ degrees of freedom, as follows:

$$MSTR = \frac{SSTR}{k-1}$$

where k is the number of levels being sampled from.

For our example, since there are three populations we are sampling from (the three different brands of golf balls that correspond to three different factor levels), $k = 3$, and since we collected a total of twelve observations, $n = 12$.

Then the grand mean for all the observations can be calculated as follows:

$$\bar{y}_G = \frac{\sum y}{n} = \frac{285 + 290 + 271 + 288 + 292 + 272 + 279 + 283 + 291 + 269 + 295 + 273}{4 + 4 + 4} = 282.33$$

Since $n_1 = n_2 = n_3 = 4$, we have that

$$SSTR = 4(283.50 - 282.33)^2 + 4(281.50 - 282.33)^2 + 4(282.00 - 282.33)^2$$

$$\approx 5.48 + 2.76 + 0.44$$

$$= 8.68$$

So, $MSTR = \dfrac{SSTR}{k-1} = \dfrac{8.68}{3-1} \approx 4.34.$

Now the mean square error is defined as

$$MSE = \frac{SSE}{n-k},$$

where $SSE = \Sigma_{i=1}^{k}(n_i - 1)s_i^2 = (n_1 - 1)s_1^2 + (n_2 - 1)s_2^2 + \cdots + (n_k - 1)s_k^2$, and s_j^2 is the sample variance of the jth group. For our example:

$$s_1^2 \approx 73.67$$

$$s_2^2 \approx 69.67$$

$$s_3^2 \approx 166.67$$

So, $SSE = (4 - 1) \cdot 73.67 + (4 - 1) \cdot 69.67 + (4 - 1) \cdot 166.67 \approx 930.03$.

SSE can be used to estimate the variability within each of the k populations. Then:

$$MSE = \frac{SSE}{n - k} = \frac{930.03}{12 - 3} \approx 103.34$$

And finally:

$$F = \frac{MSTR}{MSE} = \frac{4.34}{103.34} \approx 0.04$$

We can now compare the value of this test statistic with the value obtained from the F-table ($\alpha = 0.05$). The degrees of freedom for the numerator is $k - 1 = 3 - 1 = 2$, and the degrees of freedom for the denominator is $n - k = 12 - 3 = 9$. Using the F-table in Table 4 of Appendix A, this gives $f = 4.26$. And since 0.042 does not fall in the rejection region as established by $f = 4.26$, we do not have enough evidence to reject the null hypothesis and accept the alternative hypothesis. Therefore, we *cannot* claim that the population mean driving distance for any pair of these three different brands of golf balls is significantly different from the others.

One way to summarize all the different calculations that are needed to conduct a one-way ANOVA is to put the calculations in the form of a table. The general form of a one-way ANOVA table illustrates how the variability is partitioned based on the treatment and the error component, along with their appropriate degrees of freedom, as illustrated in Table 9.3.

TABLE 9.3

General Form of a One-Way ANOVA Table

Source	Sum of Squares	Degrees of Freedom	Mean Square	F	p-value
Treatment variation	SSTR	$k - 1$	MSTR	MSTR/MSE	—
Error variation	SSE	$n - k$	MSE	—	—
Total	SSTR + SSE	$n - 1$	—	—	—

TABLE 9.4

A One-Way ANOVA Table for the Golf Ball Example

Source	Sum of Squares	Degrees of Freedom	Mean Square	F
Treatment variation	8.68	2	4.34	0.04
Error variation	930.03	9	103.34	
Total	938.71	11		

For our last example, the one-way ANOVA table (without the respective *p*-value) is given in Table 9.4.

9.4 Model Assumptions

As with most analyses that we have considered thus far, there are some model assumptions for a one-way ANOVA that need to be considered before any relevant inferences can be made. Because a one-way ANOVA is just a special case of a regression analysis, the ANOVA model assumptions are equivalent to those model assumptions that we considered regarding the distribution of the error component as in a regression analysis. This is because the one-way ANOVA model, $y_{i,j} = \mu + \tau_j + \varepsilon_{i,j}$, is represented by a fixed component $(\mu + \tau_j)$, and a random component $(\varepsilon_{i,j})$.

The assumptions for a one-way ANOVA are as follows:

1. The samples from each of the k populations are independent and have been selected at random.
2. The dependent variable is normally distributed in each population. Because a one-way ANOVA is a special case of a regression analysis that has a fixed component and a random component, this is the same as saying that the error component is normally distributed.
3. The variance of the dependent variable is constant for each population (also referred to as homogeneity of variance). Again, since a one-way ANOVA is a special case of a regression analysis, this is the same as saying that the error component has constant variance.

The first assumption (independence) is not always easy to check. If the samples are collected in such a way that each observation is independent of all the other observations, then we can be relatively sure that this assumption has not been violated. Typically, if a random sample is collected, then one can be relatively sure that the independence assumption has not been violated. However, if there is any reason to believe that the independence assumption could have been violated, then there are some exploratory techniques that can be done to see if this is an issue. For instance, if the data are collected over time and each observation depends on another, it would clearly be a violation of the independence assumption.

One of the more common ways to check the normality and constant variance assumptions for a one-way ANOVA model is to use residual plots and formal tests, just like we did with regression analysis. Residual plots can be useful in determining whether some of the one-way ANOVA model assumptions may have been violated. The next section will provide some details on how to use MINITAB to conduct various exploratory analyses.

Also, just as with regression analysis, there may be some situations, such as when small sample sizes are drawn from each of the k different populations, where it may be difficult to determine if the model assumptions are reasonable based solely on the exploratory plots. For such cases, there are formal tests that can be used. These formal tests address the assumptions of constant variance and normality of the error component.

9.5 Assumption of Constant Variance

There are two formal tests that can be used to determine if the population variances for each of the k levels are equal; these are Bartlett's test (1937) and Levene's test (1960). For both of these tests, the null and alternative hypotheses for comparing the true but unknown variance from k populations is

$$H_0: \quad \sigma_1^2 = \sigma_2^2 = \ldots = \sigma_k^2$$

$$H_A: \quad \text{At least two population variances are unequal.}$$

For samples taken from k populations that are normally distributed, the test statistic for Bartlett's test is found by taking the sum of the natural logarithm for each of the individual variances and subtracting it from the natural logarithm of the average variance. When equal sample sizes are taken from each population, the test statistic is as follows:

$$B = \frac{3k(n-1)^2 \left[k \cdot \ln(\bar{s}^2) - \sum_{j=1}^{k} \ln\left(s_j^2\right) \right]}{3kn - 2k + 1}$$

where k is the number of different populations being sampled from, n is the sample size taken from each population (since we have equal sample sizes from each of the k populations, this means that $n = n_1 = n_2 = \ldots n_k$), s_j^2 is the sample variance for sample j, $\bar{s}^2 = \sum_{j=1}^{k} s_j^2/k$ is the average of all k sample variances, and ln is the natural logarithm.

For unequal sample sizes, the test statistic becomes quite a bit more complicated:

$$B = \frac{(n-k)\ln(\bar{s}^2) - \sum_{j=1}^{k}(n_j - 1)\ln\left(s_j^2\right)}{1 + \frac{1}{3(k-1)}\left[\sum_{j=1}^{k}\frac{1}{(n_j - 1)} - \frac{1}{n-k}\right]}$$

where k is the number of populations being sampled from, n_j is the sample size for sample j, n is the total number of observations collected across all samples $(n = n_1 + n_2 + \cdots + n_k)$, s_j^2 is the sample variance for sample j, and $\bar{s}^2 = \sum_{j=1}^{k}(n_j - 1)s_j^2/(n-k)$ is a weighted average of the sample variances for all k populations.

Notice that in rejecting the null hypothesis and accepting the alternative hypothesis for Bartlett's test, this means that there are at least two variances that are unequal. Because we want to check for constant variance, we would hope to find that there *is not* enough evidence to reject the null hypothesis and accept the alternative hypothesis. For the assumption of constant variance to hold true, this means that we do not want the test statistic to fall in the rejection region. If we *cannot* claim that there are at least two variances that are unequal (in other words, we cannot accept the alternative hypothesis), then we have reason to believe that the assumption of constant variance has not been violated.

For both equal and unequal sample sizes, we compare the appropriate test statistic with the critical value of the χ^2 distribution with $k - 1$ degrees of freedom (See Table 5 in Appendix A). The χ^2 distribution is similar to the F-distribution, but its shape is determined by $k - 1$ degrees of freedom, as illustrated in Figure 9.4.

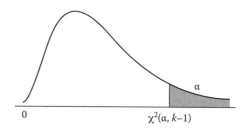

FIGURE 9.4
χ^2 distribution with $k - 1$ degrees of freedom.

Notice that when considering *equal* sample sizes, n is the sample size taken from *each* population, and when considering *unequal* sample sizes, n is the total number of observations collected across *all* samples.

Example 9.2

We will use Bartlett's test on the golf ball data in Table 9.2 as a formal test for constant variance. Because we have equal sample sizes for each of the populations we are sampling from, we can use Bartlett's test statistic for equal sample sizes. For this set of data we have the following descriptive statistics:

$n = 4$ (there are four observations from each population)

$k = 3$ (we are sampling from three different populations)

$$\left.\begin{array}{l} s_1^2 = 73.67 \\[2mm] s_2^2 = 69.67 \\[2mm] s_3^2 = 166.67 \end{array}\right\} \quad \text{(sample variances for the samples from each population)}$$

$$\bar{s}^2 = \frac{\sum\limits_{j=1}^{3} s_j^2}{k} = \frac{73.67 + 69.67 + 166.67}{3} \approx 103.34$$

And $\Sigma_{j=1}^{3} \ln\left(s_j^2\right) = \ln\left(s_1^2\right) + \ln\left(s_2^2\right) + \ln\left(s_3^2\right) = \ln(73.67) + \ln(69.67) + \ln(166.67) \approx 13.66$.

So we can calculate Bartlett's test statistic as follows:

$$B = \frac{3(3)(4-1)^2[3 \cdot \ln(103.34) - 13.66]}{3(3(4)) - 2(3) + 1} \approx 0.66$$

In comparing this to the value of the $\chi^2 = 5.9915$ distribution ($\alpha = 0.05$) for $k - 1 = 3 - 1 = 2$ degrees of freedom, we do not have enough evidence to suggest that at least two variances are different because the test statistic $B = 0.66$ does not fall in the rejection region. Therefore, we have reason to believe that the assumption of constant variance has not been violated.

Another test, Levene's test as modified by Brown and Forsythe (1974), can also be used for detecting unequal variances. The basic idea behind Levene's modified test is to use the absolute value of the difference between the observations in each treatment group from the median of the treatment. It then tests whether or not the mean of these differences is equal for all of the k treatments.

The test statistic is the same as the F-statistic for a one-way ANOVA, but the analysis is applied to the absolute value of the difference between the observation and the treatment median for each of the k treatments. If the average differences are equal for all of the treatments, this suggests that the variances of the observations in all of the treatment groups will be the same.

Thus, the test statistic for Levene's test is the familiar F-test using the absolute value of the difference between the observation and the treatment median for each of the k treatments as the response variable:

$$F = \frac{MSTR}{MSE}$$

which is compared to the f_α distribution with $(k-1, n-k)$ degrees of freedom.

Example 9.3

Let's use Levene's test on the golf ball data in Table 9.2 to test for constant variance. In order to do this, we first need to transform the response variable by taking the absolute value of the difference for each observation and the median for every observation in the three different levels of treatment. In other words:

$$|y_{i,j} - \tilde{y}_j|$$

where $y_{i,j}$ is the ith observation from sample j, and $\tilde{y}_j$ is the median for sample j.

The medians for the response variables for each of the three different brands of golf balls are as follows:

$$\tilde{y}_1 = 286.50$$

$$\tilde{y}_2 = 281.00$$

$$\tilde{y}_3 = 282.00$$

The transformed response data are then found by taking the absolute value of the difference between each individual observation and the median for each treatment, as in Table 9.5.

TABLE 9.5

Difference between Median and Response Value for Each Observation in Each Treatment

Mean	$k = 1$	$k = 2$	$k = 3$
	1.5	11	9
	3.5	9	13
	15.5	2	13
	1.5	2	9
	5.5	6	11

For instance, the first observation for Brand 1 is 285 yards, and the median of Brand 1 is 286.50. Therefore, the entry in the first row for column $k = 1$ is as follows:

$$|285 - 286.50| = 1.5$$

The means of the transformed variables for each treatment level are 5.5, 6, and 11, respectively (in other words, these are the means for the columns $k = 1$, $k = 2$, and $k = 3$ in Table 9.5).

To find the value of MSTR for the transformed data, we need to first take the average of the column means as follows:

$$\bar{y}_G = \frac{5.5 + 6 + 11}{3} = 7.5$$

So,

$$SSTR = 4(5.5 - 7.5)^2 + 4(6 - 7.5)^2 + 4(11 - 7.5)^2 = 16 + 9 + 49 = 74$$

$$MSTR = \frac{SSTR}{k-1} = \frac{74}{3-1} = 37$$

The variances for each of the treatments for the transformed variables are

$$s_1^2 \approx 45.33$$

$$s_2^2 = 22.00$$

$$s_3^2 \approx 5.33$$

Then $SSE = (4-1) \cdot 45.33 + (4-1) \cdot 22 + (4-1) \cdot 5.33 \approx 217.98$, and $MSE = \frac{217.98}{12-3} \approx 24.22$. Hence, $F = \frac{MSTR}{MSE} = \frac{37}{24.22} \approx 1.53$. Now, compare to $f_{0.05} = 4.26$ for $\alpha = 0.05$, which suggests that the assumption of constant variance has not been violated.

Typically, Bartlett's test is used when the data come from populations that have normal distributions since Bartlett's test can be very sensitive to deviations from normality. Levene's test tends to be preferred for nonconstant variance because it can be used when the data come from populations that have continuous distributions, but they do not necessarily have to be normal distributions. Because Levene's test considers the distances between the observations and their sample medians rather than their sample means, this makes the test more appropriate for smaller sample sizes and where outliers may be present.

9.6   Normality Assumption

Recall in Chapter 6 that we used the *Ryan–Joiner test* (Ryan and Joiner, 1976) to compare the unknown distribution of the error component with a normal distribution to see if they differ in shape. We compared the residuals with data from a normal distribution to create a test statistic that serves to measure the correlation between the residuals and data that come from a normal distribution. If this correlation measure is close to 1.0, then we have a greater amount of certainty that the unknown distribution of the error component is approximately normally distributed. Recall that the Ryan–Joiner test statistic is calculated as follows:

$$RJ = \frac{\sum\limits_{i=1}^{n} \hat{\varepsilon}_i \cdot z_i}{\sqrt{s^2(n-1)\sum\limits_{i=1}^{n} z_i^2}}$$

where $\hat{\varepsilon}_i$ is the residual for observation i, z_i are the normal scores for the ith residual (see Section 6.8 for a description of how to calculate normal scores), n is the sample size, and s^2 is the variance of the residuals. This test statistic is then compared to a table of critical values that have been established based on the desired level of significance and the sample size (see Table 3 in Appendix A).

Recall that the Ryan–Joiner test statistic represents a measure of correlation between the residuals and data obtained from a normal distribution, and this statistic is compared to the critical value for a given level of significance using the table given in Table 3 of Appendix A, which defines the smallest correlation between the residuals and data obtained from a normal distribution.

Example 9.4

For the data regarding the driving distance by brand of golf ball given in Table 9.2, Table 9.6 gives the residuals and normal scores.

Then, $\sum \hat{\varepsilon}_i \cdot z_i = 93.2711$, $s^2 = 84.55$, and $\sum z_i^2 = 9.85136$. So the test statistic is

$$R = \frac{\sum\limits_{i=1}^{n} \hat{\varepsilon}_i \cdot z_i}{\sqrt{s^2(n-1)\sum\limits_{i=1}^{n} z_i^2}} = \frac{93.2711}{\sqrt{84.55(11)(9.85136)}} \approx 0.974$$

TABLE 9.6

Residuals and Normal Scores for the Golf Ball Data Given in Table 9.2

Driving Distance	Brand	Residual	Normal Score
285	1	1.5	0.00000
290	1	6.5	0.53618
271	1	−12.5	−1.11394
288	1	4.5	0.31192
292	2	10.5	1.11394
272	2	−9.5	−0.79164
279	2	−2.5	−0.31192
283	2	1.5	0.00000
291	3	9.0	0.79164
269	3	−13.0	−1.63504
295	3	13.0	1.63504
273	3	−9.0	−0.53618

Table 3 in Appendix A gives the critical values for rejecting the null hypothesis. Comparing this test statistic to the critical value that falls between 0.9180 and 0.9383 for $n = 10$ and $n = 15$ with $\alpha = 0.05$ suggests that there is not enough evidence to imply that the error component does not come from a normal distribution (there is not enough evidence to reject the null hypothesis). This is because our test statistic is greater than the smallest correlation defined by the corresponding critical values. Thus, we can infer that the assumption that the error component is normally distributed does not appear to have been violated.

Recall that the assumption of normality would be rejected if the value of the Ryan–Joiner test statistic was less than the critical value for a given level of significance and sample size, because this would imply that the correlation between the residual values and the values obtained from a normal distribution is less than the minimum amount that is specified by the critical value.

9.7 Using MINITAB for One-Way ANOVAs

As you can see, it can be quite cumbersome to perform all of the various one-way ANOVA calculations by hand, so we will now use MINITAB to perform a one-way ANOVA. Using software to run an ANOVA allows for a graphical assessment of the model assumptions for a one-way ANOVA in addition to performing the formal tests of some of the assumptions that were described in the last section.

Using MINITAB for a one-way ANOVA is fairly straightforward and requires putting all the collected data into either stacked or unstacked form. In using unstacked form, we can enter the data collected from each population into separate columns, where each column consists of the data for each factor (similar to how the golf ball data are presented in Table 9.2). In using stacked form, the data are entered using only two columns, where one column consists of the response variable and the other column describes the levels of the factor. For our example, the response variable would be the driving distance in yards, and the levels of the factor for our example will be Brand 1, Brand 2, and Brand 3 to denote the three different brands of golf balls. In stacked form, the first two columns in our MINITAB worksheet are presented in Figure 9.5.

To run a one-way ANOVA using MINITAB, select **Stat, ANOVA, One-way**, as illustrated in Figure 9.6.

We can now run a one-way ANOVA routine in MINITAB using driving distance as the response and brand as the factor, as presented in the dialog box in Figure 9.7. This gives the MINITAB printout in Figure 9.8.

Notice that we arrived at approximately the same values for the SSTR, SSE, MSTR, and MSE as we did when we performed the calculations by hand (the slight differences are due to round-off error). Based on the *p*-value highlighted in Figure 9.8, we can make the similar conclusion in that we do

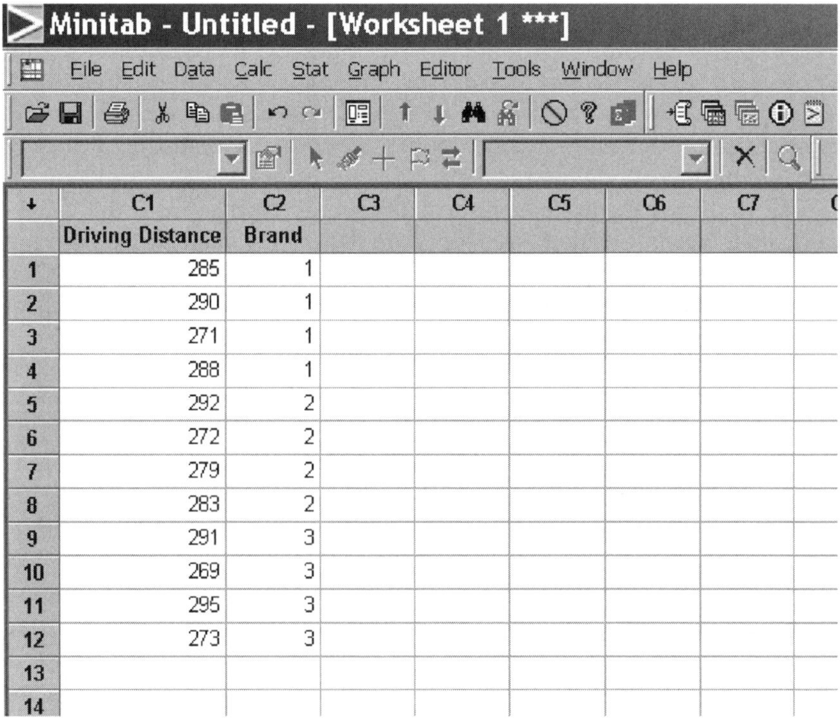

	C1 Driving Distance	C2 Brand
1	285	1
2	290	1
3	271	1
4	288	1
5	292	2
6	272	2
7	279	2
8	283	2
9	291	3
10	269	3
11	295	3
12	273	3
13		
14		

FIGURE 9.5
MINITAB worksheet with the golf ball data entered in stacked form.

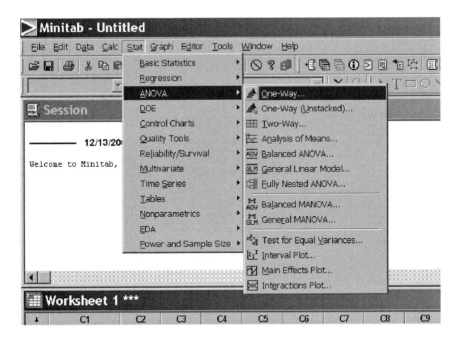

FIGURE 9.6
MINITAB commands for a one-way ANOVA.

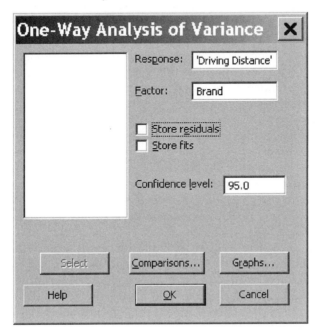

FIGURE 9.7
MINITAB dialog box for a stacked one-way ANOVA.

One-way ANOVA: Driving Distance versus Brand

```
Source   DF    SS    MS      F       P
Brand     2     9     4    0.04   0.959
Error     9   930   103
Total    11   939
```

```
S = 10.17    R-Sq = 0.92%    R-Sq(adj) = 0.00%
```

```
                            Individual 95% CIs For Mean Based on
                            Pooled StDev
Level   N    Mean   StDev   ----+---------+---------+---------+-----
1       4  283.50    8.58        (---------------*---------------)
2       4  281.50    8.35   (---------------*---------------)
3       4  282.00   12.91   (---------------*---------------)
                            ----+---------+---------+---------+-----
                            273.0      280.0      287.0      294.0
```

```
Pooled StDev = 10.17
```

FIGURE 9.8
MINITAB printout for one-way ANOVA of driving distance versus brand of golf ball.

not have enough evidence to suggest that any of the population mean driving distances for the three different brands of golf balls are significantly different from each other (since $p > .05$). Furthermore, on the bottom portion of the MINITAB printout, there is a collection of confidence intervals of the population means for each of these different brands of golf balls. Based on the width and position of the confidence intervals, you can see that they are not much different from each other since all of these intervals are overlapping.

We can also use MINITAB to determine if the assumptions for the ANOVA have been reasonably met. Similar to multiple regression analysis, checking the distribution of the error component for a one-way ANOVA can be done using MINITAB by opening the **Graphs** dialog box within the one-way ANOVA box, as in Figure 9.9. As an additional check on the assumption of constant variance, we can also check the box plots of data in the same dialog box.

The box plot in Figure 9.10 illustrates the variability for each of the different brands of golf balls.

The histogram of the residuals in Figure 9.11 indicates whether or not the distribution of the error term is approximately normal.

We can also graphically check the assumption of constant variance by plotting the residual values versus the fitted values, as in Figure 9.12. The residuals should appear to be scattered randomly about the 0 line and there should be no obvious patterns or clusters. However, with small sample sizes this plot is usually not too revealing because the observations will tend to cluster based on the number of levels of the factor.

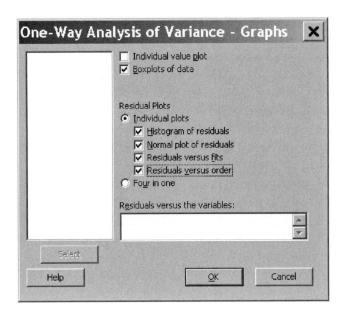

FIGURE 9.9
MINITAB dialog box to plot a box plot and check model assumptions.

The normal probability plot, as presented in Figure 9.13, should appear to be approximately linear if the distribution of the error component is normally distributed.

Finally, the residuals versus the order of the observations can be plotted as in Figure 9.14. This graph can be used to assess whether the values of the residuals

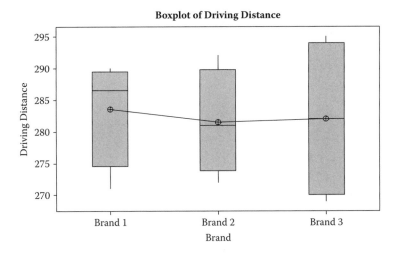

FIGURE 9.10
Box plot of driving distance by brand of golf ball.

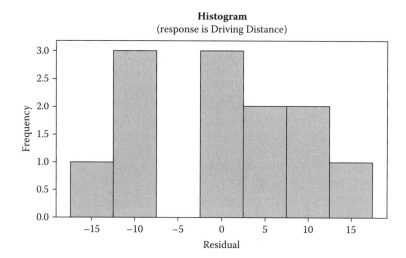

FIGURE 9.11
Histogram of residuals.

for any of the observations differ based on if the data were collected sequentially. If the pattern of this line diagram does not appear to be random, we could have reason to believe that the samples are not independent of each other.

Because we have a small sample size, it can be a challenge to use residual plots to assess whether the assumptions for a one-way ANOVA have been

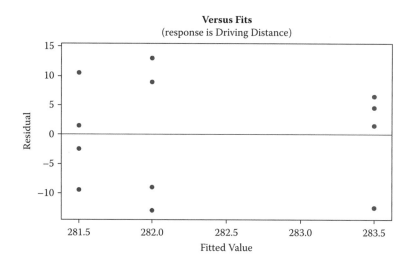

FIGURE 9.12
Residual versus fitted values.

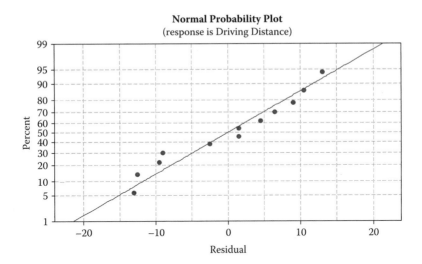

FIGURE 9.13
Normal probability plot.

violated. However, we can also use MINITAB to conduct the Ryan–Joiner test of normality, and Bartlett's and Levene's tests of constant variance.

Instead of generating a normal probability plot directly from the one-way ANOVA dialog box to check whether we are sampling from a normal distribution, we can use MINITAB to conduct the Ryan–Joiner test of normality.

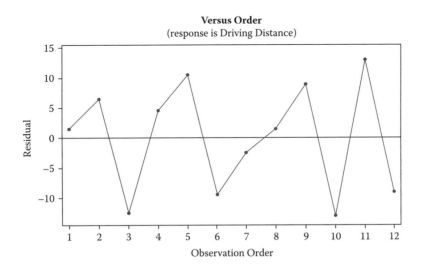

FIGURE 9.14
Residuals versus order plot.

To do this, we have to first save the residuals obtained from running the one-way ANOVA, as illustrated in Figure 9.15. To store the residuals using MINITAB, simply check the box on the one-way ANOVA dialog box to store the residuals. We can then perform the Ryan–Joiner test under **Basic Statistics, Normality Test**, as in Figure 9.16.

Selecting **Ryan–Joiner** in the dialog box in Figure 9.17 gives the MINITAB printout in Figure 9.18. Figure 9.18 gives the normal probability plot along with the value of the Ryan–Joiner test statistic (*RJ*) in the box on the right of the graph. Since the *p*-value for the Ryan–Joiner test is greater than .05 ($p > .100$), this suggests that there is not enough evidence at the .05 significance level to imply that the assumption of the normality of the error component has been violated.

MINITAB can also perform Bartlett's and Levene's tests. Using the **Test for Equal Variance** tab in the ANOVA menu, as in Figure 9.19, gives the MINITAB dialog box in Figure 9.20.

Figure 9.21 gives the test statistics, confidence intervals for the standard deviations for each level of the factor, and *p*-values for both Bartlett's and Levene's tests.

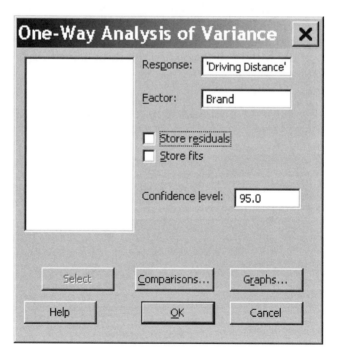

FIGURE 9.15
MINITAB dialog box for a one-way ANOVA with the residuals stored.

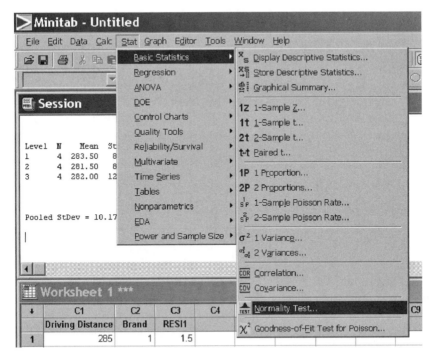

FIGURE 9.16
MINITAB commands to run a normality test.

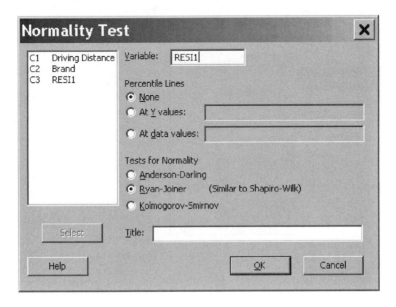

FIGURE 9.17
MINITAB dialog box to run a normality test.

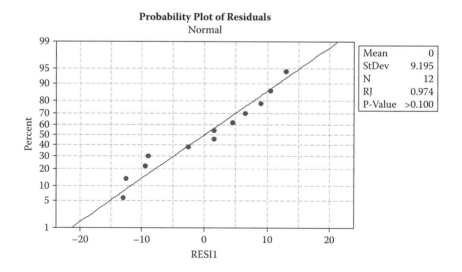

FIGURE 9.18
MIINTAB normality plot including the Ryan–Joiner test.

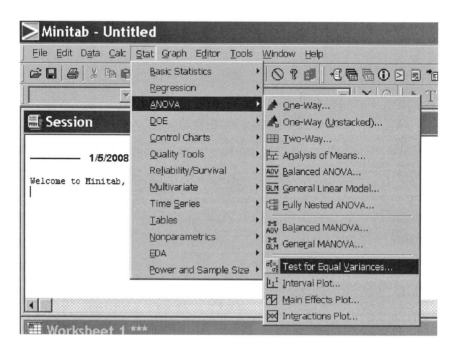

FIGURE 9.19
MINITAB commands to run Bartlett's and Levene's tests.

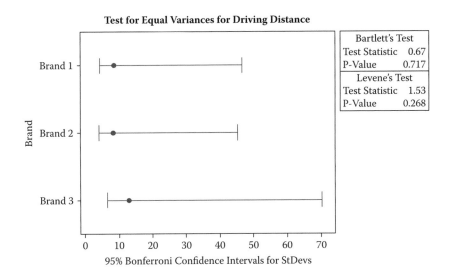

FIGURE 9.20
MINITAB dialog box to test for equal variances.

Test for Equal Variances for Driving Distance

Bartlett's Test	
Test Statistic	0.67
P-Value	0.717

Levene's Test	
Test Statistic	1.53
P-Value	0.268

95% Bonferroni Confidence Intervals for StDevs

FIGURE 9.21
MINITAB output for Bartlett's and Levene's tests.

In the box on the right-hand portion of Figure 9.21, both Barlett's and Levene's tests report the values of respective test statistics and p-values greater than .05. These p-values suggest that we do not have enough evidence to believe that the assumption of constant variance has been violated. In order to use MINITAB to test for equal variances, the data have to be in stacked form.

Example 9.5

The data set "Fertilizer" in Table 9.7 consists of the weekly growth rate of grass for six different brands of lawn fertilizer. Suppose that we want to test whether there is a difference in the weekly growth rates of grass between the six different brands of fertilizer. The six columns in Table 9.7 represent the weekly growth rates of the grass in millimeters.

Notice that for this data set, the samples sizes from each of the populations are not all equal to each other. By running a one-way ANOVA in unstacked form using MINITAB, the results are presented in Figure 9.22.

Notice in Figure 9.22 that the p-value is .000. This suggests that we can reject the null hypothesis and accept the alternative hypothesis. In other words, at least two population brands of lawn fertilizer have significantly different mean growth rates.

Also, as you can see from the confidence intervals for each brand of fertilizer given in the bottom portion of the MINITAB printout in Figure 9.22, the growth rate for Brand 2 appears to be significantly higher than those for all of the other brands of fertilizer since the confidence interval for Brand 2 does not overlap the confidence intervals for any of the other brands.

The box plot and the graphs to assess the model assumptions for a one-way ANOVA are in Figures 9.23 to 9.26.

Note that if you use the unstacked option, there is no residual versus order plot. So if you have reason to believe that the order in which the data were collected may affect the independence assumption, you would want to put the data into MINITAB using stacked form so you could plot the residual values versus the order.

TABLE 9.7

Weekly Growth Rate of Grass (in millimeters) for Six Different Brands of Fertilizer

Brand 1	Brand 2	Brand 3	Brand 4	Brand 5	Brand 6
26	35	26	25	26	27
31	36	27	34	30	27
34	42	28	37	32	28
35	47	29	38	34	29
	49	30	39	35	30
	50	34	42	42	
	52		45		
	54				

One-Way ANOVA: Brand 1, Brand 2, Brand 3, Brand 4, Brand 5, Brand 6

```
Source   DF       SS      MS      F       P
Factor    5   1450.9   290.2  10.25   0.000
Error    30    849.4    28.3
Total    35   2300.3

S = 5.321    R-Sq = 63.08%   R-Sq(adj) = 56.92%

                              Individual 95% CIs For Mean Based on
                              Pooled StDev
Level      N     Mean   StDev  -------+---------+---------+---------+--
Brand 1    4   31.500   4.041        (-------*-------)
Brand 2    8   45.625   7.190                                (----*-----)
Brand 3    6   29.000   2.828   (-----*------)
Brand 4    7   37.143   6.414                   (-----*-----)
Brand 5    6   33.167   5.382            (-----*------)
Brand 6    5   28.200   1.304  (------*------)
                              -------+---------+---------+---------+--
                                 28.0      35.0      42.0      49.0

Pooled StDev = 5.321
```

FIGURE 9.22
MINITAB printout for a one-way ANOVA testing the mean difference in the growth rate of six different brands of fertilizer.

Although we have a larger sample size than we did for the golf ball example, it is still difficult to assess whether or not the one-way ANOVA assumptions may have been violated by solely relying on the residual plots. Thus, we may want to conduct some formal tests of the assumptions.

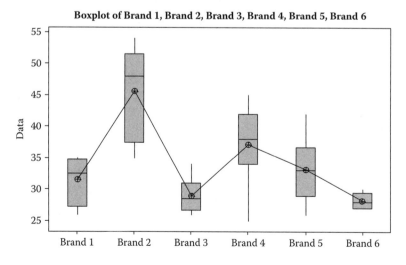

FIGURE 9.23
Box plots of growth rate for the six different brands of fertilizer.

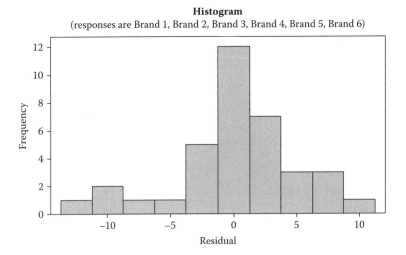

FIGURE 9.24
Histogram of the residuals.

The Ryan–Joiner test can be used to see if the normality assumption is believed to hold true. The graph of this plot is given in Figure 9.27.

The *p*-value for the Ryan–Joiner test is greater than our predetermined level of significance of .05, which suggests that there is not enough evidence to imply that the assumption of the normality of the error component has been violated.

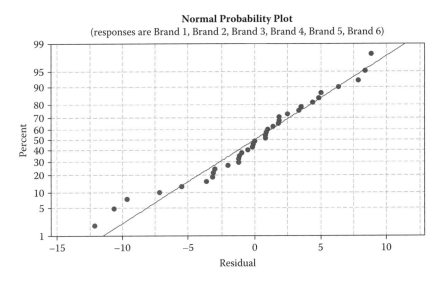

FIGURE 9.25
Normal probability plot of the residuals.

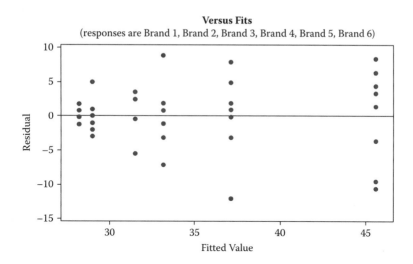

FIGURE 9.26
Residual versus fitted values plot.

Now using Bartlett's and Levene's tests for constant variance, the MINITAB output is presented in Figure 9.28.

The box on the right-hand side of Figure 9.28 gives the results for Bartlett's and Levenes's tests. For Barlett's test, the p-value of .040 suggests that the variance is not the same for at least two brands of lawn fertilizer. However, Levene's tests report a p-value greater than .05, which suggests that we do

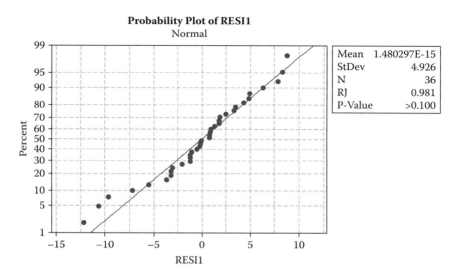

FIGURE 9.27
Ryan–Joiner test of the normality assumption.

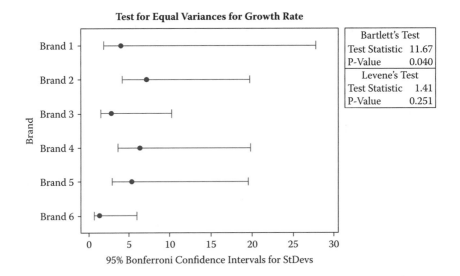

FIGURE 9.28
MINITAB output for Bartlett's and Levene's tests of constant variance.

not have enough evidence to imply that the assumption of constant variance has been violated. When the results from Bartlett's and Levene's tests disagree with each other, you may want to lean toward using Levene's test over Bartlett's test. This is because Bartlett's test is much more sensitive to outliers and extreme values than Levene's test, and for small samples, outliers can have a more pronounced effect.

As we have seen, when using small sample sizes ($n < 30$) from each of the populations we are sampling from, it can be difficult to assess whether or not the model assumptions have been reasonably met, or whether the model assumptions appear to have been violated. In Chapter 10, we will describe alternative statistical methods and techniques that can be used when considering small sample sizes or when we have reason to believe that the model assumptions have not been met.

9.8 Multiple Comparison Techniques

Once we perform a one-way ANOVA and accept the alternative hypothesis and reject the null hypothesis (and we believe that our model assumptions have been reasonably met), we can conclude that at least two population means are significantly different from each other. For instance, for the fertilizer data given in Table 9.7, we found that the mean growth rate of grass was different for at least two brands of fertilizer. There may be instances when we also may want to identify which specific pair (or pairs) of means are different from each other, and we also may want to estimate the magnitude and direction of such a difference or differences. We will use *multiple comparison techniques* to make such assessments.

In order to rank and compare a pairwise collection of means, we will use *Tukey's multiple comparison method*. This method is based on the studentized range distribution (Table 6 in Appendix A). The studentized range distribution can be described by its degrees of freedom as represented by two parameters: κ (kappa) and ν (nu), as presented in Figure 9.29.

We can find the value of q_α for a given level of significance of the studentized range distribution with parameters $\kappa = k$ and $\nu = n - k$, whereas before, k is the number of populations being sampled from and n is the total number of all observations.

Once we find the value of q_α, we can create confidence intervals for the difference in any pair of means, $u_l - \mu_m$, where l and m are the indices of the two individual populations we are interested in comparing. To calculate a confidence interval for the difference between a pair of means for a *balanced* one-way ANOVA, we use the formula

$$(\bar{x}_l - \bar{x}_m) \pm \frac{q_\alpha}{\sqrt{n}} \cdot \sqrt{MSE}$$

where n represents the number of observations sampled from *each* population $(n = n_1 = n_2 = \ldots = n_k)$, $\bar{x}_l$ and $\bar{x}_m$ are the sample means for populations l and m, respectively, and $\sqrt{MSE}$ is the root mean square error.

This confidence interval represents the range of values for the difference in the means for populations l and m. If this confidence interval does not contain the value 0, then we can say that the two means from populations l and m are significantly different from each other. Otherwise, if the interval contains the value of 0, we cannot claim that the two means are different.

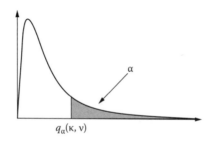

FIGURE 9.29
Studentized range-distribution with κ and ν degrees of freedom.

Example 9.6

Suppose we are interested in comparing the mean grade point averages (on a scale of 0.00–4.00) for the population of entering college freshmen who come from three different academic tracks in high school: liberal arts (or general studies), vocational and technical, or college prep. The data set "High School GPA" in Table 9.8 consists of the grade point averages (GPAs) for a random sample of twenty-one students from the three different academic tracks.

Given $\alpha = 0.05$, we want to determine if there is a difference in the population mean GPA for entering college freshmen based on the type of high school track.

The descriptive statistics for each of the three levels of the factor are given in Table 9.9.

The one-way ANOVA table found using MINITAB is given in Table 9.10.

Since the *p*-value is less than the predetermined level of significance, the mean grade point average for entering college freshmen is significantly different for at least two types of high school tracks.

Because our sample size is the same for each of the three levels of our factor, we can now use Tukey's multiple comparison technique for a balanced one-way ANOVA to determine which pairs of means are significantly different from each other, as well as the estimated magnitude of such differences.

TABLE 9.8

High School Grade Point Averages for a Sample of Twenty-One Students from Three Different High School Tracks

Liberal Arts	College Prep	Vocational/Technical
2.05	3.78	1.38
1.89	2.75	2.65
2.25	3.03	2.87
3.19	3.53	3.25
2.10	3.16	3.49
2.54	2.95	3.36
3.04	3.51	2.57

TABLE 9.9

Descriptive Statistics for the College Grade Point Average Data Given in Table 9.8

Statistic	Liberal Arts	College Prep	Vocational/ Technical
Mean	2.437	3.244	2.796
Standard deviation	0.506	0.370	0.717
Variance	0.256	0.137	0.515

TABLE 9.10

One-Way ANOVA Table for Comparing If There Is a Difference in the Mean College Grade Point Average for the Population of Students Coming from Three Different High School Tracks

Source	SS	df	MS	F	p-value
Treatment variation	2.290	2	1.145	3.78	0.043
Error variation	5.451	18	0.303		
Total	7.740	20			

Based on three factors and a total sample size of 21, the value for $q_{0.05} = 3.61$. The confidence interval that compares the entering GPA for the college-prep track subtracted from that for the liberal arts track can be calculated as follows:

$$(2.437 - 3.244) \pm \frac{3.61}{\sqrt{7}} \cdot \sqrt{0.303} = -0.807 \pm 0.751 = (-1.56, -0.06)$$

Therefore, we are 95% confident that for the population of students who take the liberal arts track in high school, their average entering GPAs are between 0.06 and 1.56 points less than the GPAs for the population of students who take the college-prep track in high school.

Similar calculations for the liberal arts track subtracted from the vocational/technical track, and college-prep track subtracted from the vocational/technical track are, respectively:

$$(-0.39, 1.11)$$

$$(-1.20, 0.30)$$

However, since both of these confidence intervals contain the value of 0, there is not a significant difference between the mean GPA for these given pairs of high school tracks.

We can also use a modification of Tukey's multiple comparison technique to calculate confidence intervals for the difference between any pair of means when we have an unbalanced one-way ANOVA, which is when the samples taken from the different populations are not all the same size:

$$(\bar{x}_l - \bar{x}_m) \pm \frac{q_\alpha}{\sqrt{2}} \cdot \sqrt{MSE} \cdot \sqrt{\left(\frac{1}{n_l}\right) + \left(\frac{1}{n_m}\right)}$$

where $\sqrt{MSE}$ is the root mean square error, n_l is the sample size from population l, and n_m is the size of the sample from population m. Similarly, the value of q_α for a given level of significance of the studentized range-distribution with parameters $\kappa = k$ and $\nu = n - k$, where k is the number of populations being sampled from and n is the total number of all observations.

Generally, multiple comparison techniques should only be performed following a one-way ANOVA when there is enough evidence to accept the alternative hypothesis and reject the null hypothesis, and we can conclude that at least two means are significantly different from each other.

Example 9.7

Now let's return to the example where we are interested in determining if there is a difference in the weekly growth rates of grass for the six different brands of lawn fertilizer. Recall this set of data was given in Table 9.7.

Given $\alpha = 0.05$, suppose that we want to determine if there is a significant difference in the growth rates for the different brands of lawn fertilizer, and we want to be able to determine which particular brands are significantly different from each other, as well as estimating the magnitude of such differences.

The descriptive statistics for each of the six brands of fertilizer (or levels of the factor) are as given in Table 9.11

The one-way ANOVA table is given in Table 9.12.

Since the p-value is less than the predetermined level of significance, we can accept the alternative hypothesis and reject the null hypothesis, and thus conclude that the population mean weekly growth rates for the different brands of lawn fertilizer are significantly different for at least two different brands of fertilizer. Furthermore, as was demonstrated earlier, we have reason to believe that the model assumptions have been reasonably met.

TABLE 9.11

Descriptive Statistics for the Fertilizer Data in Table 9.7

Statistic	Brand 1	Brand 2	Brand 3	Brand 4	Brand 5	Brand 6
Mean	31.50	45.62	29.00	37.14	33.17	28.20
Standard deviation	4.04	7.19	2.83	6.41	5.38	1.30
Variance	16.33	51.70	8.00	41.14	28.97	1.70

TABLE 9.12

One-Way ANOVA to Determine If There Is a Difference in the Mean Weekly Growth Rates of Grass across Six Different Brands of Fertilizer

Source	SS	df	MS	F	p-value
Treatment variation	1450.9	5	290.2	10.25	0.000
Error variation	849.4	30	28.3		
Total	2300.3	35			

We can now use Tukey's multiple comparison technique adjusted for different sample sizes to determine whether there is a difference in the weekly growth rate between the six different brands of lawn fertilizer.

Based on six factors and a total sample of size 36, the value for $q_{0.05} = 4.30$. The confidence interval comparing the mean weekly growth rate for Brand 1 subtracted from that for Brand 2 is calculated as follows:

$$(45.62 - 31.50) \pm \frac{4.30}{\sqrt{2}} \sqrt{28.3} \cdot \sqrt{\frac{1}{8} + \frac{1}{4}} = 14.12 \pm 9.91 = (4.21, \ 24.03)$$

Interpreting this confidence interval says that in subtracting Brand 1 from Brand 2, we are 95% confident that the difference in the population mean weekly growth rate will fall in this interval. In other words, we can claim that there is a significant difference in the weekly growth rates between Brand 1 and Brand 2. Specifically, Brand 2 has a significantly higher growth rate than Brand 1. This is because the difference, Brand 2 – Brand 1, is greater than 0. Since the confidence interval for Brand 1 subtracted from Brand 2 does not contain the value of 0, we can conclude that the mean weekly growth rates for Brands 1 and 2 are significantly different from each other. Furthermore, we can estimate that the magnitude of the difference between these two means to be between 4.21 and 24.03 mm.

If we take Brand 1 subtracted from Brand 4, for example, then there is no difference between the mean weekly growth rates for the difference because the confidence interval (–4.50, 15.78) contains the value of 0 (see Exercise 2).

9.9   Using MINITAB for Multiple Comparisons

It is very easy to use MINITAB to perform Tukey's multiple comparison technique for both balanced and unbalanced one-way ANOVAs. This can be done by selecting the **Comparison** tab on the one-way ANOVA dialog box, to give the multiple comparison dialog box presented in Figure 9.30.

Then checking the Tukey's family error rate box and entering the desired level of significance (by default a value of 5, which assumes a .05 level of significance), we get the printout in Figure 9.31 in addition to the one-way ANOVA printout.

The printout in Figure 9.31 gives the confidence intervals for each pair of differences between the mean growth rates for all of the specified brands of fertilizer. For example, Figure 9.32 is the portion of the last MINITAB printout that displays Brand 1 subtracted from all of the other brands.

Notice that similar to our hand calculations, the confidence interval for Brand 1 subtracted from Brand 2 is (4.218, 24.032). Thus, by using Tukey's

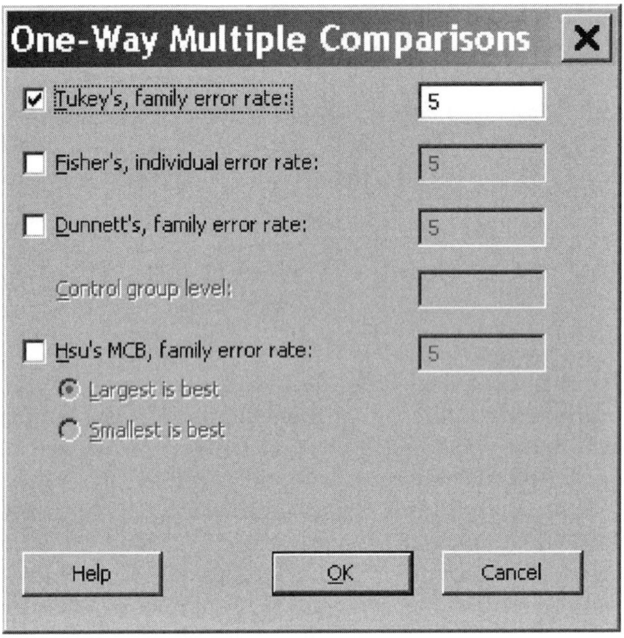

FIGURE 9.30
MINITAB dialog box for running multiple comparison techniques.

multiple comparison method, we can identify and compare the individual pairs of means that are significantly different from each other, and we can also distinguish which specific brands of fertilizer are generating a higher (or lower) growth rate, along with a range of values that represent the magnitude of the difference.

9.10   Power Analysis and One-Way ANOVA

Recall that a power analysis allows you to determine how large a sample size is needed to make reasonable inferences and how likely it is that the statistical test you are considering will be able to detect a difference of a given size. Also recall that the power of a test is the probability of correctly rejecting the null hypothesis when it is, in fact, false. For example, a power value of 0.80 would suggest that there is an 80% chance of correctly rejecting the null hypothesis and that there is a 20% chance of failing to detect a difference if such a difference actually exists. Sample size and power analysis

```
Tukey 95% Simultaneous Confidence Intervals
All Pairwise Comparisons among Levels of Brand

Individual confidence level = 99.51%

Brand = Brand 1 subtracted from:

Brand      Lower   Center   Upper    --------+---------+---------+---------+-
Brand 2    4.218   14.125   24.032                         (-----*------)
Brand 3  -12.943   -2.500    7.943           (------*------)
Brand 4   -4.498    5.643   15.783                 (------*------)
Brand 5   -8.777    1.667   12.110              (------*-------)
Brand 6  -14.153   -3.300    7.553           (------*------)
                                     --------+---------+---------+---------+-
                                          -15         0        15        30


Brand = Brand 2 subtracted from:

Brand      Lower   Center   Upper    --------+---------+---------+---------+-
Brand 3  -25.362  -16.625   -7.888   (-----*-----)
Brand 4  -16.855   -8.482   -0.109       (----*-----)
Brand 5  -21.196  -12.458   -3.721    (-----*-----)
Brand 6  -26.648  -17.425   -8.202   (-----*-----)
                                     --------+---------+---------+---------+-
                                          -15         0        15        30


Brand = Brand 3 subtracted from:

Brand      Lower   Center   Upper    --------+---------+---------+---------+-
Brand 4   -0.858    8.143   17.144                 (-----*-----)
Brand 5   -5.174    4.167   13.507              (-----*-----)
Brand 6  -10.597   -0.800    8.997           (-----*------)
                                     --------+---------+---------+---------+-
                                          -15         0        15        30


Brand = Brand 4 subtracted from:

Brand      Lower   Center   Upper    --------+---------+---------+---------+-
Brand 5  -12.977   -3.976    5.025           (-----*-----)
Brand 6  -18.416   -8.943    0.530       (-----*-----)
                                     --------+---------+---------+---------+-
                                          -15         0        15        30


Brand = Brand 5 subtracted from:

Brand      Lower   Center   Upper    --------+---------+---------+---------+-
Brand 6  -14.763   -4.967    4.830           (------*-----)
                                     --------+---------+---------+---------+-
                                          -15         0        15        30
```

FIGURE 9.31

MINITAB printout of Tukey's multiple comparisons between the six different brands of fertilizer.

go hand in hand, and one way to increase the power of a test is to increase the number of observations in your sample.

Sample size estimation and power analysis are very important aspects of any empirical study because if the sample size is too small, the experiment may lack precision, and thus not allow the investigator to make a reasonable inference. Also, if it is costly to obtain a large sample size, it can be a waste of resources if a smaller sample can adequately detect a given difference.

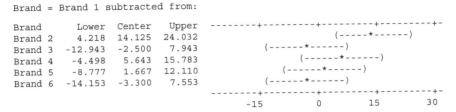

FIGURE 9.32
MINITAB printout comparing Brand 1 subtracted from all of the other brands.

Recall in Chapter 4 that we used MINITAB to determine the sample size needed to achieve an adequate level of power, or to determine the power level for a given sample size for some basic statistical inferences. In order to conduct a power analysis, we had to specify two of three values, namely, the sample size, difference, or power. Similarly, for a one-way ANOVA, we also need to specify two of the following: the power, the sample size (the number of observations from each level of the factor), or the difference between the smallest and largest factor means. However, with a one-way ANOVA, we have different levels of a factor, so we also need to specify the number of levels of the factor and provide an estimate of the standard deviation (typically this can be estimated by using the root mean square error).

Example 9.8

Suppose we want to determine if the sample we collected for our golf ball experiment has at least 80% power, to detect a difference of approximately 10 yards difference between the means among the different brands. From our one-way ANOVA, we found that the MSE was approximately 103, so the estimate of our standard deviation would be approximately 10. Our sample size from each factor level is 4, so the power of our analysis would be entered into MINITAB as illustrated in Figure 9.33. The results of this power analysis are presented in Figure 9.34.

Thus, in order to detect a 10-yard difference between the smallest and largest factor level means, a sample size of 4 from each level only achieves about 17% power. Hence, our one-way ANOVA is underpowered, which means that we are not likely to detect a difference of this size if one does exist.

To determine how large of a sample we would need to achieve 80% power for a difference of 10 yards, we would need to sample at least twenty-one observations from each factor level, as described in Figure 9.35.

Similarly, if we wanted to find a larger difference, say 50 yards, then we would need at least three observations from each level, as presented in Figure 9.36.

As Figure 9.36 illustrates, larger differences require smaller sample sizes to achieve adequate power, whereas smaller differences require larger sample sizes to achieve adequate power. This is because larger differences are easier to see than smaller differences.

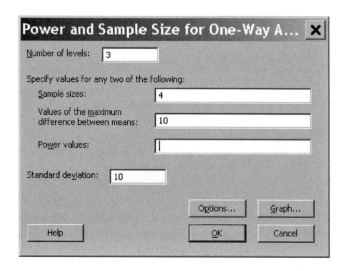

FIGURE 9.33
MINITAB dialog box for a power analysis for a one-way ANOVA.

Power and Sample Size

One-Way ANOVA

Alpha = 0.05   Assumed standard deviation = 10   Number of Levels = 3

SS Means	Sample Size	Power	Maximum Difference
50	4	0.173282	10

The sample size is for each level.

FIGURE 9.34
MINITAB printout for the golf ball data in Table 9.2 with a sample size of 4 for each level with three different levels and a difference of 10 yards.

Power and Sample Size

One-Way ANOVA

Alpha = 0.05   Assumed standard deviation = 10   Number of Levels = 3

SS Means	Sample Size	Target Power	Actual Power	Maximum Difference
50	21	0.8	0.814770	10

The sample size is for each level.

FIGURE 9.35
Power analysis to find a difference of 10 yards with 80% power.

Power and Sample Size

One-Way ANOVA

Alpha = 0.05 Assumed standard deviation = 10 Number of Levels = 3

SS Means	Sample Size	Target Power	Actual Power	Maximum Difference
1250	3	0.8	0.988251	50

The sample size is for each level.

FIGURE 9.36
Power analysis to find a difference of 50 yards with 80% power at three levels.

In conducting a power analysis for an unbalanced one-way ANOVA, a conservative approach would use the largest sample size from a given level.

Exercises

1. a. Complete the following ANOVA table:

Source	SS	df	MS	F
Treatment variation		2		
Error variation	1,580			
Total	2,586	114		

b. From the ANOVA table, determine the sample size and the number of levels of the factor.

c. Based on the ANOVA table, is there enough evidence to suggest that there is a significant difference in the means for the given populations? Justify your answer.

2. For the data in Table 9.7, which gives the weekly growth rate of grass treated with six different brands of fertilizer, find a 95% confidence interval for the difference between the population growth rate for Brand 1 subtracted from that for Brand 4.

3. The data set "Miles per Gallon" in Table 9.13 gives the average highway mileage per gallon (MPG) of gasoline for a random sample of minivans, mid-size cars, and SUVs by make.

a. Conduct a one-way ANOVA to determine whether the mean highway MPG differs among minivans, mid-size cars, and SUVs.

b. What assumptions need to be met in order to make a valid inference about the difference in the mean MPG among minivans, mid-size cars, and SUVs? Using both exploratory analyses and

TABLE 9.13

"Miles Per Gallon" Data Set

Minivan	Mid-size	SUV	Make
20	25	24	Brand 1
19	25	24	Brand 1
19	25	24	Brand 1
19	24	23	Brand 1
18	24	22	Brand 1
18	24	22	Brand 1
18	22	22	Brand 2
18	22	21	Brand 2
18	21	21	Brand 2
18	21	21	Brand 2
18	21	21	Brand 2
18	19	20	Brand 2
18	19	20	Brand 3
18	18	20	Brand 3
17	18	19	Brand 3
17	18	17	Brand 3
17	17	17	Brand 3
17	17	17	Brand 3

formal tests, establish whether these assumptions have been reasonably met.

c. Do the data provide sufficient evidence to indicate that there is a difference in the mean MPG among minivans, mid-size cars, and SUVs? Justify your response.

d. Estimate the magnitude and direction of the difference in means for the different models of vehicles.

e. Approximately what sample size would be needed to achieve 80% power with a difference of 2 MPG, 4 MPG, and 8 MPG?

4. Using the "Miles per Gallon" data set:

a. Conduct a one-way ANOVA to determine whether the mean MPG differs among brands.

b. What assumptions need to be met in order to make a valid inference about the difference in the mean MPG by brand? Using both exploratory analyses and formal tests, establish whether these assumptions have been reasonably met.

c. Do the data provide sufficient evidence to indicate that there is a difference in the mean MPG by brand of vehicle? Justify your response.

d. Estimate the magnitude and direction of the difference in means for the different brands of vehicles.

e. Approximately what sample size would be needed to achieve 80% power with a difference of 2 MPG, 4 MPG, and 8 MPG?

5. Greenhouse gases are substances in the atmosphere, such as water vapor and carbon dioxide, that can trap heat near the earth's surface, and thus can contribute to global warming. One of the main causes of greenhouse gas comes from burning fossil fuels in cars and trucks. The U.S. Department of Energy and the Environmental Protection Agency have a website (http://www.fueleconomy.gov/feg/byclass.htm) that gives the estimated greenhouse gas emissions based on the make and type of vehicle.

The data set "Automobile" contains the greenhouse gases (in tons/year) by make and type of vehicle in terms of full fuel-cycle estimates. Full fuel-cycle estimates consider all the steps in the use of a fuel for a vehicle, from its production and refining to distribution and final use.

a. Conduct a one-way ANOVA ($\alpha = 0.10$) to determine whether there is a difference in the amount of greenhouse gases produced based on the *type* of the vehicle.

b. Conduct a one-way ANOVA ($\alpha = 0.10$) to determine whether there is a difference in the amount of greenhouse gases produced based on the *make* of the vehicle.

c. What assumptions need to be met in order to make a valid inference about the difference in the mean greenhouse gas emissions among the different *types* of vehicles? Have these assumptions been reasonably met? Justify your response.

d. What assumptions need to be met in order to make a valid inference about the difference in the mean greenhouse gas emissions among the different *makes* of vehicles? Have these assumptions been reasonably met? Justify your response.

e. Given a difference of 1.5 tons/year, what size sample is needed from each of the different makes of vehicles to ensure 0.80 power?

f. Based on the sample data collected, are you able to determine which makes and types of vehicles have the best and worst emissions of greenhouse gases? Justify your answer.

g. Is there any particular make of vehicle that has the best emission of greenhouse gases? Explain.

6. Is there a difference by state in number of pupils per teacher? The data set "Student Teacher Ratio" (http://nces.ed.gov/ccd/districtsearch/) contains a random sample of the number of pupils per teacher from some schools in the six New England states: Massachusetts (MA),

Connecticut (CT), Vermont (VT), Maine (ME), Rhode Island (RI), and New Hampshire (NH).

a. Identify the factor(s) and factor levels for this experiment.

b. Determine if there is a difference in the mean number of pupils per teacher by state.

c. Justify whether you believe the model assumptions were reasonably met. Use both exploratory plots and formal analyses to validate the model assumptions.

d. In verifying the assumption of constant variance, do you think Bartlett's or Levene's test is more appropriate? Explain why.

e. Based on your analysis, can you make a reasonable inference regarding whether there is a difference in the number of pupils per teacher by state? Explain.

7. Using MINITAB, verify that performing a one-way ANOVA on the transformed response variables as described in Example 3 gives the same values for *MSTR*, *MSE*, and *F* as were found in doing the calculations by hand.

References

Bartlett, M. 1937. Properties of sufficiency and statistical tests. *Proceedings of the Royal Society* A160:268–82.

Brown, M., and Forsythe, A. 1974. Robust tests for the equality of variances. *Journal of the American Statistical Association* 69:364–67.

Levene, H. 1960. *Contributions to probability and statistics: Essays in honor of Harold Hotelling*, ed. I. Olkin, 278–92. Stanford, CA: Stanford University Press.

Ryan, T., and Joiner, B. 1976. *Normal probability plots and tests for normality*. Technical report. University Park, PA: The Pennsylvania State University.

10

Other Topics

10.1 Introduction

Throughout the last few chapters we have introduced some of the more common topics in applied statistical inference, such as multiple regression and one-way analysis of variance (ANOVA). However, these topics only scratch the surface of the number of different topics that can be used to make an inference about an unknown population parameter of interest. In this chapter, we will describe some more advanced topics, such as two-way ANOVA, nonparametric statistics, and basic time series analysis.

10.2 Two-Way Analysis of Variance

Recall that a one-way ANOVA compares the means of a variable that is classified by only one single factor. However, there may be situations where you may want to compare the means of a variable that is classified by *two* factors.

Example 10.1

Suppose the manager of a manufacturing company is interested in evaluating whether there is a difference in the average number of defective products that are produced at three different manufacturing plants using five different types of machines. The data provided in Table 10.1 consist of a random sample of the number of defective products that were produced by plant and by machine at two different collection times.

We are interested in testing whether there is a difference in the mean number of defective products that are produced by machine and by plant. We can describe one factor that represents the five different levels that correspond to the five different types of machines (this can be referred to as the row factor), and another factor that represents the three levels that correspond to the three different plants (this can be referred to as the column factor).

TABLE 10.1

Number of Defective Products Produced
by Machine and by Plant Collected at
Random at Two Different Times

	Plant 1	Plant 2	Plant 3
Machine 1	12	18	19
	15	13	17
Machine 2	16	22	21
	19	15	10
Machine 3	12	10	8
	12	9	8
Machine 4	25	20	20
	24	21	19
Machine 5	18	19	26
	10	8	9

In a fashion similar to that of a one-way ANOVA, let α describe the effect of the different levels of the machine (the row factor), and let β describe the effect of the different levels of the plant (column factor). We can then model the average number of defective items for the ith row and the jth column by using the following regression equation:

$$y_{i,j,k} = \mu + \alpha_i + \beta_j + (\alpha\beta)_{i,j} + \varepsilon_{i,j,k}$$

where $y_{i,j,k}$ is the kth observation at level α_i and level β_j, μ represents the mean common to all populations, $(\alpha\beta)_{i,j}$ represents the interaction between the row factor α at level i and the column factor β at level j, and $\varepsilon_{i,j,k}$ represents the error component that corresponds to the kth observation of the row factor at level i and the column factor at level j.

A *two-way analysis of variance* can be used to test whether there is a significant difference between the mean number of defective products based on the type of machine or the plant where it was manufactured. A two-way ANOVA can also test whether or not there is an *interaction* or dependency between the row and column factors. For our example, this would represent a dependency between the machine and the plant. For instance, an interaction could occur if there is a particular machine at a certain plant that is producing a larger number of defects than can be explained by the type of machine and the plant on their own.

The appropriate null and alternative hypotheses for the row effect, the column effect, and the interaction effect corresponding to the five machines and three plants would be as follows:

$$H_0 : \alpha_1 = \alpha_2 = \alpha_3 = \alpha_4 = \alpha_5$$

H_A : At least two population machine (row) effects are not equal.

$H_0 : \beta_1 = \beta_2 = \beta_3$

H_A : At least two plant (column) population effects are not equal.

H_0 : There is no interaction between the row and column factors.

H_A : There is an interaction between the row and column factors.

A two-way ANOVA also requires using the F-distribution with the following test statistics for each of the row, column, and interaction effects, along with the respective degrees of freedom for the numerator and denominator:

$$F = \frac{MSRow}{MSE} \qquad d.f. = [r-1, \; rc(m-1)]$$

$$F = \frac{MSColumn}{MSE} \qquad d.f. = [c-1, \; rc(m-1)]$$

$$F = \frac{MSInteraction}{MSE} \qquad d.f. = [(r-1)(c-1), \; rc(m-1)]$$

where r is the number of row factors, c is the number of column factors, and m is the number of replicates in each cell.

In order to calculate the mean squares for the row, column, and interaction effects, first we need to find the appropriate sums of squares and mean squares for the row, column, and interaction. In Table 10.1, notice that there are two observations (or replicates) for each machine and plant. In order to find the sum of squares for the rows and the columns, it is usually easier first to take the average of the observations in each cell, as illustrated in Table 10.2.

To find the sum of squares for the rows, we can use the following formula:

$$SSRow = mc \sum_{i=1}^{r} (\bar{R}_i - \bar{y})^2$$

TABLE 10.2

Average of the Observations in Each Cell from the Data in Table 10.1

	Plant 1	Plant 2	Plant 3	Row Mean ($\bar{R}_i$)
Machine 1	$\frac{12+15}{2} = 13.5$	$\frac{18+13}{2} = 15.5$	$\frac{19+17}{2} = 18.0$	$\frac{13.5+15.5+18}{3} = 15.6667$
Machine 2	17.5	18.5	15.5	17.1667
Machine 3	12.0	9.5	8.0	9.8333
Machine 4	24.5	20.5	19.5	21.5000
Machine 5	14.0	13.5	17.5	15.0000
Column average	16.3000	15.5000	15.7000	15.8333

where m is the number of observations (or replicates) in each cell (in this example $m = 2$), c is the number of column factors (in this example $c = 3$ because there are three plants), and r is the number of row factors (in this example $r = 5$ because there are five types of machines). The statistic $\bar{R}_i$ is the mean of the ith row, and $\bar{y}$ is the mean of all of the observations.

To find the sum of squares for the row factor, first we need to find the mean of all of the observations. For the data in Table 10.2, this can be calculated by adding up all the row means and dividing by 5, as follows:

$$\bar{y} = \frac{15.6667 + 17.1667 + 9.8333 + 21.500 + 15.0000}{5} = 15.8333$$

The squared difference between each of the five row means and the total mean of all the observations is

$$(\bar{R}_1 - \bar{y})^2 = (15.6667 - 15.8333)^2 = 0.0278$$

$$(\bar{R}_2 - \bar{y})^2 = (17.1667 - 15.8333)^2 = 1.7780$$

$$(\bar{R}_3 - \bar{y})^2 = (9.8333 - 15.8333)^2 = 36.0000$$

$$(\bar{R}_4 - \bar{y})^2 = (21.5000 - 15.8333)^2 = 32.1115$$

$$(\bar{R}_5 - \bar{y})^2 = (15.0000 - 15.8333)^2 = 0.6944$$

Then summing all of these squared differences gives

$$\sum_{i=1}^{5} (\bar{R}_i - \bar{y})^2 = 0.0278 + 1.7780 + 36.0000 + 32.1115 + 0.6944 = 70.612$$

Therefore:

$$SSRow = mc \sum_{i=1}^{5} (\bar{R}_i - \bar{y})^2 = (2)(3)(70.612) \approx 423.672$$

Thus:

$$MSRow = \frac{SSRow}{r-1} = \frac{423.672}{5-1} = 105.918$$

Similarly, to find the sum of squares and the mean square for the column factor:

$$SSColumn = mr \sum_{i=1}^{c} (\bar{C}_i - \bar{y})^2 = 3.467$$

where m is the number of observations in each cell, r is the number of rows, c is the number of columns, $\bar{C}_i$ is the mean of the ith column, and $\bar{y}$ is the mean of all of the observations so,

TABLE 10.3

Sum of the Observations in Each Cell Squared from the Data Presented in Table 10.1

	Plant 1	Plant 2	Plant 3
Machine 1	$(12 + 15)^2 = 729$	$(18 + 13)^2 = 961$	$(19 + 17)^2 = 1296$
Machine 2	$(16 + 19)^2 = 1225$	$(22 + 15)^2 = 1369$	$(21 + 10)^2 = 961$
Machine 3	$(12 + 12)^2 = 576$	$(10 + 9)^2 = 361$	$(8 + 8)^2 = 256$
Machine 4	$(25 + 24)^2 = 2401$	$(20 + 21)^2 = 1681$	$(20 + 19)^2 = 1521$
Machine 5	$(18 + 10)^2 = 784$	$(19 + 8)^2 = 729$	$(26 + 9)^2 = 1225$

$$MSColumn = \frac{SSColumn}{c-1} = \frac{3.467}{3-1} = 1.734$$

In order to test for an interaction between the row and column factor, we need to find the sum of squares for the interaction term. This is done by summing the squares of the individual observations in each of the cells, dividing by the number of observations in each cell, and then subtracting the sum of squares for both the row and column factors, and finally subtracting the average of all the values squared. This can be represented by the following formula:

$$SSInteraction = \frac{\text{Sum of squares of all the cell values}}{m}$$

$$- SSRow - SSColumn - \frac{\left(\Sigma_{i=1}^{n} y_i\right)^2}{n}$$

For our example, we first sum up the values of the observations in each cell, and then square them, as illustrated in Table 10.3.

Then summing all of these values in Table 10.3, we get a total of 16,075, and then dividing this value by $m = 2$ gives 8037.500.

To find *SSInteraction*, we subtract from this value *SSRow* and *SSColumn*, and since

$$\left(\sum_{i=1}^{30} y_i\right)^2 = (475)^2 = 225625$$

$$\frac{\left(\Sigma_{i=1}^{30} y_i\right)^2}{30} = 7520.833$$

then *SSInteraction* = 8037.500 − 423.672 − 3.467 − 7520.833 = 89.528, so:

$$MSInteraction = \frac{SSInteraction}{(c-1)(r-1)} = \frac{89.528}{(3-1)(5-1)} = 11.191$$

TABLE 10.4

Square of the Difference between the Observations and the Cell Means

	Plant 1	Plant 2	Plant 3
Machine 1	$(12 - 13.5)^2 = 2.25$	$(18 - 15.5)^2 = 6.25$	$(19 - 18)^2 = 1$
	$(15 - 13.5)^2 = 2.25$	$(13 - 15.5)^2 = 6.25$	$(17 - 18)^2 = 1$
Machine 2	$(16 - 17.5)^2 = 2.25$	$(22 - 18.5)^2 = 12.25$	$(21 - 15.5)^2 = 30.25$
	$(19 - 17.5)^2 = 2.25$	$(15 - 18.5)^2 = 12.25$	$(10 - 15.5)^2 = 30.25$
Machine 3	$(12 - 12)^2 = 0$	$(10 - 9.5)^2 = 0.25$	$(8 - 8)^2 = 0$
	$(12 - 12)^2 = 0$	$(9 - 9.5)^2 = 0.25$	$(8 - 8)^2 = 0$
Machine 4	$(25 - 24.5)^2 = 0.25$	$(20 - 20.5)^2 = 0.25$	$(20 - 19.5)^2 = 0.25$
	$(24 - 24.5)^2 = 0.25$	$(21 - 20.5)^2 = 0.25$	$(19 - 19.5)^2 = 0.25$
Machine 5	$(18 - 14)^2 = 16$	$(19 - 13.5)^2 = 30.25$	$(26 - 17.5)^2 = 72.25$
	$(10 - 14)^2 = 16$	$(8 - 13.5)^2 = 30.25$	$(9 - 17.5)^2 = 72.25$

To find SSE, we subtract the means from the value of the observations in each cell, square them, and then add them all together, as illustrated in Table 10.4.

$$SSE = \sum_{i=1}^{n} (y_i - \text{cell mean})^2$$

Then, summing up all of these values gives $SSE = 347.50$ and

$$MSE = \frac{SSE}{(c \cdot r)(m - 1)} = \frac{347.50}{(3 \cdot 5)(2 - 1)} = 23.167$$

Thus, the test statistics for the row, column, and interaction effect can be determined as follows:

$$F = \frac{MSRow}{MSE} = \frac{105.918}{23.167} = 4.572$$

$$F = \frac{MSColumn}{MSE} = \frac{1.734}{23.167} = 0.075$$

$$F = \frac{MSInteraction}{MSE} = \frac{11.191}{23.167} = 0.483$$

We then need to compare each of the values of these test statistics to the respective *F*-distributions with the appropriate significance level and degrees of freedom.

For a significance level of $\alpha = 0.05$, the row factor will have $r - 1 = 5 - 1 = 4$ degrees of freedom for the numerator and $r \cdot c (m - 1) = (5)(3)(2 - 1) = 15$ degrees of freedom for the denominator. Then we compare the test statistic $F = 4.572$ to the value $f(4,15) = 3.06$ that defines the rejection region, as in Figure 10.1.

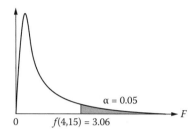

FIGURE 10.1
Rejection region for the row factor for a two-way ANOVA.

Since $F = 4.52 > f = 3.06$, as Figure 10.1 illustrates, we can accept the alternative hypothesis that at least two population machine (row) effects are different from each other.

Similarly, for the column factor, with $k - 1 = 3 - 1 = 2$ degrees of freedom for the numerator and 15 degrees of freedom for the denominator, we would compare the value of $f(2,15) = 3.68$ to the value of the test statistic $F = 0.075$, as in Figure 10.2.

From Figure 10.2, since $F = 0.075 < f = 3.68$, we do not have enough evidence to suggest that the plant (column) effects are different from each other.

And finally, for the interaction with $(r - 1)(c - 1) = (4)(2) = 8$ degrees of freedom for the numerator and 15 degrees of freedom for the denominator, we would compare the value of $f(8,15) = 2.64$ to the test statistic $F = 0.483$, as in Figure 10.3.

Similarly, from Figure 10.3, since $F = 0.483 < f = 2.64$, we do not have enough evidence to suggest that there is an interaction between the two factors.

Thus, our two-way ANOVA reveals that the only significant differences are in the mean number of defective products produced by machine, and not by plant or the interaction between plant and machine.

One thing to keep in mind when running a two-way ANOVA is that in addition to the model assumptions (which also happen to be the same as for a multiple regression analysis), the model is constructed such that there are only two factors that can impact the number of defects. Similar to a one-way ANOVA, we are assuming that there are no other factors that could impact the number

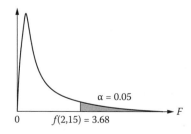

FIGURE 10.2
Rejection region for the column factor for a two-way ANOVA.

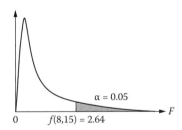

FIGURE 10.3
Rejection region for the interaction for a two-way ANOVA.

of defects that are produced. If you have reason to believe that there are more than two factors, such as the temperature and the time at which the defects were obtained, you may want to reconsider using a two-way ANOVA.

The analysis just described is called a *balanced two-way ANOVA*. This is because the same number of replicates are in each cell. Furthermore, in order to test for an interaction, there must be at least two observations in each cell. There are more sophisticated methods that can be used to handle unbalanced two-way ANOVAs, but they are beyond the scope of this book and can be found in some more advanced texts on multivariate analysis.

10.3   Using MINITAB for a Two-Way ANOVA

Using MINITAB for a two-way ANOVA for the data given in Table 10.1 requires that the data be entered in three separate columns: one column represents the response (the number of defects), one column represents the row factor (machine), and one column represents the column factor (plant). A portion of how the data are to be entered in MINITAB is illustrated in Figure 10.4.

The MINITAB commands for running a two-way ANOVA are found under the **ANOVA** command from the **Stat** menu bar, as illustrated in Figure 10.5. This gives the dialog box presented in Figure 10.6, where the response, row factor, and column factor need to be specified. This gives the MINITAB printout illustrated in Figure 10.7. Notice that the values for *SS*, *MS*, and *F* match those that were calculated by hand for the row, column, and interaction factors. Also, in interpreting the highlighted *p*-values for the row factor, column factor, and interaction, we see that the only significant effect is the row factor (machine).

We can also use MINITAB to check the usual regression model assumptions by selecting **Graphs** on the **Two-Way ANOVA** dialog box to get the dialog box illustrated in Figure 10.8.

The usual residual plots are given in Figures 10.9 to 10.11.

We can also use MINITAB to provide a box plot of the mean number of defects with respect to the machine and plant factors by checking the **Boxplots of data** box in Figure 10.8 to create the box plots by row and column factor, as illustrated in Figure 10.12.

↓	C1	C2	C3	C4	
	Defects	**Plant**	**Machine**		
1	12	1	1		
2	15	1	1		
3	16	1	2		
4	19	1	2		
5	12	1	3		
6	12	1	3		
7	25	1	4		
8	24	1	4		
9	18	1	5		
10	10	1	5		
11	18	2	1		
12	13	2	1		
13	22	2	2		
14	15	2	2		

FIGURE 10.4
Format for entering data into MINITAB in order to run a two-way ANOVA.

We can also run the Ryan–Joiner test of the normality assumption of the error component by storing the residuals to get the normality plot with the results of the Ryan–Joiner tests, as presented in Figure 10.13.

MINITAB can also be used to test for constant variance by using Bartlett's and Levene's tests, but there are some restrictions on the data that will be described in more detail in Exercise 2.

MINITAB also provides a main effects plot and an interaction plot. In a *main effects plot* the points correspond to the means of the response variable at the different levels for both the row and the column factor.

To create a main effects plot using MINITAB, first select the **ANOVA** command from the **Stat** menu and then select **Main Effects Plot**, as illustrated in Figure 10.14. This brings up the main effects dialog box, where the response and the row and column factors need to be specified, as illustrated in Figure 10.15.

The main effects plot is given in Figure 10.16, where the reference line drawn is the mean of all the observations. The row and column effects are

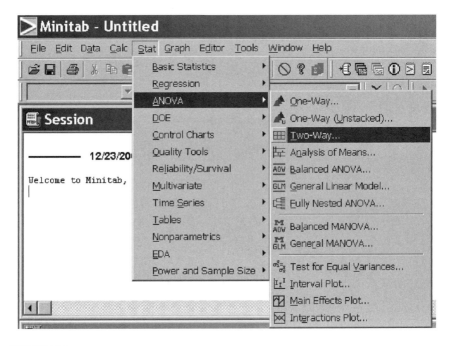

FIGURE 10.5
MINITAB commands for a two-way ANOVA.

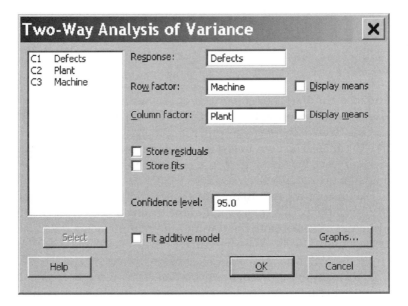

FIGURE 10.6
MINITAB dialog box for a two-way ANOVA.

Two-Way ANOVA: Defects versus Machine, Plant

Source	DF	SS	MS	F	P
Machine	4	423.667	105.917	4.57	0.013
Plant	2	3.467	1.733	0.07	0.928
Interaction	8	89.533	11.192	0.48	0.850
Error	15	347.500	23.167		
Total	29	864.167			

S = 4.813   R-Sq = 59.79%   R-Sq(adj) = 22.26%

FIGURE 10.7
MINITAB printout for a two-way ANOVA for the data in Table 10.1.

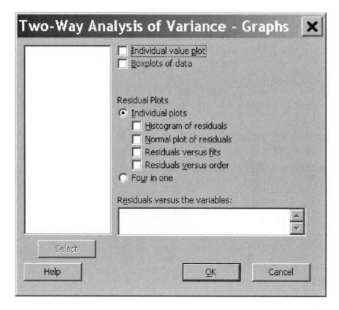

FIGURE 10.8
Dialog box for graphs for a two-way ANOVA.

represented by the differences between the points, which represent the means for each level of each factor, and the reference line, which is the mean of all of the observations. In Figure 10.16, the effect of the different machines on the number of defective products produced is quite large compared to the number of defective products produced at the different plants. Notice that there is more variability from the mean of all the observations in the means in the main effects plot by machine versus the variability in the means by plant.

Similarly, MINITAB can also graph an interaction plot. An *interaction plot* consists of the mean values for each level of a row factor with respect to the

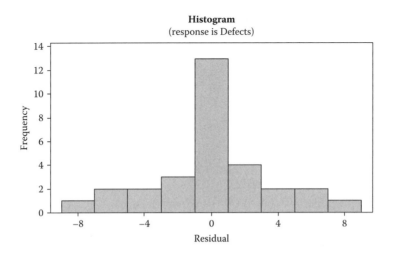

FIGURE 10.9
Histogram of the residuals.

column factor. Recall that an interaction occurs when the row and column factors affect each other. An interaction plot depicts potential interactions as non-parallel lines.

The dialog box for an interaction plot is given in Figure 10.17.

Figure 10.18 gives the interaction plot for the data in Table 10.1. This plot shows the mean number of defects versus the plant for each of the five different types of machines.

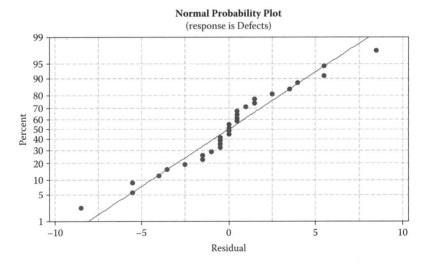

FIGURE 10.10
Normal probability plot of the residuals.

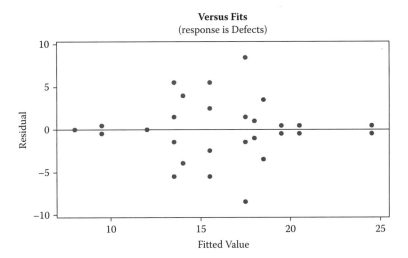

FIGURE 10.11
Residual versus fitted values.

An interaction would occur if the number of defects produced on any given machine depends on a given plant (or plants). If such an interaction were present, it would show up in the interaction plot as non-parallel lines. If there is no interaction, the interaction plot would primarily consist of parallel lines because the row and column factors do not depend on each other. If an interaction is present, then it does not make sense to conduct individual

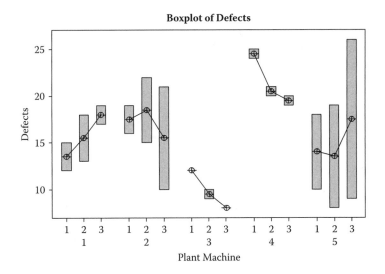

FIGURE 10.12
Box plot of the mean number of defects by plant and by machine.

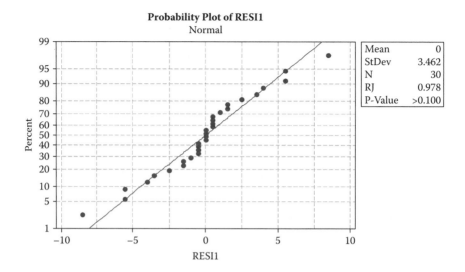

FIGURE 10.13
Ryan–Joiner test of the normality assumption.

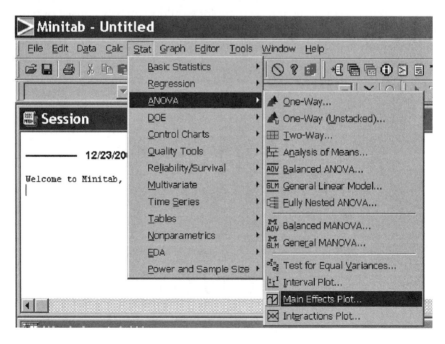

FIGURE 10.14
MINITAB commands to create a main effects plot.

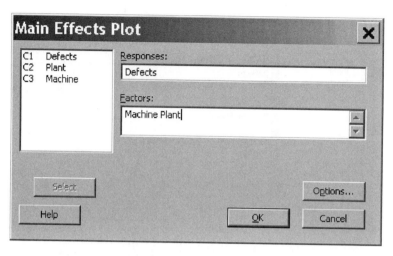

FIGURE 10.15
MINITAB dialog box for a main effects plot.

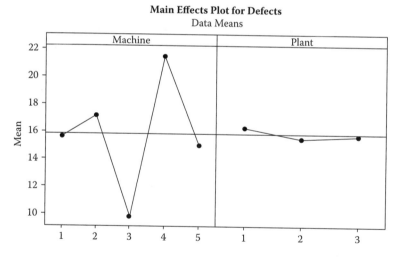

FIGURE 10.16
Main effects plot for the two-way ANOVA using the data in Table 10.1.

tests for the row and column effects because an interaction suggests that differences in the row factor depend on the levels of the column factor (or vice
versa). If the row and column effects are dependent on each other, then testing them individually would not be appropriate

Notice that the interaction plot in Figure 10.18 does not reveal any obvious non-parallel lines among machines by plant. This supports the formal

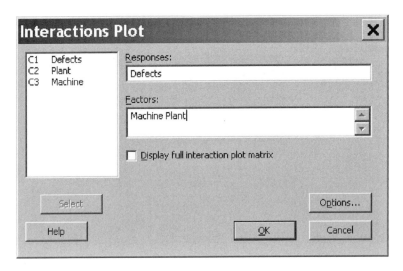

FIGURE 10.17
MINITAB dialog box for an interaction plot.

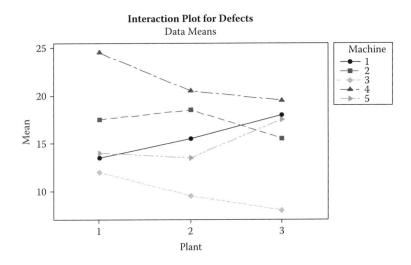

FIGURE 10.18
Interaction plot for the two-way ANOVA using the data in Table 10.1.

analysis, which suggests that there may not be a significant interaction between the row (machine) and column (plant) factors.

Example 10.2

The data set "Scores" in Table 10.5 presents the scores on a standardized achievement test for a random sample of thirty sixth-grade boys and thirty

TABLE 10.5

Scores on a Standardized Achievement
Test for a Random Sample of Thirty
Sixth-Grade Boys and Thirty Sixth-Grade
Girls at a Certain Middle School

	Boys	Girls
One week	65	78
	78	65
	84	81
	82	55
	76	72
	69	70
	81	68
	95	84
	87	80
	90	75
Two weeks	78	62
	84	78
	86	75
	81	76
	69	81
	79	83
	82	84
	98	89
	90	90
	86	70
Three weeks	85	85
	89	78
	91	82
	94	86
	82	85
	78	81
	90	83
	85	89
	87	91
	91	90

sixth-grade girls at a certain middle school. The test scores are presented by gender and by the length of time that was spent preparing for the test.

We want to know whether there is a difference in the population mean score received on the standardized test by gender and by the amount of time that was spent preparing for the test. We would use a two-way ANOVA if we had reason to believe that these are the only two factors along with a

Two-Way ANOVA: Exam Score versus Time Preparing, Gender

```
Source            DF      SS        MS       F      P
Time Preparing     2   876.10   438.050    7.88   0.001
Gender             1   355.27   355.267    6.39   0.014
Interaction        2    82.23    41.117    0.74   0.482
Error             54  3001.00    55.574
Total             59  4314.60
```

```
S = 7.455  R-Sq = 30.45%   R-Sq(adj) = 24.01%
```

FIGURE 10.19
MINITAB printout for the two-way ANOVA testing if there is a difference in the mean examination score based on time for preparation and gender.

potential interaction effect that could impact the scores received on the standardized examination.

Thus, our null and alternative hypotheses would be as follows:

$H_0 : \alpha_1 = \alpha_2 = \alpha_3$

H_A : There is a difference in the population mean test score by time preparing.

$H_0 : \beta_1 = \beta_2$

H_A : There is a difference in the population mean score based on gender.

H_0 : There is no interaction between time preparing and gender.

H_A : There is an interaction between the time preparing and gender.

Using MINITAB to run this analysis gives the results of the two-way ANOVA
that are presented in Figure 10.19.

There does not appear to be an interaction effect between time preparing
for the examination and gender ($p = .482$). Therefore, it makes sense to look
at the individual row and column effects. Based on the p-values from the
MINITAB printout in Figure 10.19, there does appear to be a significant difference in the mean examination score based on the amount of time spent
preparing for the examination($p = .001$) and by gender ($p = .014$). These row
and column effects can also be seen in the box plot in Figure 10.20.

By visually inspecting the graph in Figure 10.20, it appears that there may be
a difference by gender because for each of these sets of box plots, males appear
to score higher on the examination than females. Also, there appears to be a
difference in the scores based on the amount of time that was spent in preparing for the examinations because the scores are higher with more time spent.

To check the two-way ANOVA model assumptions, a four-in-one residual
plot is given in Figure 10.21, and the results of the Ryan–Joiner test are presented in Figure 10.22.

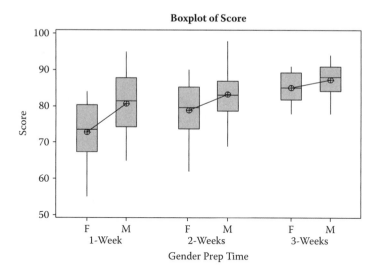

FIGURE 10.20
Box plot of exam score based on time preparing for the exam and gender.

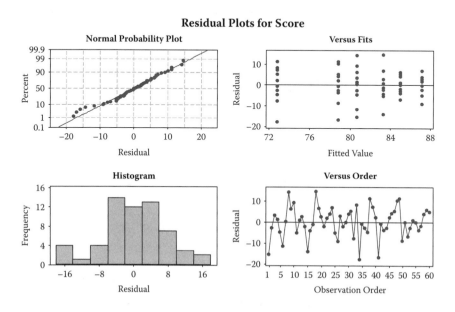

FIGURE 10.21
Residual plots for the two-way ANOVA from Figure 10.19.

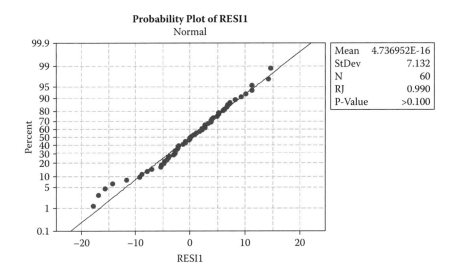

FIGURE 10.22
Ryan–Joiner test of the assumption that the residuals are normally distributed.

The graphs of the residual plots in Figure 10.21 suggest that the model assumptions appear to hold true. The results of the Ryan–Joiner test presented in Figure 10.22 also support the claim that the assumption of normality of the error component does not appear to have been violated.

The main effects plot and the interaction plot also support these conclusions, as represented in Figures 10.23 and 10.24.

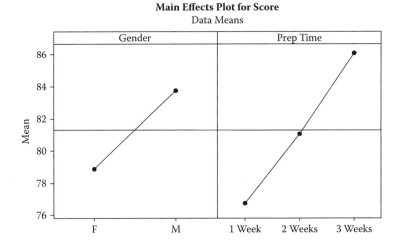

FIGURE 10.23
Main effects plot for scores on a standardized achievement test based on gender and amount of time preparing for the test.

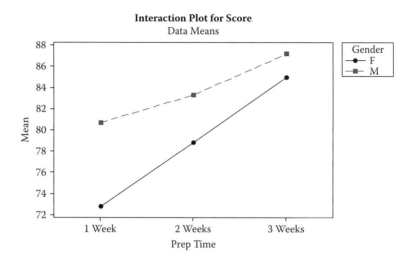

FIGURE 10.24

Interaction plot for scores on a standardized achievement test based on gender and amount of time preparing for the test.

The plot in Figure 10.23 suggests that there could be both a row (preparation) and column (gender) effect, whereas the interaction plot in Figure 10.24 suggests that there does not appear to be an interaction because the mean values by gender do not appear to be affected by the time spent preparing for the examination.

Example 10.3

The data set "Airline Data," as provided in Table 10.6, presents a sample of the flight delay times in minutes by airline and by whether the delay was caused by the airline carrier or the weather. We want to determine if there is

TABLE 10.6

Flight Delay Times (in minutes) by Airline and by Delay Caused by the Carrier or the Weather

	Airline 1	Airline 2	Airline 3	Airline 4
Airline carrier	257	3,850	1,260	1650
	229	1,542	3,300	208
	451	3,501	1,521	710
	164	1,980	205	198
Weather	285	125	102	180
	270	250	76	12
	480	167	109	59
	195	102	111	102

Two-way ANOVA: Delay versus Carrier/WX, Airline

```
Source        DF         SS         MS      F       P
Carrier/WX     1   10581150   10581150  24.67   0.000
Airline        3    6579552    2193184   5.11   0.007
Interaction    3    7561374    2520458   5.88   0.004
Error         24   10291982     428833
Total         31   35014059


S = 654.9     R-Sq = 70.61%   R-Sq(adj) = 62.03%
```

FIGURE 10.25
MINITAB printout for flight delay data from Table 10.6.

a difference in the population mean delay time based on the airline carrier and the type of delay, and whether there is an interaction between the airline carrier and the type of delay.

Running a two-way ANOVA using MINITAB, with the type of delay as the row factor and the airline carrier as the column factor, gives the printout in Figure 10.25. Figure 10.26 gives a four-way plot of the residuals, and Figure 10.27 gives the box plot by airline and type of delay. Figure 10.28 gives the results of the Ryan–Joiner test. Figures 10.29 and 10.30 give the main effects and interaction plots.

The two-way ANOVA in Figure 10.25 suggests that there is a significant interaction effect. The interaction plot in Figure 10.30 illustrates the

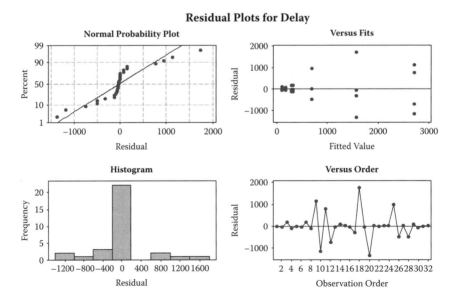

FIGURE 10.26
Four-way residual plots for the airline data from Table 10.6.

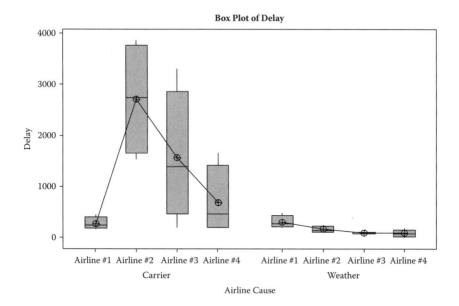

FIGURE 10.27
Box plot of airline delay data by airline and type of delay.

interaction between the type of delay and airline. An interaction may be present because some of the lines in Figure 10.30 that represent the different airlines are non-parallel to each other. For instance, the graph representing the mean delay time for Airline 1 is perpendicular to the graph representing

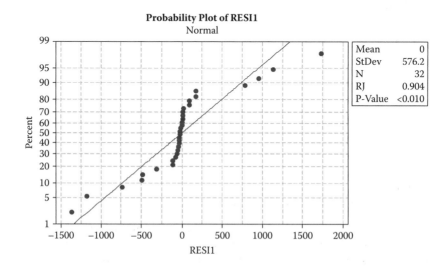

FIGURE 10.28
Ryan–Joiner test of the normality assumption for the airline data given in Table 10.6.

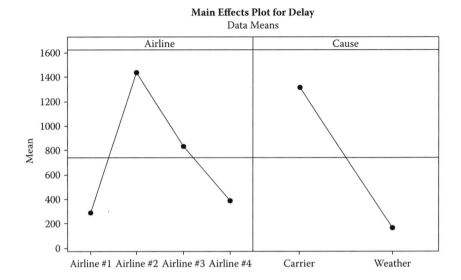

FIGURE 10.29
Main effects plot for the flight delay data by airline and type of delay.

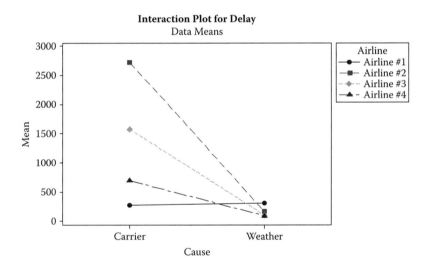

FIGURE 10.30
Interaction plot for the flight delay data by airline and the type of delay.

the mean delay times for Airlines 2 and 3. What the interaction plot suggests is that there is an interaction between the airline carrier and the type of delay. In other words, the cause of the delay depends upon the different airline carriers. Because there is an interaction effect, it is not appropriate to interpret the individual row and column effects. Even though the p-values and the main effects plot in Figure 10.29 suggest a row effect (type of delay) and column effect (airline), because of the interaction effect, the interpretation of the row and column effect test is meaningless.

However, we need to be cautious of our interpretations because the results of the Ryan–Joiner test, illustrated in Figure 10.28, suggest that the assumption that the error component is normally distributed may have been violated. Therefore, any inferences that we made based on our analysis could be suspect because the assumption of normality of the error component for the two-way ANOVA may not have been met.

In some instances (such as in the last example), it can be very difficult to assess whether the underlying model assumptions have been met. When this is the case, there are other statistical techniques, called nonparametric statistics, in which the underlying model assumptions are more relaxed than what we have seen thus far.

10.4   Nonparametric Statistics

All the topics that we have studied up to now are considered *parametric methods of statistical inference*. This is because underlying many of these techniques there has been a collection of model assumptions. For instance, in order to make reasonable inferences from a multiple regression analysis, we are assuming that the error component is normally distributed, has constant variance, and that the error terms are independent of each other. Also, in many of the analyses we have seen thus far, the presence of outliers can often have a profound impact on any inferences we make.

Nonparametric statistics refers to a set of statistical methods and procedures that do not rely on such strong model assumptions, and such methods and procedures tend to be more resilient to outliers and other extreme values. Often nonparametric methods are referred to as distribution-free tests because they can be performed without reliance on a specific shape of the distribution, such as the shape of a normal curve. However, nonparametric tests do rely on some basic assumptions, but such assumptions are more relaxed than parametric tests.

You may think that since you do not need to have such strict distributional assumptions with nonparametric methods, these methods would obviously be preferred over the standard parametric methods. However, this is not usually the case because nonparametric methods rely on some basic distributional assumptions. Furthermore, nonparametric methods are not as powerful as parametric methods when all the model assumptions have been met. Thus, in practice, nonparametric methods tend to be used when there are gross violations of the model assumptions, as is often the case when there are a significant number of outliers or extreme values that could impact the analysis or when smaller sample sizes are used.

10.5 Wilcoxon Signed-Rank Test

The first nonparametric test that we will be considering is the *Wilcoxon signed-rank test*. This test is similar to a *t*-test for comparing a population mean against a specific hypothesized value. However, the difference between the Wilcoxon signed-rank test and the one-sample *t*-test is that the Wilcoxon signed-rank test is testing the population *median* against some hypothesized value, whereas the one-sample *t*-test is testing the population *mean* against some hypothesized value.

The Wilcoxon signed-rank test relies on the assumption that the sample is drawn from a population that has a *symmetric distribution*, but no specific shape of the distribution is required. Recall in Chapter 4 that for a one-sample *t*-test with a small sample size ($n < 30$), we assumed the population we sampled from was approximately normally distributed. If the population we sampled from did come from a normal distribution, then we can use the Wilcoxon signed-rank test as an alternative to a one-sample *t*-test. This is because for the Wilcoxon signed-rank test, we are only assuming that the underlying population has a symmetric distribution, and this does not necessarily imply that the underlying population has the specific shape of the normal distribution. The illustrations presented in Figure 10.31 show two different distributions that are symmetric but not normally distributed.

Example 10.4

Consider the data set in Table 10.7, which consists of a sample of the final examination grades for ten students selected at random from the population of students in an introductory statistics course. Because we have such a small sample size, we may be unsure whether the population we are sampling

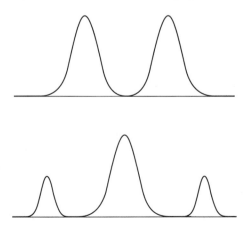

FIGURE 10.31
Example of two different distributions that are symmetric but not normally distributed.

TABLE 10.7

Final Examination Grades for a Sample of Ten
Students in an Introductory Statistics Course

80	90	71	79	75
68	83	89	66	58

from has the shape of a normal distribution, so we will use the Wilcoxon
signed-rank test instead of a one-sample *t*-test.

Suppose we were interested in determining whether the true population
median final exam grade is greater than 70. The difference between using a
one-sample *t*-test and the Wilcoxon signed-rank test is that we will be looking at whether the true but unknown population median (versus the population mean, as with a *t*-test) is greater than 70. Thus, the appropriate null and
alternative hypotheses would be

$$H_0 : \tilde{\mu}_0 = 70$$

$$H_A : \tilde{\mu}_0 > 70$$

where $\tilde{\mu}_0$ is the true but unknown median being tested under the null
hypothesis.

To conduct the Wilcoxon signed-rank test, we first need to rank each observation in the data set according to the distance the observation is from the median
that we are testing under the null hypothesis (which in this case is $\tilde{\mu}_0 = 70$).
We then take the difference between each observation and the hypothesized
median, take the absolute value of this difference, rank the absolute value of
this difference, and then attach the sign of the difference to this rank. Table 10.8
provides a summary of how this is done.

TABLE 10.8

Ranks and Signed Ranks for the Data in Table 10.7

Grade	Signed Difference D	$|D|$	Rank of $|D|$	Signed Rank
80	80 – 70 = 10	10	6	6
90	90 – 70 = 20	20	10	10
71	71 – 70 = 1	1	1	1
79	79 – 70 = 9	9	5	5
75	75 – 70 = 5	5	4	4
68	68 – 70 = –2	2	2	–2
83	83 – 70 = 13	13	8	8
89	89 – 70 = 19	19	9	9
66	66 – 70 = –4	4	3	–3
58	58 – 70 = –12	12	7	–7

If there are ties for the ranks, then the rank for each of the tied values will be assigned the mean of the ranks. Now, by taking the sum of the absolute values of the ranks in column 4 of Table 10.8, we get

$$6 + 10 + 1 + 5 + 4 + 2 + 8 + 9 + 3 + 7 = 55$$

The logic behind the Wilcoxon signed-rank test is that if the null hypothesis were true, then we would expect that the sum of the signed positive ranks and the sum of the negative signed ranks to both equal about half of the sum of all the ranks. So for the current example, since the sum of the absolute value of all the ranks is 55, and if the sum of the positive ranks is significantly greater than 27.5 (with some predetermined level of significance), then we would reject the null hypothesis and accept the alternative hypothesis, which implies that the median examination score is significantly greater than 70.

From Table 10.8 we can see that the sum of the positive ranks is $6 + 10 + 1 + 5 + 4 + 8 + 9 = 43$, which is quite a bit greater than 27.5. This value of the sum of the positive ranks, $W^+ = 43$, is used to calculate the test statistic that represents the sum of the positive ranks. The sampling distribution of the sum of the positive ranks is approximately normally distributed with the following test statistic:

$$Z_w^+ = \frac{W^+ - \dfrac{n(n+1)}{4} - \dfrac{1}{2}}{\sqrt{\dfrac{n(n+1)(2n+1)}{24}}}$$

where W is the sum of the positive ranks and n is the sample size.

Thus, for a sample size of $n = 10$, the value of Z_w^+ is

$$Z_w^+ = \frac{W^+ - \dfrac{n(n+1)}{4} - \dfrac{1}{2}}{\sqrt{\dfrac{n(n+1)(2n+1)}{24}}} = \frac{43 - \dfrac{10(11)}{4} - \dfrac{1}{2}}{\sqrt{\dfrac{10(11)(21)}{24}}} \approx 1.53$$

In order to reject the null hypothesis and accept the alternative hypothesis, the value of the test statistic Z_w^+ would have to be greater than the value of z that defines the rejection region for a given level of significance and the specified direction of the alternative hypothesis. For a significance level of $\alpha = 0.05$ and a right-tailed test, we would compare $Z_w^+ = 1.53$ to the value $z = 1.645$, as illustrated in Figure 10.32.

For our example, since $Z_w^+ = 1.53$ does not fall in the rejection region as described in Figure 10.32, we do not have enough evidence to reject the null hypothesis and accept the alternative. In other words, we do not have enough evidence to claim that the median final examination grade is significantly greater than 70.

Because the sampling distribution of the test statistic for a Wilcoxon signed-rank test is approximately normally distributed, we can easily find the corresponding p-value, as illustrated in Figure 10.33, by finding the probability that $z > Z_w^+$.

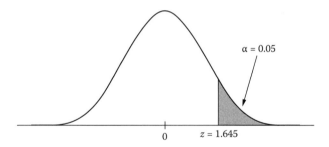

FIGURE 10.32
Rejection region for a one-sided test of whether the population median is greater than 70 for the examination data in Table 10.7.

Similarly, since the *p*-value ($p = .063$) is greater than our predetermined level of significance ($\alpha = 0.05$), we cannot infer that the median final examination grade is significantly greater than 70. If we were to test the alternative hypothesis $H_A : \tilde{\mu}_0 \neq 70$, the *p*-value would be $2(0.063) = 0.1260$.

If we wanted to test the alternative hypothesis $H_A : \tilde{\mu}_0 < 70$, we could sum up the negative ranks as follows:

$$W^- = 2 + 3 + 7 = 12$$

The test statistic would be calculated as follows:

$$Z_w^- = \frac{W^- - \frac{n(n+1)}{4} - \frac{1}{2}}{\sqrt{\frac{n(n+1)(2n+1)}{24}}} = \frac{12 - \frac{10(11)}{4} - \frac{1}{2}}{\sqrt{\frac{10(11)(21)}{24}}} \approx -1.63$$

The corresponding *p*-value is the probability that $z > Z_w^-$, which would be $0.5000 + 0.4485 = 0.9485$. Notice that calculating the *p*-value for a left-tailed test requires finding the probability that $z > Z_w^-$. The reason for this is because of how we describe the test statistic as using either the sum of the positive or the sum of the negative ranks.

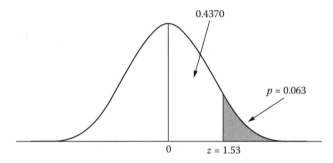

FIGURE 10.33
Corresponding *p*-value for the test statistic $Z_w = 1.53$.

10.6   Using MINITAB for the Wilcoxon Signed-Rank Test

We can use MINITAB to perform the Wilcoxon signed-rank test for the data in Table 10.7 by selecting **1-Sample Wilcoxon** in the **Nonparametric** tab, as illustrated in Figure 10.34.

We can test ($\alpha = 0.05$) whether the median score is greater than 70 using the Wilcoxon signed-rank test dialog box, as illustrated in Figure 10.35. This gives the MINITAB printout in Figure 10.36.

Similar to what we found by doing the calculations by hand, the *p*-value is greater than .05, and therefore, we do not have enough evidence to suggest that the median grade is greater than 70. It is important to note that even when performing a left-tailed test where we would use the sum of the negative ranks, MINITAB will always report the value of the Wilcoxon statistic as the sum of

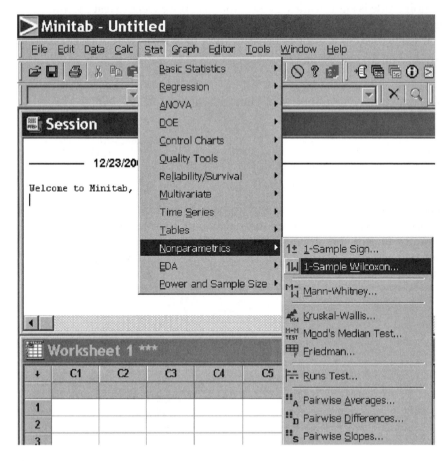

FIGURE 10.34
MINITAB commands for the Wilcoxon signed-rank test.

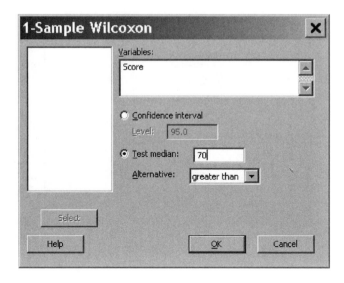

FIGURE 10.35
MINITAB dialog box for the Wilcoxon signed-rank test.

Wilcoxon Signed Rank Test: Score

```
Test of median = 70.00 versus median > 70.00

              N for   Wilcoxon             Estimated
         N    Test   Statistic      P       Median
Score   10     10        43.0   0.063        75.50
```

FIGURE 10.36
MINITAB printout for the Wilcoxon signed-rank test of whether or not the median final examination grade is greater than 70.

the positive ranks. However, in determining the *p*-value, MINITAB calculates it based on the sum of the negative ranks, as was described in the last section.

Even though the Wilcoxon signed-rank test is nonparametric, this does not mean that there are absolutely *no* underlying distributional assumptions. In fact, for this particular test it is assumed that the population being sampled from has a symmetric distribution but no specific shape of the distribution is required. We can run an exploratory check on this by plotting a histogram of the data, as in Figure 10.37.

As we have seen before, with small sample sizes it is difficult to assess by eye the shape of the underlying distribution. When this is the case, we can also use the Ryan–Joiner test as a formal check on the normality of the distribution of the data, as illustrated in Figure 10.38. The Ryan–Joiner test can be used to compare the distribution of the sample data to a normal distribution,

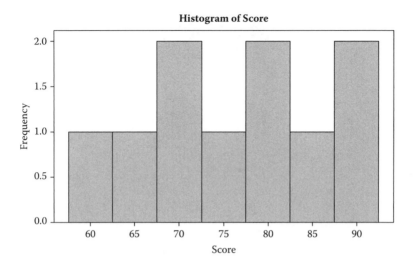

FIGURE 10.37
Histogram of scores on a statistics final examination for a random sample of ten students.

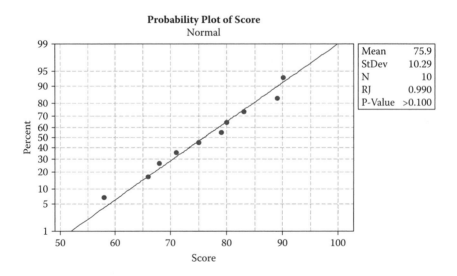

FIGURE 10.38
Ryan–Joiner test comparing the distribution of the sample data with a normal distribution.

and since the *p*-value is greater than .100, we have reason to believe that the data come from a population that is normally distributed. The skewness can also be calculated to help assess if the distribution is symmetric (see Exercise 9 in Chapter 3).

Example 10.5

The data set in Table 10.9 gives the amount spent (in dollars) per month on monthly grocery bills for a random sample of eighteen consumers at a local supermarket.

Suppose we are interested in determining whether the median amount that consumers spend on their monthly grocery bill is less than $500 ($\alpha = 0.05$).

We can first graph an exploratory histogram, as in Figure 10.39, to see if the data are roughly symmetrical, and run a Ryan–Joiner test, as in Figure 10.40, to see if the underlying population is normally distributed.

Notice that for this example the assumption the underlying population is normally distributed appears to hold true.

By running the Wilcoxon signed-rank test in MINITAB in Figure 10.41, we get the results in Figure 10.42.

Since the p-value is not less than the predetermined level of significance of .05, we cannot accept the alternative hypothesis and reject the null hypothesis. Thus, we cannot claim that the population median amount spent on monthly grocery bills is less than $500.

The Wilcoxon signed-rank test is similar to a t-test of the mean (this was covered in Section 4.4). Now suppose that we look at what happens when we run a t-test on the same set of data, as illustrated in Figure 10.43.

TABLE 10.9

Amount Spent per Month (in dollars) on Grocery Bills for a Random Sample of Eighteen Consumers

283	528	359	483	387	398	298	559	592
499	512	475	389	358	529	527	542	496

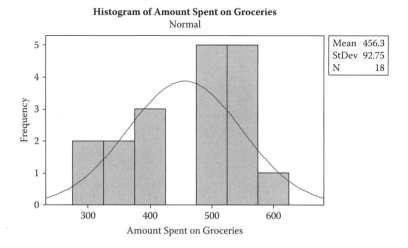

FIGURE 10.39
Histogram of monthly cost for groceries for a sample of eighteen consumers.

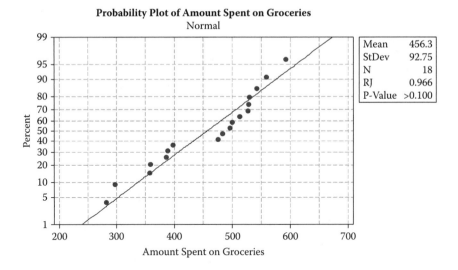

FIGURE 10.40
Ryan–Joiner test of the grocery data in Table 10.9.

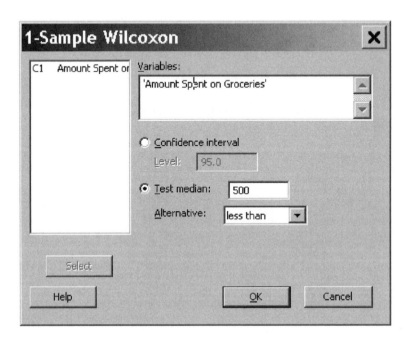

FIGURE 10.41
MINITAB dialog box to test if the population median spent on monthly grocery bills is less than $500.

Wilcoxon Signed Rank Test: Amount Spent on Groceries

```
Test of median = 500.0 versus median < 500.0

                            N for    Wilcoxon          Estimated
                        N    Test    Statistic    P     Median
Amount Spent on Groceries 18   18       54.0   0.088     458.0
```

FIGURE 10.42
MNITAB output for the Wilcoxon signed-rank test of whether the true population median spent on monthly grocery bills is less than $500.

One-Sample T: Amount Spent on Groceries

```
Test of mu = 500 vs < 500

                                              95% Upper
Variable                   N Mean StDev SE Mean  Bound   T      P
Amount Spent on Grocerie  18 456.3  92.8   21.9  494.4 -2.00 0.031
```

FIGURE 10.43
One-sample t-test of whether the true population mean amount spent on monthly grocery bills is less than $500.

Notice for the one-sample t-test that the p-value in Figure 10.43 is less than .05, whereas the p-value for the Wilcoxon signed-rank test presented in Figure 10.42 is not less than .05. Aside from the fact that the Wilcoxon signed-rank test is testing medians and the t-test is testing means, one reason why parametric methods tend to be preferred over nonparametric methods is because parametric methods are more powerful than nonparametric methods, and thus you are more likely to detect an effect if one exists. For this example, the Ryan–Joiner test in Figure 10.40 suggests that the sample comes from a population that has a normal distribution. Because parametric tests are more powerful, we would be more likely to find an effect (if one exists) when using a parametric test than a nonparametric test.

If you have reason to believe that the underlying assumption of normality has not been violated, you may prefer to use parametric methods versus nonparametric methods. Even though an exploratory analysis of your data may suggest that the underlying distributions are not exactly what they should be, as illustrated in Figure 10.40, using nonparametric methods may not provide any significant results because these methods have less power than their equivalent parametric statistical methods when the assumptions have been reasonably met.

10.7   Kruskal–Wallis Test

The Kruskal–Wallis test is a nonparametric alternative to a one-way ANOVA. Recall that a one-way ANOVA relies on the assumption that the populations sampled from are normally distributed (see Section 9.4). The Kruskal–Wallis test can be used for samples collected from populations that have the same shape with similar variance. However, unlike the Wilcoxon signed-rank test, the Kruskal–Wallis test does not require that the distributions are normal or even symmetric; the only requirement is that the distributions for all of the populations sampled from have a similar shape and variance. Similar to the Wilcoxon signed-rank test, the test statistic for the Kruskal–Wallis test is also based on ranks.

The null and alternative hypotheses for the Kruskal–Wallis test are as follows:

H_o : The populations have the same mean ranks

H_A : At least two popuations have different mean ranks

The basic idea behind the Kruskal–Wallis test is that if the null hypothesis were true, then we would expect the mean rank for each of the k samples to be approximately the same. If there is a large amount of variability among the ranks of the k populations being sampled from, then we would expect that we have enough evidence to reject the null hypothesis and accept the alternative. Thus, we could conclude that at least two of the populations we are sampling from have different mean ranks.

The only assumptions underlying the Kruskal–Wallis test is that the populations we are sampling from are independent, with similar shape and variance. There is no need for the population distributions to have any specific shape, such as the shape of a normal distribution.

Example 10.6

Consider the "GPS" data set in Table 10.10, which shows the price (in dollars) for a particular brand of a portable GPS navigation system from three different regions of the country.

We are interested in determining if there is a difference in the mean ranks of the prices based on the region of the country. To use the Kruskal–Wallis test, we first need to rank the data from *all three samples combined*. This can be very tedious to do and will take some time to do by hand, especially when you have a large sample. Table 10.11 gives the rank-ordered data from Table 10.10.

Notice in Table 10.11 that if two or more observations have the same rank, then we need to find the mean of the rank and assign it to each of the tied values. For example, the highlighted portion in Table 10.11 illustrates that for a GPS price of $279 we have a tie because there are two observations that share this same price. Instead of ranking one with a 5 and the other with a 6, the ranks assigned to each of these observations is the average of 5.5.

TABLE 10.10

Price (in dollars) of a Portable GPS
Navigation System by Region of
the Country

North	West	South
285	292	288
297	272	269
299	298	275
290	289	285
291	291	279
299	288	280
302	290	281
295	279	289
293	283	273

TABLE 10.11

Rank-Ordered Data Where the Highlighted Data
Represent Ties

North	Rank	West	Rank	South	Rank
285	10.5	292	20	288	12.5
297	23	272	2	269	1
299	25.5	298	24	275	4
290	16.5	289	14.5	285	10.5
291	18.5	291	18.5	279	5.5
299	25.5	288	12.5	280	7
302	27	290	16.5	281	8
295	22	279	5.5	289	14.5
293	21	283	9	273	3

Once we obtain the ranks for each of the individual observations, we can then find the total and mean of the ranks for each of the k different samples, as presented in Table 10.12. The total rank is found by adding up the ranks separately for each of the three different regions, and the mean of each rank is found by taking the mean of the ranks for each of the three different regions.

If there are no ties among the ranks, the Kruskal–Wallis test is based on the following test statistic, which has a χ^2 distribution with $k-1$ degrees of freedom. The test statistic below represents the variability in the rank positions between the k different samples:

$$H = \frac{12}{n(n+1)} \sum_{j=1}^{k} \frac{R_j^2}{n_j} - 3(n+1)$$

where n is the total sample size, n_j is the sample size for group j, and R_j^2 is the rank squared for the jth group.

TABLE 10.12

Total and Average Rank for Each of
the Three Regions

	Total Rank	Average Rank
North	189.5	$189.5/9 = 21.06$
West	122.5	$122.5/9 = 13.61$
South	66	$66/9 = 7.33$

If there are ties, the test statistic gets adjusted as follows:

$$H_{adj} = \frac{H}{1 - \dfrac{\sum_{j=1}^{n}(t_j^3 - t_j)}{(n^3 - n)}}$$

where t_j is the number of ties in the jth rank. The test statistic adjusted for ties also has the χ^2 distribution with $k - 1$ degrees of freedom.

For our example, the value of H would be

$$H = \frac{12}{27(28)}\left[\frac{189.5^2}{9} + \frac{122.5^2}{9} + \frac{66^2}{9}\right] - 3(27 + 1) = 13.48$$

To calculate the test statistic adjusted for ties, we would have to find $\sum_{j=1}^{n}(t_j^3 - t_j)$ for all of the ties in the sample. This value is found by counting up the number of ties for each rank. For our example, since there are 0 ties for the first rank, the value $(t_1^3 - t_1) = (0^3 - 0) = 0$. However, notice that there are two values that tied for the fifth rank, then $(t_5^3 - t_5) = (2^3 - 2) = 6$. Similarly, there are two values tied for the 10th, 12th, 14th, 16th, 18th, and 25th positions. By summing up all the values, if $(t_j^3 - t_j)$, we would get

$$\sum_{j=1}^{n}\left(t_j^3 - t_j\right) = 6 + 6 + 6 + 6 + 6 + 6 + 6 = 42$$

Thus, the value of H_{adj} would be

$$H_{adj} = \frac{H}{1 - \dfrac{\sum\left(t_j^3 - t_j\right)}{(n^3 - n)}} = \frac{13.48}{1 - \dfrac{42}{(27^3 - 27)}} \approx 13.51$$

Now by comparing this value to the value of the χ^2 distribution with 2 degrees of freedom $(\chi^2_{0.05} = 5.9915)$, we see that H_{adj} falls in the rejection region, as illustrated in Figure 10.44.

This suggests that we can reject the null hypothesis and accept the alternative hypothesis. In other words, at least two of the regions have population mean ranks that are significantly different from each other.

A typical rule of thumb in using the Kruskal–Wallis test is to make sure that there are at least five observations sampled from each population. If less

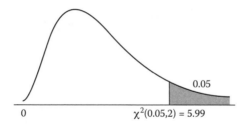

FIGURE 10.44

χ^2 distribution with 2 degrees of freedom, where $H_{adj} = 13.51$ falls in the rejection region.

than five observations are obtained from any of the different populations being sampled from, then the value of the test statistic H (or H_{adj}) may not have the χ^2 distribution, and thus any inference could be suspect.

10.8 Using MINITAB for the Kruskal–Wallis Test

To use MINITAB for the Kruskal–Wallis test, we need to enter the data as two columns: one column for the response variable (which for this example would be the price of the GPS navigation systems) and the other column for the factor (which for this example describes the region of the country). Figure 10.45 shows what a portion of this data set would look like in MINITAB.

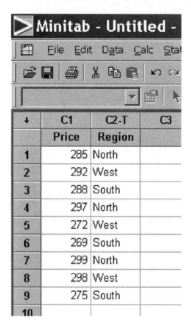

FIGURE 10.45

Data for GPS navigation systems by price and region.

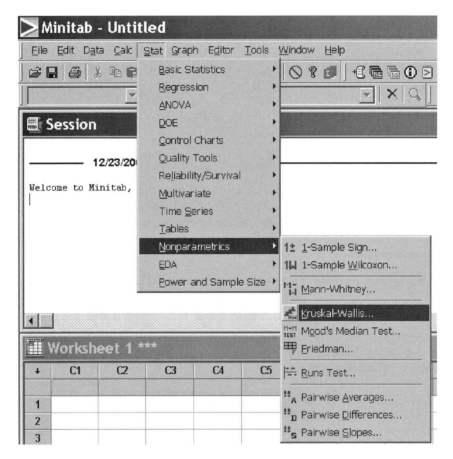

FIGURE 10.46
MINITAB commands to run a Kruskal–Wallis test.

The MINITAB commands for a Kruskal–Wallis test are given in Figure 10.46. This brings up the dialog box for a Kruskal–Wallis test in Figure 10.47.

Notice that the response represents the price of the navigation systems, and the factor is the region. The MINITAB printout is given in Figure 10.48.

The highlighted portion of Figure 10.48 gives the value of the test statistic H, the test statistic adjusted for ties H_{adj}, along with the corresponding p-values. Also notice in the MINITAB printout in Figure 10.48 that the value of the test statistic H is only slightly different when it is adjusted for ties, H_{adj}. Recall that ties occur when the observations are ranked such that there are two or more observations that have the same rank.

The p-value of .001 suggests that we can reject the null hypothesis and accept the alternative hypothesis. Thus, we can claim that the mean ranks for

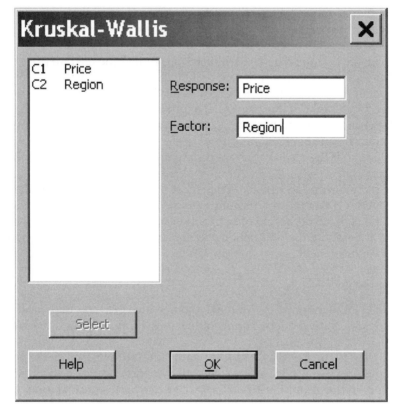

FIGURE 10.47
MINITAB dialog box for the Kruskal–Wallis test.

Kruskal–Wallis Test: Price versus Region

```
Kruskal-Wallis Test on Price

Region    N  Median  Ave Rank      Z
North     9   295.0      21.1   3.27
South     9   280.0       7.3  -3.09
West      9   289.0      13.6  -0.18
Overall  27              14.0

H = 13.48 DF = 2 P = 0.001
H = 13.51 DF = 2 P = 0.001 (adjusted for ties)
```

FIGURE 10.48
MINITAB printout for the Kruskal–Wallis test of whether the price for GPS navigation systems is different by region.

at least two of the populations we are sampling from (which represent the price for GPS systems in different regions of the country) are significantly different from each other.

Also notice that the Kruskal–Wallis test generated the median price of the GPS for each of the three different regions. Even though MINITAB reports the median for each factor, we cannot claim that the medians are different; only the mean ranks of the factors are significantly different from each other. Similarly, unlike an ANOVA, we also cannot make any claims regarding the population means being significantly different between the factors; we can only claim that the mean ranks are different between the different levels of the factor.

One reason why the Kruskal–Wallis test may be preferred over the one-way ANOVA is that the only assumptions are that the data are sampled from populations that have a similar distribution and variance. Thus, the distributions of the underlying populations do not have to be symmetric or normal; the only requirement is that they are similar.

Example 10.7

Table 10.13 gives the data set "Mutual Funds," which provides the year-to-date rate of return for a random sample of low-, medium-, and high-risk mutual funds.

If we were to run a one-way ANOVA as in Figure 10.49, based on the p-value we would find that we do not have enough evidence to suggest that there is a difference in the mean year-to-date rate of return for the mutual funds based on the degree of risk.

However, upon further investigation, we may find that the distributional assumptions for the one-way ANOVA are likely to have been violated, as illustrated in Figures 10.50 to 10.52.

TABLE 10.13

Year-to-Date Returns for a Sample of Thirty Different Mutual Funds Based on Risk

Low Risk	Medium Risk	High Risk
3.47	5.17	18.24
2.84	4.69	17.63
6.78	11.52	10.22
5.24	7.63	54.24
4.98	5.68	18.72
5.22	7.23	–15.24
3.81	8.82	10.54
4.43	7.44	–32.45
5.19	6.78	19.25
8.54	9.54	10.20

One-Way ANOVA: Rate of Return versus Risk

```
Source  DF    SS    MS     F      P
Risk    2     188   94    0.54   0.591
Error   27    4735  175
Total   29    4923
```

```
S = 13.24    R-Sq = 3.82%    R-Sq(adj) = 0.00%
```

```
                              Individual 95% CIs For Mean Based on
                              Pooled StDev
Level     N   Mean  StDev   ------+---------+---------+---------+---
HIGH      10  11.13 22.78                 (--------------*-------------)
LOW       10   5.05  1.65   (-------------*--------------)
MEDIUM    10   7.45  2.09      (-------------*--------------)
                            ------+---------+---------+---------+---
                               0.0       6.0      12.0      18.0
```

```
Pooled StDev = 13.24
```

FIGURE 10.49
One-way ANOVA results for the year-to-date rate of return versus risk for mutual funds.

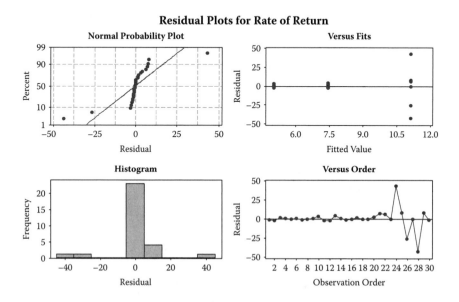

FIGURE 10.50
Four-in-one residual plots for a one-way ANOVA for the mutual fund data in Table 10.13.

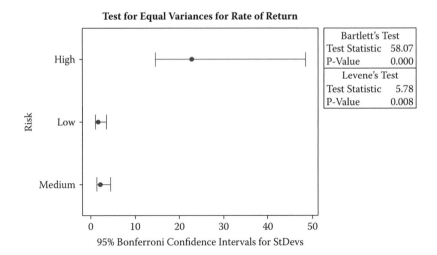

FIGURE 10.51
Bartlett's and Levene's tests for equal variances for the mutual fund data in Table 10.13.

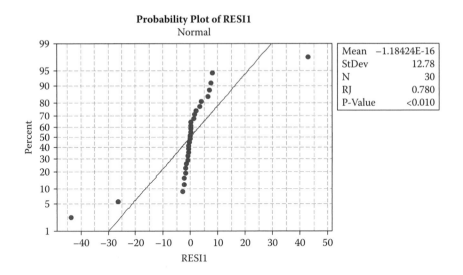

FIGURE 10.52
Ryan–Joiner test for normality for the mutual fund data in Table 10.13.

Because the one-way ANOVA assumptions appear to have been violated, we can run the Kruskal–Wallis test to see if there is a difference in the average ranks for the year-to-date rate of returns based on degree of risk. The results for the Kruskal–Wallis test are given in Figure 10.53.

The results in Figure 10.53 suggest that at least two populations have different average ranks, and thus we would conclude that the ranks of the

Kruskal–Wallis Test: Rate of Return versus Risk

```
Kruskal-Wallis Test on Rate of Return

Risk       N  Median  Ave Rank      Z
HIGH      10  14.085      21.2    2.51
LOW       10   5.085       9.3   -2.75
MEDIUM    10   7.335      16.1    0.24
Overall   30              15.5

H = 9.27   DF = 2   P = 0.010
H = 9.27   DF = 2   P = 0.010   (adjusted for ties)
```

FIGURE 10.53
MINITAB printout for the Kruskal–Wallis test for the difference in ranks between the year-to-date rates of return by risk.

year-to-date rates of return are not all the same, which contradicts what we found when using a one-way ANOVA.

However, even with the Kruskal–Wallis test we are still assuming that the populations we are sampling from have a similar shape and variance. Looking at the box plot in Figure 10.54, we may believe that the variability between the populations is not the same, since clearly there is more variability in the risky mutual funds than in the funds with low or moderate risk. Thus, even when using nonparametric methods that have weaker assumptions, there are still some basic distributional assumptions that if violated may lend to unreliable inferences.

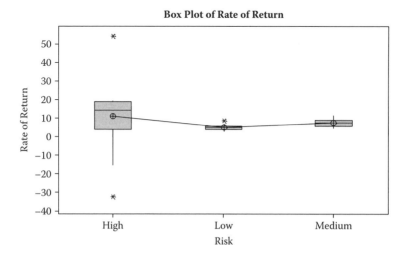

FIGURE 10.54
Box plot of year-to-date rate of return for mutual funds based on degree of risk.

As we have seen with small sample sizes, it can be difficult to determine whether or not the basic distributional assumptions for a one-way ANOVA have been met. As a general rule of thumb, if you can, use the parametric methods! When the assumptions are met, your analysis will have more power and you are more likely to detect an effect if there is one. If you have a small sample or if you have reason to believe that the distributional assumptions for a parametric analysis are grossly violated, then consider using the nonparametric methods. But even when using nonparametric methods, be aware that there are still some assumptions that need to hold true in order to make meaningful inferences.

10.8   Basic Time Series Analysis

Up to now, we have only discussed statistical methods that can be used for data that are collected at a single point in time. However, there may be circumstances when you want to analyze data that are collected over a period of time. We call this type of data *time series* data because it is data that are measured on the same variable sequentially over time.

Example 10.8

Suppose we are interested in how the enrollment patterns at a given university change over time. Such data would be considered time series data because the same variable, namely, enrollment, is measured sequentially over time. Table 10.14 gives the total graduate and undergraduate fall enrollments for Central Connecticut State University for the years 1997 through 2006.

To visualize how the trend in enrollment changes over time, we can graph what is called a *time series plot*. A time series plot is basically a scatter plot where the enrollment is plotted against the chronological order of the observations and the ordered pairs are then connected by straight lines.

MINITAB can easily be used to create a time series plot by first entering the enrollment data in *chronological order*, then selecting **Time Series** and then **Time Series Plot** under the **Stat** menu, as illustrated in Figure 10.55.

This then gives the dialog box in Figure 10.56. We will select a simple time series plot. We then need to select the variable (which needs to be in chronological order), as presented in Figure 10.57.

TABLE 10.14

Graduate and Undergraduate Fall Enrollments for Central Connecticut State University for the Years 1997 through 2006

Year	1997	1998	1999	2000	2001	2002	2003	2004	2005	2006
Total	11,625	11,686	11,903	12,252	12,368	12,642	12,131	12,320	12,315	12,144

Source: http://web.ccsu.edu/oira/data/factbook/200809/enrollment/Historical.htm

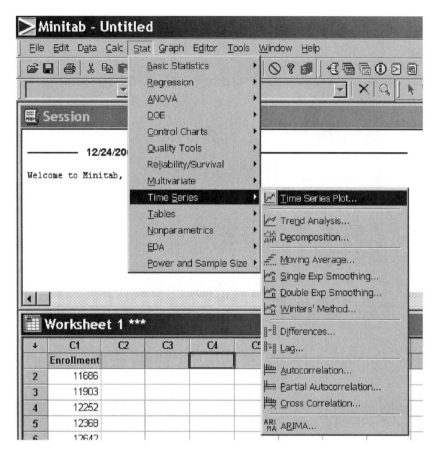

FIGURE 10.55
MINITAB commands to draw a time series plot.

Then by selecting the **Time/Scale** tab, we can specify the format in which the data were entered in. For our example, the data were entered chronologically by year, as illustrated in Figure 10.58. Also notice that the years start at 1997.

This gives the time series plot of the enrollment data by year, as illustrated in Figure 10.59.

One way to describe time series data is by *trends*. One type is a *long-term trend*. A long-term trend would suggest whether the series tends to increase or decrease over a period of time. We can also look to determine if there is any *seasonal trend* in the time series. A seasonal trend could be described by a repeating pattern that seems to reoccur during a certain interval of time. Another trend that could be observed is a *cyclical trend*. A cyclical trend would be a pattern that oscillates about a long-term trend.

Because we cannot describe every possible type of trend, there is also a *random trend*, which consists of any trend that may not have been observed.

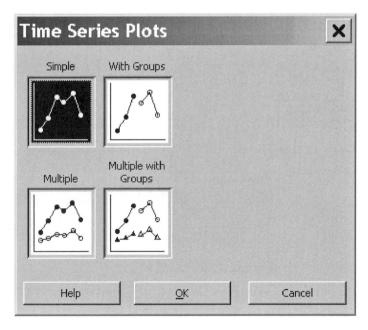

FIGURE 10.56
MINITAB dialog box to draw a simple time series plot.

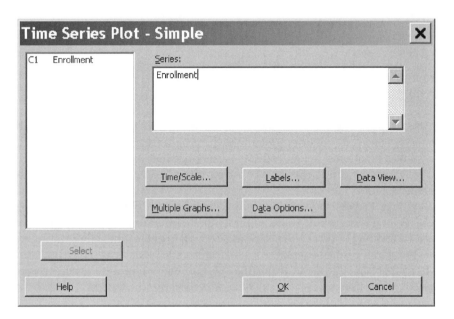

FIGURE 10.57
MINITAB dialog box to draw a simple time series plot.

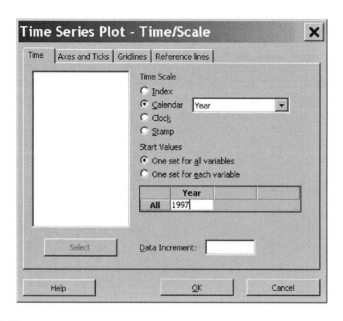

FIGURE 10.58
MINITAB dialog box to specify the chronological ordering of the data.

It is what is left over after the long-term, seasonal, and cyclical trends have been accounted for.

Notice in Figure 10.59 that there appears to be an increasing long-term trend. In other words, the enrollments at the institution appear to be increasing over time.

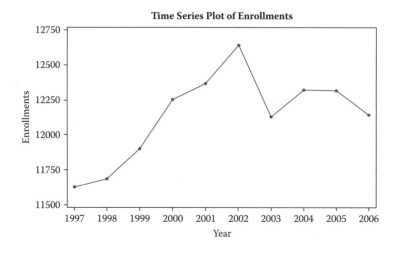

FIGURE 10.59
Time series plot of enrollment by year for the data in Table 10.14.

Regression Analysis: Enrollment versus Time

```
The regression equation is
Enrollment = 11770 + 67.1 Time

Predictor      Coef SE Coef        T       P
Constant    11769.6    176.8    66.57   0.000
Time          67.09    28.49     2.35   0.046

S = 258.794    R-Sq = 40.9%      R-Sq(adj) = 33.6%

Analysis of Variance

Source           DF        SS       MS       F       P
Regression        1    371348   371348    5.54   0.046
Residual Error    8    535796    66975
Total             9    907144
```

FIGURE 10.60
Regression analysis of enrollments over time for the data in Table 10.14.

We can use simple forecasting techniques to analyze time series data. This basically entails modeling observable trends or patterns in the data and then using such patterns to predict future observations.

One way to see if there is a long-term trend is to do a *trend analysis*. This entails fitting a regression model to the time series data, and then using such a model to extrapolate for future time periods.

A *linear forecast model* uses a linear model to represent the relationship between the response variables of interest at time *t*, as follows:

$$y_t = \beta_0 + \beta_1 t_t + \varepsilon_t$$

where y_t is the response variable at time *t*, β_1 represents the average change from one time period to the next, and *t* represents a given time period.

For the time series data given in Table 10.14 we can perform a linear forecast by running a regression analysis. This is done by creating time periods and using them as the predictor variable and the enrollments as the response variable and running a basic linear regression analysis, as seen in Figure 10.60.

We can then use the estimated model from this regression analysis to forecast what the enrollments will be for future years. We can also use MINITAB to perform a trend analysis by selecting **Trend Analysis** under **Stat** and **Time Series** to get the trend analysis dialog box presented in Figure 10.61.

We can use the model to generate forecasts for future observations by checking the **Generate forecasts** box and then indicating the **Number of forecasts**. The number of forecasts will be the same scale as the time series data. Since for our example the time is in years, we will use the model to forecast the enrollments for the next 3 years. This will generate the trend analysis graph presented in Figure 10.62.

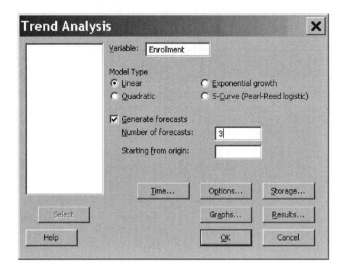

FIGURE 10.61
MINITAB dialog box to perform a trend analysis for the data in Table 10.14.

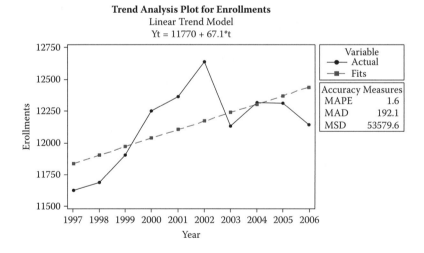

FIGURE 10.62
MINITAB output for a linear trend analysis for the change in enrollments over time.

Notice in Figure 10.62 that the linear trend model is the same as the linear regression model presented in Figure 10.60. The trend analysis plot not only provides the time series plot, but also provides the equation of the line of best fit along with the line of best fit for the forecast (or estimated) values. Also, you will notice that there is a set of accuracy measures in the box on the right-hand side of Figure 10.62. These accuracy measures can be used to assess how well the model fits the data as they compare the observed values

TABLE 10.15

Enrollment and Fitted Values for the Data Given in Table 10.14

Time (in years) t	Enrollment y_t	Fitted Value $\hat{y}_t$	$\left\| \left(\dfrac{y_t - \hat{y}_t}{y_t} \right) \right\|$
1	11,625	11836.691	0.0182100
2	11,686	11903.782	0.0186361
3	11,903	11970.873	0.0057022
4	12,252	12037.964	0.0174695
5	12,368	12105.055	0.0212601
6	12,642	12172.145	0.0371662
7	12,131	12239.236	0.0089223
8	12,320	12306.327	0.0011098
9	12,315	12373.418	0.0047437
10	12,144	12440.509	0.0244161

of the time series data with the forecast values. These measures can be very useful for comparing different models.

The first measure, MAPE, stands for the *mean absolute percentage error*. This measure gives the average percentage error by finding the average of the absolute value of the difference between each observation and the fitted value multiplied by 100. The smaller this value is, the better the fit of the forecast model. It can be calculated by using the following equation:

$$MAPE = \frac{\sum\limits_{t=1}^{n} \left\| \left(\dfrac{y_t - \hat{y}_t}{y_t} \right) \right\|}{n} \cdot 100$$

where y_t is the value of the observation at time t, $\hat{y}_t$ is the fitted value at time t, and n is the sample size.

Table 10.15 gives the observations and the corresponding fitted values for the data in Table 10.14.

Then to calculate MAPE:

$$MAPE = \frac{\sum\limits_{t=1}^{n} \left\| \left(\dfrac{y_t - \hat{y}_t}{y_t} \right) \right\|}{n} \cdot 100 = \frac{0.157636}{10} \cdot 100 \approx 1.58$$

Similarly, the MAD, the *mean absolute deviation*, measures the accuracy of the fitted model by taking the average difference between the observations and the fitted value, and is calculated as follows:

$$MAD = \frac{\sum\limits_{t=1}^{n} \left| y_t - \hat{y}_t \right|}{n}$$

where y_t is the value of the observation at time t, $\hat{y}_t$ is the fitted value at time t, and n is the sample size. Using the calculations in Table 10.15, we can calculate the MAD as follows:

$$MAD = \frac{\sum\limits_{t=1}^{n} |y_t - \hat{y}_t|}{n} = \frac{1921.02}{10} = 192.10$$

Similar to the MAPE, the smaller the value of the MAD, the better the fit of the fitted model.

Another measure is the MSD, or the *mean squared deviation*. This measure gives the average of the squared deviations between the observed and fitted values, and it is found using the following formula:

$$MSD = \frac{\sum\limits_{t=1}^{n} (y_t - \hat{y}_t)^2}{n}$$

where y_t is the value of the observation at time t, $\hat{y}_t$ is the fitted value at time t, and n is the sample size. Using the calculations in Table 10.15, we can calculate the MSD as follows:

$$MSD = \frac{\sum\limits_{t=1}^{n} (y_t - \hat{y}_t)^2}{n} = \frac{535796}{10} = 53579.60$$

Because the long-term trend may be nonlinear, we can also use MINITAB to fit a quadratic model and forecast the enrollments for the next 3 years, as can be seen in Figure 10.63. This gives the quadratic trend model presented in Figure 10.64.

Notice that the accuracy measures in Figure 10.64 for the quadratic model are smaller than those for the linear model in Figure 10.62. This suggests that the quadratic model fits the data better than the linear model because the deviations between the observed and fitted values are smaller for the quadratic model. Also, using this model we can forecast the enrollments for the next 3 years.

One limitation to conducting a trend analysis on time series data is that a simple trend analysis like the one just illustrated does not account for seasonal trends. Seasonal trends occur when a repeating pattern is observed over a specific interval of time. We can use *decomposition* to break down a time series analysis into linear and seasonal trends.

There are two types of decomposition models: additive and multiplicative. An additive model represents linear and season trends added together to form the model, and it can be represented by the following equation:

$$y = \text{Linear Trend} + \text{Seasonal Trend} + \varepsilon$$

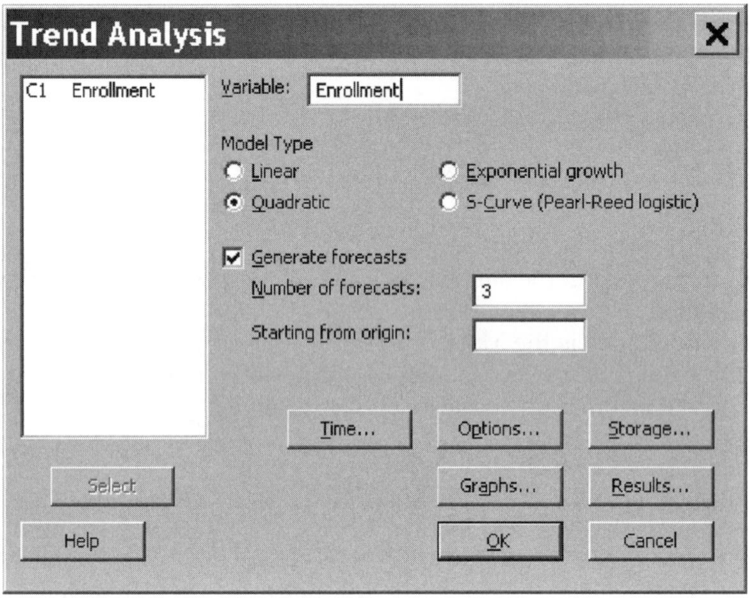

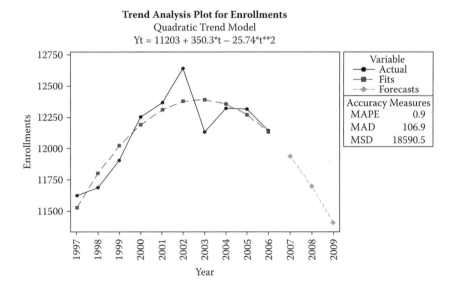

FIGURE 10.63
MINITAB dialog box for a quadratic trend analysis.

FIGURE 10.64
MINITAB quadratic trend analysis for the data in Table 10.14.

TABLE 10.16

Annual Government Hires Rate as a Percent of Total Employment for the Years 2001 through 2006

Year	Jan	Feb	Mar	Apr	May	Jun	Jul	Aug	Sep	Oct	Nov	Dec
2001	1.7	1.3	1.2	1.2	1.8	2.1	2.0	2.8	2.2	1.5	1.2	0.9
2002	1.4	1.0	1.0	1.2	1.6	2.1	1.8	2.5	2.5	1.6	1.2	1.0
2003	1.3	1.0	1.0	1.1	1.4	1.8	1.8	2.1	2.2	1.3	1.2	1.0
2004	1.3	1.1	1.0	1.0	1.4	1.9	1.7	2.5	2.5	1.4	1.2	0.9
2005	1.4	1.0	1.1	1.1	1.6	2.0	1.8	2.7	2.4	1.5	1.3	0.9
2006	1.5	1.2	1.2	1.2	1.8	2.4	2.0	3.1	2.8	1.5	1.2	1.0

Source: http://data.bls.gov/cgi-bin/surveymost?jt. Taken from the job openings and turnover survey, form JTU90000000HIR.

A multiplicative model represents a seasonal pattern that depends on the level of the data; thus, as more observations are obtained, the seasonal pattern increases, and it can be represented by the following equation:

$$y = (\text{Linear Trend}) \cdot (\text{Seasonal Trend}) \cdot \varepsilon$$

Example 10.9

Suppose we are interested in whether the size of the government workforce changes over time. We could obtain a measure on the number of government hires as a percent of the total employed workforce and look to see if this measure changes over time. Thus, the hires rate can be calculated as follows:

$$\text{Hires Rate} = \frac{\text{Number of hires duirng month}}{\text{Total number of employees}} \cdot 100$$

The data set "Government Hires Rate" in Table 10.16 provides the number of annual government hires as a percent of total employment for each month for the years 2001 through 2006.

One way to determine if the government's hiring rate is increasing over time is to plot the hires rate versus time, as in Figure 10.65.

For the graph in Figure 10.65, it appears that the hires rate peaks during the summer months, and this seasonal pattern seems to repeat during each of the given years.

Because the seasonal pattern does not appear to be increasing with time, we will consider a decomposition using an additive model. Using MINITAB, we can select **Decomposition** under the **Time Series** tab on the **Stat** menu to get the decomposition dialog box illustrated in Figure 10.66.

We need to specify the variable, the type of model (in this case we are considering an additive model), and the components of the model (which in this case include the trend and the seasonal components). We also need to specify

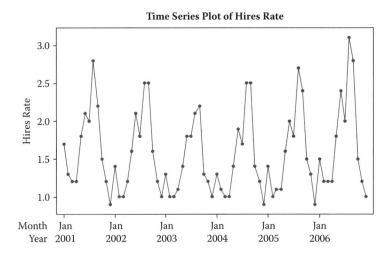

FIGURE 10.65
Time series plot of hires rate as a function of time for the data in Table 10.16.

the **Seasonal length**. For our example, since the data were collected monthly, we will use a seasonal length of 12.

We can also generate forecasts to predict the hires rate for the next 3 years, or 36 months. We get three different decomposition plots: a seasonal analysis as presented in Figure 10.67, a component analysis in Figure 10.68,

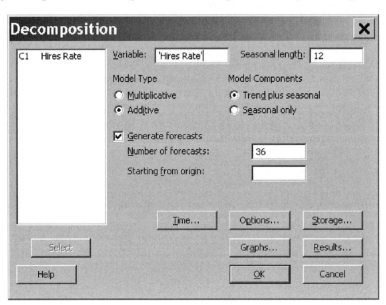

FIGURE 10.66
MINITAB dialog box for a time series decomposition.

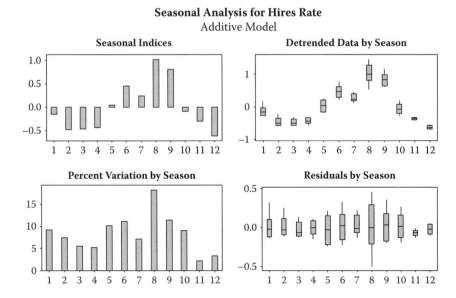

FIGURE 10.67
Seasonal analysis plot for the hires rate data from Table 10.16.

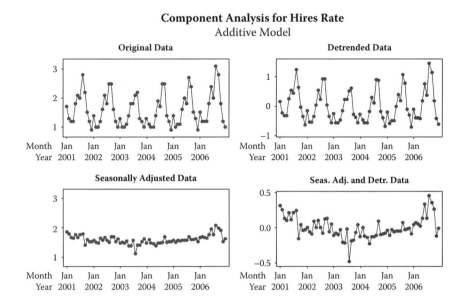

FIGURE 10.68
Component analysis plot for the hires rate data from Table 10.16.

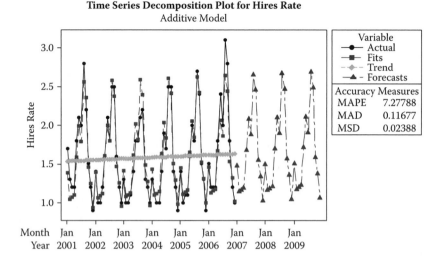

FIGURE 10.69
Time series decomposition plot for the hires rate data from Table 10.16.

and a time series decomposition plot in Figure 10.69. These plots can be used to visualize how to separate time series data into linear and seasonal components.

The seasonal analysis plot in Figure 10.67 shows the percent variation by season relative to the sum of variation by season along with box plots of the data and the residuals by seasonal period. Notice that the month of August shows the most variability in the hires rate.

The component analysis in Figure 10.68 shows the original time series data, a time series of the detrended data (the detrended data include only the seasonal and error components), a time series of the seasonally adjusted data (this includes the trend and error components only), and a time series of the seasonally adjusted and detrended data (the error component only). The plot of the seasonally adjusted data represents the graph of the time series data if there is no seasonal effect.

The time series decomposition plot in Figure 10.69 shows the time series with the fitted trend line along with the forecast.

From these graphs, we can see that there does not appear to be an increase in the hires rate, as the seasonally adjusted data are not showing such a trend. Thus, by partitioning out the seasonal trend, we can visualize whether a linear trend exists.

Exercises

1. Verify that $SSColumn = 3.467$ and $MSColumn = 1.733$ from Table 10.2.

2. Using the data given in Table 10.1 and MINITAB, run a test for equal variances. Notice that only Bartlett's test result is displayed. This is because MINITAB will only display Levene's test if two or more cells have replicates and at least one of these cells has three or more replicates. Now using the data in Table 10.5, use MINITAB to run a test for equal variances. Notice that both Bartlett's and Levene's tests are reported. This is because there are more than three replicates in each cell.

3. Using the data in Table 10.5, describe the row and column factors, and find the values for r, c, m, and n. From the MINITAB printout in Figure 10.19, identify $SSRow$, $SSColumn$, $SSInteraction$, $MSRow$, $MSColumn$, and $MSInteraction$.

4. The data set "Freshmen" consists of a random sample of 350 first-year grade point averages for a given class of freshmen at a university.

   a. Using the Wilcoxon signed-rank test, determine if the *median* grade point average for the population of first-year freshmen is less than 2.00.

   b. Use a one-sample t-test to determine if the *mean* grade point average for the population of first-year freshmen is less than 2.00.

   c. Comment on the difference in using these two methods. Which method do you believe is more appropriate for the given set of data?

5. a. Using the data in Table 10.9, calculate by hand the value of the test statistic Z_w to determine if the population median amount that consumers spend on their monthly grocery bill is less than $500. Remember when doing a left-tailed Wilcoxon signed-rank test that the sum of the negative ranks is used.

   b. Find, by hand, the p-value, and compare it to the p-value that was found in Figure 10.42.

6. The Wilcoxon signed-rank test can also be used as a nonparametric alternative to a paired t-test (see Exercise 21 in Chapter 4). Using the "Weight Loss" data given in Table 4.10, with MINITAB run a Wilcoxon signed-rank test to determine if the median difference in weight loss for the program participants is significantly greater than 0. Were your results different from running the Wilcoxon signed-rank test compared to the paired t-test? Justify your answer.

7. a. Using the data set "Automobile," run a one-way ANOVA and a Kruskal–Wallis test for the difference in the highway miles per gallon based on the make of the car. Which method would you prefer? Why?

    b. Using the data set "Automobile," run a one-way ANOVA and a Kruskal–Wallis test for the difference in the highway miles per gallon based on the type of the car. Which method would you prefer? Why?

8. Is there a difference in the mean number of hours of television that students and faculty members watch per week by gender? The data set "TV" in Table 10.17 consists of the number of hours of television watched per week for a sample of twenty-four faculty members and twenty-four students by gender.

TABLE 10.17

"TV" Data Set Showing the Number of Hours
Watched for Male and Female Faculty and Students

	Males	**Females**
Faculty	22	20
	18	15
	9	10
	17	5
	35	4
	25	3
	18	4
	22	9
	13	8
	4	12
	6	15
	15	20
Students	28	18
	24	22
	30	15
	12	14
	32	16
	21	32
	28	18
	29	19
	31	17
	22	20
	8	18
	0	5

a. Using MINITAB, run a two-way ANOVA to determine if there is a difference between the number of hours of television watched between faculty members and students by gender ($\alpha = 0.05$).

b. Check any relevant model assumptions and comment on what you see.

c. Create a main effects plot and an interaction plot and comment on whether these plots support your findings.

9. a. Using a two-way ANOVA on the data set "Miles Per Gallon" and MINITAB, determine if there is a difference in the average miles per gallon based on the make and model of the vehicle.

b. Check any relevant model assumptions and comment on what you see.

c. Create a main effects plot and an interaction plot and comment on whether these plots support your findings.

10. The data set "Mortgage Rates" in Table 10.18 gives the contract rate on 30-year-fixed, conventional home mortgages compiled by the Federal Reserve from the years 1972 through 2006 (source: http://www.federalreserve.gov/releases/h15/data/Annual/ H15_MORTG_NA.txt).

a. Using MINITAB, plot a basic time series of the average yearly interest rate for 30-year-fixed, conventional home mortgages.

b. Run a regression analysis of the average yearly interest rate over time.

c. Using MINITAB, conduct a linear trend analysis and forecast the average interest rate over the next 15 years.

d. Using MINITAB, conduct a quadratic trend analysis and forecast the average interest rate over the next 15 years.

e. Comment on whether the linear or quadratic model best fits this set of time series data.

11. The data set "Housing Sales" (source: http://www.census.gov/ const/www/newressalesindex_excel.html) gives the number of houses sold (in thousands) by month between January 2000 and December 2006 in the United States.

a. Using MINITAB, plot a basic time series plot of the housing sales for the years 2000 through 2005.

b. Run a regression analysis of the housing sales over time.

c. Using MINITAB, conduct a linear trend analysis and forecast the housing sales over the next 5 years.

d. Using MINITAB, conduct a quadratic trend analysis and forecast the housing sales over the next 5 years.

e. Comment on whether the linear or quadratic model best fits this set of time series data.

f. Do you think there is a seasonal effect on housing sales? If so, explain.

TABLE 10.18

Contract Rate on Thirty-Year
Fixed-Rate Mortgages

Date	Mortgage Rate
1972	7.38
1973	8.04
1974	9.19
1975	9.04
1976	8.86
1977	8.84
1978	9.63
1979	11.19
1980	13.77
1981	16.63
1982	16.08
1983	13.23
1984	13.87
1985	12.42
1986	10.18
1987	10.2
1988	10.34
1989	10.32
1990	10.13
1991	9.25
1992	8.4
1993	7.33
1994	8.35
1995	7.95
1996	7.8
1997	7.6
1998	6.94
1999	7.43
2000	8.06
2001	6.97
2002	6.54
2003	5.82
2004	5.84
2005	5.86
2006	6.41

Appendix A

TABLE 1

Standard Normal Table

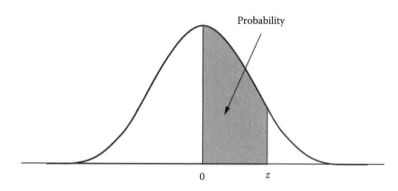

z	.00	.01	.02	.03	.04	.05	.06	.07	.08	.09
0.0	0.0000	0.0040	0.0080	0.0120	0.0160	0.0199	0.0239	0.0279	0.0319	0.0359
0.1	0.0398	0.0438	0.0478	0.0517	0.0557	0.0596	0.0636	0.0675	0.0714	0.0754
0.2	0.0793	0.0832	0.0871	0.0909	0.0948	0.0987	0.1026	0.1064	0.1103	0.1141
0.3	0.1179	0.1217	0.1255	0.1293	0.1331	0.1368	0.1406	0.1443	0.1480	0.1517
0.4	0.1554	0.1591	0.1628	0.1664	0.1700	0.1736	0.1772	0.1808	0.1844	0.1879
0.5	0.1915	0.1950	0.1985	0.2019	0.2054	0.2088	0.2123	0.2157	0.2190	0.2224
0.6	0.2258	0.2291	0.2324	0.2357	0.2389	0.2421	0.2454	0.2486	0.2518	0.2549
0.7	0.2580	0.2611	0.2642	0.2673	0.2703	0.2734	0.2764	0.2793	0.2823	0.2852
0.8	0.2881	0.2910	0.2939	0.2967	0.2995	0.3023	0.3051	0.3078	0.3106	0.3133
0.9	0.3159	0.3186	0.3212	0.3238	0.3264	0.3289	0.3315	0.3340	0.3365	0.3389
1.0	0.3413	0.3438	0.3461	0.3485	0.3508	0.3531	0.3554	0.3577	0.3599	0.3621
1.1	0.3643	0.3665	0.3686	0.3708	0.3729	0.3749	0.3770	0.3790	0.3810	0.3830
1.2	0.3849	0.3869	0.3888	0.3907	0.3925	0.3943	0.3962	0.3980	0.3997	0.4015
1.3	0.4032	0.4049	0.4066	0.4082	0.4099	0.4115	0.4131	0.4147	0.4162	0.4177
1.4	0.4192	0.4207	0.4222	0.4236	0.4251	0.4265	0.4278	0.4292	0.4306	0.4319
1.5	0.4332	0.4345	0.4357	0.4370	0.4382	0.4394	0.4406	0.4418	0.4429	0.4441
1.6	0.4452	0.4463	0.4474	0.4485	0.4495	0.4505	0.4515	0.4525	0.4535	0.4545
1.7	0.4554	0.4564	0.4573	0.4582	0.4591	0.4599	0.4608	0.4616	0.4625	0.4633
1.8	0.4641	0.4648	0.4656	0.4664	0.4671	0.4678	0.4686	0.4693	0.4699	0.4706

(continued)

TABLE 1 (CONTINUED)

Standard Normal Table

z	.00	.01	.02	.03	.04	.05	.06	.07	.08	.09
1.9	0.4713	0.4719	0.4726	0.4732	0.4738	0.4744	0.4750	0.4756	0.4761	0.4767
2.0	0.4772	0.4778	0.4783	0.4788	0.4793	0.4798	0.4803	0.4808	0.4812	0.4817
2.1	0.4821	0.4826	0.4830	0.4834	0.4838	0.4842	0.4846	0.4850	0.4854	0.4857
2.2	0.4861	0.4865	0.4868	0.4871	0.4875	0.4878	0.4881	0.4884	0.4887	0.4890
2.3	0.4893	0.4896	0.4898	0.4901	0.4904	0.4906	0.4909	0.4911	0.4913	0.4916
2.4	0.4918	0.4920	0.4922	0.4925	0.4927	0.4929	0.4930	0.4932	0.4934	0.4936
2.5	0.4938	0.4940	0.4941	0.4943	0.4945	0.4946	0.4648	0.4949	0.4951	0.4952
2.6	0.4953	0.4955	0.4956	0.4957	0.4959	0.4960	0.4961	0.4962	0.4963	0.4964
2.7	0.4965	0.4966	0.4967	0.4968	0.4969	0.4970	0.4971	0.4972	0.4973	0.4974
2.8	0.4974	0.4975	0.4976	0.4977	0.4977	0.4978	0.4979	0.4980	0.4980	0.4981
2.9	0.4981	0.4982	0.4982	0.4983	0.4984	0.4984	0.4985	0.4985	0.4986	0.4986
3.0	0.4987	0.4987	0.4987	0.4988	0.4988	0.4989	0.4989	0.4989	0.4990	0.4990
3.1	0.4990	0.4991	0.4991	0.4991	0.4992	0.4992	0.4992	0.4992	0.4993	0.4993
3.2	0.4993	0.4993	0.4994	0.4994	0.4994	0.4994	0.4994	0.4995	0.4995	0.4995
3.3	0.4995	0.4995	0.4996	0.4996	0.4996	0.4996	0.4996	0.4996	0.4996	0.4997
3.4	0.4997	0.4997	0.4997	0.4997	0.4997	0.4997	0.4997	0.4997	0.4998	0.4998
3.5	0.4998	0.4998	0.4998	0.4998	0.4998	0.4998	0.4998	0.4998	0.4998	0.4998
3.6	0.4998	0.4999	0.4999	0.4999	0.4999	0.4999	0.4999	0.4999	0.4999	0.4999
3.7	0.4999	0.4999	0.4999	0.4999	0.4999	0.4999	0.4999	0.4999	0.4999	0.4999
3.8	0.4999	0.4999	0.4999	0.4999	0.4999	0.4999	0.4999	0.5000	0.5000	0.5000
3.9	0.5000	0.5000	0.5000	0.5000	0.5000	0.5000	0.5000	0.5000	0.5000	0.5000

Abridged and adapted from *Standard Probability and Statistical Tables and Formulae* (Boca Raton: Chapman & Hall/CRC, 2000).

TABLE 2

Critical Values for the Rejection Region for the t-distribution

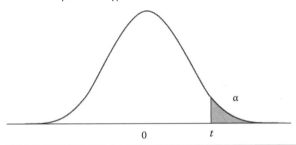

	α = 0.10	0.05	0.025	0.01	0.005	0.0025	0.001
d.f. =1	3.078	6.314	12.706	31.821	63.657	318.309	636.619
2	1.886	2.920	4.303	6.965	9.925	22.327	31.599
3	1.638	2.353	3.182	4.541	5.841	10.215	12.924

TABLE 2 (CONTINUED)

Critical Values for the Rejection Region for the t-distribution

	$\alpha = 0.10$	0.05	0.025	0.01	0.005	0.0025	0.001
4	1.533	2.132	2.776	3.747	4.604	7.173	8.610
5	1.476	2.015	2.571	3.365	4.032	5.893	6.869
6	1.440	1.943	2.447	3.143	3.707	5.208	5.959
7	1.415	1.895	2.365	2.998	3.499	4.785	5.408
8	1.397	1.860	2.306	2.896	3.355	4.501	5.041
9	1.383	1.833	2.262	2.821	3.250	4.297	4.781
10	1.372	1.812	2.228	2.764	3.169	4.144	4.587
11	1.363	1.796	2.201	2.718	3.106	4.025	4.437
12	1.356	1.782	2.179	2.681	3.055	3.930	4.318
13	1.350	1.771	2.160	2.650	3.012	3.852	4.221
14	1.345	1.761	2.145	2.624	2.977	3.787	4.140
15	1.341	1.753	2.131	2.602	2.947	3.733	4.073
16	1.337	1.746	2.120	2.583	2.921	3.686	4.015
17	1.333	1.740	2.110	2.567	2.898	3.646	3.965
18	1.330	1.734	2.101	2.552	2.878	3.610	3.922
19	1.328	1.729	2.093	2.539	2.861	3.579	3.883
20	1.325	1.725	2.086	2.528	2.845	3.552	3.850
21	1.323	1.721	2.080	2.518	2.831	3.527	3.819
22	1.321	1.717	2.074	2.508	2.819	3.505	3.792
23	1.319	1.714	2.069	2.500	2.807	3.485	3.768
24	1.318	1.711	2.064	2.492	2.797	3.467	3.745
25	1.316	1.708	2.060	2.485	2.787	3.450	3.725
26	1.315	1.706	2.056	2.479	2.779	3.435	3.707
27	1.314	1.703	2.052	2.473	2.771	3.421	3.690
28	1.313	1.701	2.048	2.467	2.763	3.408	3.674
29	1.311	1.699	2.045	2.462	2.756	3.396	3.659
30	1.310	1.697	2.042	2.457	2.750	3.385	3.646
35	1.306	1.690	2.030	2.438	2.724	3.340	3.591
40	1.303	1.684	2.021	2.423	2.704	3.307	3.551
45	1.301	1.679	2.014	2.412	2.690	3.281	3.520
50	1.299	1.676	2.009	2.403	2.678	3.261	3.496
100	1.290	1.660	1.984	2.364	2.626	3.174	3.390
∞	1.282	1.645	1.960	2.326	2.576	3.091	3.291

Abridged and adapted from *Standard Probability and Statistical Tables and Formulae* (Boca Raton: Chapman & Hall/ CRC, 2000).

TABLE 3

Approximate Critical Values for the Ryan–Joiner Test of Normality

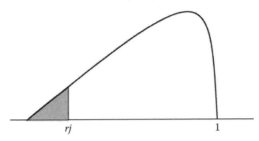

n	α = 0.10	α = 0.05	α = 0.01
4	0.8951	0.8734	0.8318
5	0.9033	0.8804	0.8320
10	0.9347	0.9180	0.8804
15	0.9506	0.9383	0.9110
20	0.9600	0.9503	0.9290
25	0.9662	0.9582	0.9408
30	0.9707	0.9639	0.9490
40	0.9767	0.9715	0.9597
50	0.9807	0.9764	0.9664
60	0.9835	0.9799	0.9710
75	0.9865	0.9835	0.9757

Source: MINITAB website: (http://www.minitab.com/uploadedFiles/Shared_Resources/ Documents/Articles/normal_probability_plots.pdf)

TABLE 4

Critical Values for the Rejection Region for the *F*-Distribution

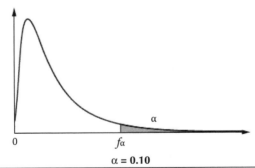

α = 0.10

df_{num} = 1	2	3	4	5	6	7	8	9	10	50	100	∞
df_{denom} = 1 39.86	49.50	53.59	55.83	57.24	58.20	58.91	59.44	59.86	60.19	62.69	63.01	63.33
2 8.53	9.00	9.16	9.24	9.29	9.33	9.35	9.37	9.38	9.39	9.47	9.48	9.49

TABLE 4 (CONTINUED)

Critical Values for the Rejection Region for the *F*-Distribution

df_{num} = 1	2	3	4	5	6	7	8	9	10	50	100	∞
3	5.54	5.46	5.39	5.34	5.31	5.28	5.27	5.25	5.24	5.23	5.15	5.14	5.13
4	4.54	4.32	4.19	4.11	4.05	4.01	3.98	3.95	3.94	3.92	3.80	3.78	3.76
5	4.06	3.78	3.62	3.52	3.45	3.40	3.37	3.34	3.32	3.30	3.15	3.13	3.10
6	3.78	3.46	3.29	3.18	3.11	3.05	3.01	2.98	2.96	2.94	2.77	2.75	2.72
7	3.59	3.26	3.07	2.96	2.88	2.83	2.78	2.75	2.72	2.70	2.52	2.50	2.47
8	3.46	3.11	2.92	2.81	2.73	2.67	2.62	2.59	2.56	2.54	2.35	2.32	2.29
9	3.36	3.01	2.81	2.69	2.61	2.55	2.51	2.47	2.44	2.42	2.22	2.19	2.16
10	3.29	2.92	2.73	2.61	2.52	2.46	2.41	2.38	2.35	2.32	2.12	2.09	2.06
11	3.23	2.86	2.66	2.54	2.45	2.39	2.34	2.30	2.27	2.25	2.04	2.01	1.97
12	3.18	2.81	2.61	2.48	2.39	2.33	2.28	2.24	2.21	2.19	1.97	1.94	1.90
13	3.14	2.76	2.56	2.43	2.35	2.28	2.23	2.20	2.16	2.14	1.92	1.88	1.85
14	3.10	2.73	2.52	2.39	2.31	2.24	2.19	2.15	2.12	2.10	1.87	1.83	1.80
15	3.07	2.70	2.49	2.36	2.27	2.21	2.16	2.12	2.09	2.06	1.83	1.79	1.76
16	3.05	2.67	2.46	2.33	2.24	2.18	2.13	2.09	2.06	2.03	1.79	1.76	1.72
17	3.03	2.64	2.44	2.31	2.22	2.15	2.10	2.06	2.03	2.00	1.76	1.73	1.69
18	3.01	2.62	2.42	2.29	2.20	2.13	2.08	2.04	2.00	1.98	1.74	1.70	1.66
19	2.99	2.61	2.40	2.27	2.18	2.11	2.06	2.02	1.98	1.96	1.71	1.67	1.63
20	2.97	2.59	2.38	2.25	2.16	2.09	2.04	2.00	1.96	1.94	1.69	1.65	1.61
25	2.92	2.53	2.32	2.18	2.09	2.02	1.97	1.93	1.89	1.87	1.61	1.56	1.52
50	2.81	2.41	2.20	2.06	1.97	1.90	1.84	1.80	1.76	1.73	1.44	1.39	1.34
100	2.76	2.36	2.14	2.00	1.91	1.83	1.78	1.73	1.69	1.66	1.35	1.29	1.20
∞	2.71	2.30	2.08	1.94	1.85	1.77	1.72	1.67	1.63	1.60	1.24	1.17	1.00

Abridged and adapted from *Standard Probability and Statistical Tables and Formulae* (Boca Raton: Chapman & Hall/ CRC, 2000).

α = 0.05

	df_{num} = 1	2	3	4	5	6	7	8	9	10	50	100	∞
df_{denom} = 1	161.4	199.5	215.7	224.6	230.2	234.0	236.8	238.9	240.5	241.9	251.8	253.0	254.3
2	18.51	19.00	19.16	19.25	19.30	19.33	19.35	19.37	19.38	19.40	19.48	19.49	19.50
3	10.13	9.55	9.28	9.12	9.01	8.94	8.89	8.85	8.81	8.79	8.58	8.55	8.53
4	7.71	6.94	6.59	6.39	6.26	6.16	6.09	6.04	6.00	5.96	5.70	5.66	5.63
5	6.61	5.79	5.41	5.19	5.05	4.95	4.88	4.82	4.77	4.74	4.44	4.41	4.36
6	5.99	5.14	4.76	4.53	4.39	4.28	4.21	4.15	4.10	4.06	3.75	3.71	3.67
7	5.59	4.74	4.35	4.12	3.97	3.87	3.79	3.73	3.68	3.64	3.32	3.27	3.23
8	5.32	4.46	4.07	3.84	3.69	3.58	3.50	3.44	3.39	3.35	3.02	2.97	2.93
9	5.12	4.26	3.86	3.63	3.48	3.37	3.29	3.23	3.18	3.14	2.80	2.76	2.71
10	4.96	4.10	3.71	3.48	3.33	3.22	3.14	3.07	3.02	2.98	2.64	2.59	2.54
11	4.84	3.98	3.59	3.36	3.20	3.09	3.01	2.95	2.90	2.85	2.51	2.46	2.40
12	4.75	3.89	3.49	3.26	3.11	3.00	2.91	2.85	2.80	2.75	2.40	2.35	2.30

(*continued*)

$df_{num}=1$	2	3	4	5	6	7	8	9	10	50	100	∞
13	4.67	3.81	3.41	3.18	3.03	2.92	2.83	2.77	2.71	2.67	2.31	2.26	2.21
14	4.60	3.74	3.34	3.11	2.96	2.85	2.76	2.70	2.65	2.60	2.24	2.19	2.13
15	4.54	3.68	3.29	3.06	2.90	2.79	2.71	2.64	2.59	2.54	2.18	2.12	2.07
16	4.49	3.63	3.24	3.01	2.85	2.74	2.66	2.59	2.54	2.49	2.12	2.07	2.01
17	4.45	3.59	3.20	2.96	2.81	2.70	2.61	2.55	2.49	2.45	2.08	2.02	1.96
18	4.41	3.55	3.16	2.93	2.77	2.66	2.58	2.51	2.46	2.41	2.04	1.98	1.92
19	4.38	3.52	3.13	2.90	2.74	2.63	2.54	2.48	2.42	2.38	2.00	1.94	1.88
20	4.35	3.49	3.10	2.87	2.71	2.60	2.51	2.45	2.39	2.35	1.97	1.91	1.84
25	4.24	3.39	2.99	2.76	2.60	2.49	2.40	2.34	2.28	2.24	1.84	1.78	1.71
50	4.03	3.18	2.79	2.56	2.40	2.29	2.20	2.13	2.07	2.03	1.60	1.52	1.45
100	3.94	3.09	2.70	2.46	2.31	2.19	2.10	2.03	1.97	1.93	1.48	1.39	1.28
∞	3.84	3.00	2.60	2.37	2.21	2.10	2.01	1.94	1.88	1.83	1.35	1.25	1.00

Abridged and adapted from *Standard Probability and Statistical Tables and Formulae* (Boca Raton: Chapman & Hall/ CRC, 2000).

$$\alpha = 0.01$$

	$df_{num}=1$	2	3	4	5	6	7	8	9	10	50	100	∞
$df_{denom}=1$	4052	5000	5403	5625	5764	5859	5928	5981	6022	6056	6303	6334	6336
2	98.50	99.00	99.17	99.25	99.30	99.33	99.36	99.37	99.39	99.40	99.48	99.49	99.50
3	34.12	30.82	29.46	28.71	28.24	27.91	27.67	27.49	27.35	27.23	26.35	26.24	26.13
4	21.20	18.00	16.69	15.98	15.52	15.21	14.98	14.80	14.66	14.55	13.69	13.58	13.46
5	16.26	13.27	12.06	11.39	10.97	10.67	10.46	10.29	10.16	10.05	9.24	9.13	9.02
6	13.75	10.92	9.78	9.15	8.75	8.47	8.26	8.10	7.98	7.87	7.09	6.99	6.88
7	12.25	9.55	8.45	7.85	7.46	7.19	6.99	6.84	6.72	6.62	5.86	5.75	5.65
8	11.26	8.65	7.59	7.01	6.63	6.37	6.18	6.03	5.91	5.81	5.07	4.96	4.86
9	10.56	8.02	6.99	6.42	6.06	5.80	5.61	5.47	5.35	5.26	4.52	4.41	4.31
10	10.04	7.56	6.55	5.99	5.64	5.39	5.20	5.06	4.94	4.85	4.12	4.01	3.91
11	9.65	7.21	6.22	5.67	5.32	5.07	4.89	4.74	4.63	4.54	3.81	3.71	3.60
12	9.33	6.93	5.95	5.41	5.06	4.82	4.64	4.50	4.39	4.30	3.57	3.47	3.36
13	9.07	6.70	5.74	5.21	4.86	4.62	4.44	4.30	4.19	4.10	3.38	3.27	3.17
14	8.86	6.51	5.56	5.04	4.69	4.46	4.28	4.14	4.03	3.94	3.22	3.11	3.00
15	8.68	6.36	5.42	4.89	4.56	4.32	4.14	4.00	3.89	3.80	3.08	2.98	2.87
16	8.53	6.23	5.29	4.77	4.44	4.20	4.03	3.89	3.78	3.69	2.97	2.86	2.75
17	8.40	6.11	5.18	4.67	4.34	4.10	3.93	3.79	3.68	3.59	2.87	2.76	2.65
18	8.29	6.01	5.09	4.58	4.25	4.01	3.84	3.71	3.60	3.51	2.78	2.68	2.57
19	8.18	5.93	5.01	4.50	4.17	3.94	3.77	3.63	3.52	3.43	2.71	2.60	2.49
20	8.10	5.85	4.94	4.43	4.10	3.87	3.70	3.56	3.46	3.37	2.64	2.54	2.42
25	7.77	5.57	4.68	4.18	3.85	3.63	3.46	3.32	3.22	3.13	2.40	2.29	2.17
50	7.17	5.06	4.20	3.72	3.41	3.19	3.02	2.89	2.78	2.70	1.95	1.82	1.70
100	6.90	4.82	3.98	3.51	3.21	2.99	2.82	2.69	2.59	2.50	1.74	1.60	1.45
∞	6.63	4.61	3.78	3.32	3.02	2.80	2.64	2.51	2.41	2.32	1.53	1.32	1.00

Abridged and adapted from *Standard Probability and Statistical Tables and Formulae* (Boca Raton: Chapman & Hall/ CRC, 2000).

TABLE 5

Critical Values for the Chi-Square Distribution

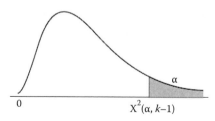

$$X^2(\alpha, k-1)$$

df	α = 0.10	0.05	0.025	0.01
1	2.7055	3.8415	5.0239	6.6349
2	4.6052	5.9915	7.3778	9.2103
3	6.2514	7.8147	9.3484	11.3449
4	7.7794	9.4877	11.1433	13.2767
5	9.2364	11.0705	12.8325	15.0863
6	10.6446	12.5916	14.4494	16.8119
7	12.0170	14.0671	16.0128	18.4753
8	13.3616	15.5073	17.5345	20.0902
9	14.6837	16.9190	19.0228	21.6660
10	15.9872	18.3070	20.4832	23.2093
11	17.2750	19.6751	21.9200	24.7250
12	18.5493	21.0261	23.3367	26.2170
13	19.8119	22.3620	24.7356	27.6882
14	21.0641	23.6848	26.1189	29.1412
15	22.3071	24.9958	27.4884	30.5779
16	23.5418	26.2962	28.8454	31.9999
17	24.7690	27.5871	30.1910	33.4087
18	25.9894	28.8693	31.5264	34.8053
19	27.2036	30.1435	32.8523	36.1909
20	28.4120	31.4104	34.1696	37.5662
21	29.6151	32.6706	35.4789	38.9322
22	30.8133	33.9244	36.7807	40.2894
23	32.0069	35.1725	38.0756	41.6384
24	33.1962	36.4150	39.3641	42.9798
25	34.3816	37.6525	40.6465	44.3141
26	35.5632	38.8851	41.9232	45.6417
27	36.7412	40.1133	43.1945	46.9629
28	37.9159	41.3371	44.4608	48.2782
29	39.0875	42.5570	45.7223	49.5879
30	40.2560	43.7730	46.9792	50.8922
31	41.4217	44.9853	48.2319	52.1914
32	42.5847	46.1943	49.4804	53.4858

(*continued*)

TABLE 5 (CONTINUED)

Critical Values for the Chi-Square Distribution

df	α = 0.10	0.05	0.025	0.01
33	43.7452	47.3999	50.7251	54.7755
34	44.9032	48.6024	51.9660	56.0609
35	46.0588	49.8018	53.2033	57.3421
36	47.2122	50.9985	54.4373	58.6192
37	48.3634	52.1923	55.6680	59.8925
38	49.5126	53.3835	56.8955	61.1621
39	50.6598	54.5722	58.1201	62.4281
40	51.8051	55.7585	59.3417	63.6907

df	α = 0.10	0.05	0.025	0.01
41	52.95	56.94	60.56	64.95
42	54.09	58.12	61.78	66.21
43	55.23	59.30	62.99	67.46
44	56.37	60.48	64.20	68.71
45	57.51	61.66	65.41	69.96
46	58.64	62.83	66.62	71.20
47	59.77	64.00	67.82	72.44
48	60.91	65.17	69.02	73.68
49	62.04	66.34	70.22	74.92
50	63.17	67.50	71.42	76.15
60	74.40	79.08	83.30	88.38
70	85.53	90.53	95.02	100.43
80	96.58	101.88	106.63	112.33
90	107.57	113.15	118.14	124.12
100	118.50	124.34	129.56	135.81
200	226.02	233.99	241.06	249.45
300	331.79	341.40	349.87	359.91
400	436.65	447.63	457.31	468.72
500	540.93	553.13	563.85	576.49
600	644.80	658.09	669.77	683.52
700	748.36	762.66	775.21	789.97
800	851.67	866.91	880.28	895.98
900	954.78	970.90	985.03	1001.63
1000	1057.72	1074.68	1089.53	1106.97
1500	1570.61	1591.21	1609.23	1630.35
2000	2081.47	2105.15	2125.84	2150.07
2500	2591.04	2617.43	2640.47	2667.43
3000	3099.69	3128.54	3153.70	3183.13
3500	3607.64	3638.75	3665.87	3697.57

df	α = 0.10	0.05	0.025	0.01
4000	4115.05	4148.25	4177.19	4211.01
4500	4622.00	4657.17	4687.83	4723.63
5000	5128.58	5165.61	5197.88	5235.57
5500	5634.83	5673.64	5707.45	5746.93
6000	6140.81	6181.31	6216.59	6257.78
6500	6646.54	6688.67	6725.36	6768.18
7000	7152.06	7195.75	7233.79	7278.19
7500	7657.38	7702.58	7741.93	7787.86
8000	8162.53	8209.19	8249.81	8297.20
8500	8667.52	8715.59	8757.44	8806.26
9000	9172.36	9221.81	9264.85	9315.05
9500	9677.07	9727.86	9772.05	9823.60
10000	1081.66	10233.75	10279.07	10331.93

Abridged and adapted from *Standard Probability and Statistical Tables and Formulae* (Boca Raton: Chapman & Hall/ CRC, 2000).

TABLE 6

Critical Values for the Studentized Range Distribution

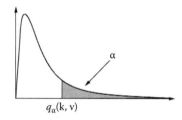

$q_\alpha(k, v)$

α = 0.10

κ = 2	3	4	5	6	7	8	9	10	11	12	13	14	15	16	17	18	19	20
v = 1 8.93	13.44	16.36	18.49	20.15	21.51	22.64	23.62	24.48	25.24	25.92	26.54	27.10	26.62	28.10	28.54	28.96	29.35	29.71
2 4.13	5.73	6.77	7.54	8.14	8.63	9.05	9.41	9.72	10.01	10.26	10.49	10.70	10.89	11.07	11.24	11.39	11.54	11.68
3 3.33	4.47	5.20	5.74	6.16	6.51	6.81	7.06	7.29	7.49	7.67	7.83	7.98	8.12	8.25	8.37	8.48	8.58	8.68
4 3.01	3.98	4.59	5.03	5.39	5.68	5.93	6.14	6.33	6.49	6.64	6.78	6.91	7.02	7.13	7.23	7.33	7.41	7.50
5 2.85	3.72	4.26	4.66	4.98	5.24	5.46	5.65	5.82	5.97	6.10	6.22	6.34	6.44	6.54	6.63	6.71	6.79	6.86
6 2.75	3.56	4.06	4.43	4.73	4.97	5.17	5.34	5.50	5.64	5.76	5.87	5.98	6.07	6.16	6.25	6.32	6.40	6.47
7 2.68	3.45	3.93	4.28	4.55	4.78	4.97	5.14	5.28	5.41	5.53	5.64	5.73	5.83	5.91	5.99	6.06	6.13	6.19
8 2.63	3.37	3.83	4.17	4.43	4.65	4.83	4.99	5.13	5.25	5.36	5.46	5.56	5.64	5.72	5.80	5.87	5.93	6.00
9 2.59	3.12	3.76	4.08	4.34	4.54	4.72	4.87	5.01	5.13	5.23	5.33	5.42	5.51	5.58	5.65	5.72	5.79	5.85
10 2.56	3.27	3.70	4.02	4.26	4.46	4.64	4.78	4.91	5.03	5.13	5.23	5.32	5.40	5.47	5.54	5.61	5.67	5.73
11 2.54	3.23	3.66	3.96	4.20	4.40	4.57	4.71	4.84	4.95	5.05	5.15	5.23	5.31	5.38	5.45	5.51	5.57	5.63
12 2.52	3.20	3.62	3.92	4.16	4.35	4.51	4.65	4.78	4.89	4.99	5.08	5.16	5.24	5.31	5.37	5.44	5.49	5.55

(continued)

TABLE 6 (CONTINUED)

Critical Values for the Studentized Range Distribution

κ = 2	3	4	5	6	7	8	9	10	11	12	13	14	15	16	17	18	19	20
13 2.50	3.18	3.59	3.88	4.12	4.30	4.46	4.60	4.72	4.83	4.93	5.02	5.10	5.18	5.24	5.31	5.37	5.43	5.48
14 2.49	3.16	3.56	3.85	4.08	4.27	4.42	4.56	4.68	4.79	4.88	4.97	5.05	5.12	5.19	5.26	5.32	5.37	5.43
15 2.48	3.14	3.54	3.83	4.05	4.23	4.39	4.52	4.64	4.75	4.84	4.93	5.01	5.08	5.15	5.21	5.27	5.32	5.38
16 2.47	3.12	3.52	3.80	4.03	4.21	4.36	4.49	4.61	4.71	4.80	4.89	4.97	5.04	5.11	5.17	5.23	5.28	5.33
17 2.46	3.11	3.50	3.78	4.00	4.18	4.33	4.46	4.58	4.68	4.77	4.86	4.93	5.00	5.07	5.13	5.19	5.24	5.29
18 2.45	3.10	3.49	3.77	3.98	4.16	4.31	4.44	4.55	4.65	4.75	4.83	4.90	4.97	5.04	5.10	5.16	5.21	5.26
19 2.45	3.09	3.47	3.75	3.97	4.14	4.29	4.42	4.53	4.63	4.72	4.80	4.88	4.95	5.01	5.07	5.13	5.18	5.23
20 2.44	3.08	3.46	3.74	3.95	4.12	4.27	4.40	4.51	4.61	4.70	4.78	4.85	4.92	4.99	5.05	5.10	5.15	5.20
24 2.42	3.05	3.42	3.69	3.90	4.07	4.21	4.34	4.44	4.54	4.63	4.71	4.78	4.85	4.91	4.97	5.02	5.07	5.12
30 2.40	3.02	3.39	3.65	3.85	4.02	4.16	4.27	4.38	4.47	4.56	4.63	4.71	4.77	4.83	4.89	4.94	4.99	5.03
40 2.38	2.99	3.35	3.60	3.80	3.96	4.10	4.21	4.32	4.41	4.49	4.56	4.63	4.69	4.75	4.81	4.86	4.90	4.95
60 2.36	2.96	3.31	3.56	3.75	3.91	4.04	4.15	4.25	4.34	4.42	4.49	4.56	4.62	4.67	4.73	4.77	4.82	4.86
120 2.34	2.93	3.28	3.52	3.71	3.86	3.99	4.10	4.19	4.28	4.35	4.42	4.48	4.54	4.60	4.65	4.69	4.74	4.78
∞ 2.33	2.90	3.24	3.48	3.66	3.81	3.93	4.04	4.13	4.21	4.28	4.35	4.41	4.47	4.52	4.57	4.61	4.65	4.69

Abridged and adapted from *Tables of Range and Studentized Range*, (Annals of Mathematical Statistics by Leon Harter, Volume 31, Issue 4, 1960, with permission of the Institute of Mathematical Statistics).

$$\alpha = 0.05$$

κ = 2	3	4	5	6	7	8	9	10	11	12	13	14	15	16	17	18	19	20
ν = 1 17.97	26.98	32.82	37.08	40.41	43.12	45.40	47.36	49.07	50.59	51.96	53.20	54.33	55.36	56.32	57.22	58.04	58.83	59.56
2 6.08	8.33	9.80	10.88	11.74	12.44	13.03	13.54	13.99	14.39	14.75	15.08	15.38	15.65	15.91	16.14	16.37	16.57	16.77
3 4.50	5.91	6.82	7.50	8.04	8.48	8.85	9.18	9.46	9.72	9.95	10.15	10.35	10.53	10.69	10.84	10.98	11.11	11.24
4 3.93	5.04	5.76	6.29	6.71	7.05	7.35	7.60	7.83	8.03	8.21	8.37	8.52	8.66	8.79	8.91	9.03	9.13	9.23
5 3.63	4.60	5.22	5.67	6.03	6.33	6.58	6.80	6.99	7.17	7.32	7.47	7.60	7.72	7.83	7.93	8.03	8.12	8.21
6 3.46	4.34	4.90	5.30	5.63	5.89	6.12	6.32	6.49	6.65	6.79	6.92	7.03	7.14	7.24	7.34	7.43	7.51	7.59
7 3.34	4.16	4.68	5.06	5.36	5.61	5.81	6.00	6.16	6.30	6.43	6.55	6.66	6.76	6.85	6.94	7.02	7.10	7.17
8 3.26	4.04	4.53	4.89	5.17	5.40	5.60	5.77	5.92	6.05	6.17	6.29	6.39	6.48	6.57	6.65	6.73	6.80	6.87
9 3.20	3.95	4.41	4.76	5.02	5.24	5.43	5.59	5.74	5.87	5.98	6.09	6.19	6.28	6.36	6.44	6.51	6.58	6.64
10 3.15	3.88	4.33	4.65	4.91	5.12	5.30	5.46	5.60	5.72	5.83	5.93	6.03	6.11	6.19	6.27	6.34	6.40	6.47
11 3.11	3.82	4.26	4.57	4.82	5.03	5.20	5.35	5.49	5.60	5.71	5.81	5.90	5.98	6.06	6.13	6.20	6.26	6.33
12 3.08	3.77	4.20	4.51	4.75	4.95	5.12	5.26	5.39	5.51	5.61	5.71	5.80	5.88	5.95	6.02	6.09	6.15	6.21
13 3.05	3.73	4.15	4.45	4.69	4.88	5.05	5.19	5.32	5.43	5.53	5.62	5.71	5.79	5.86	5.93	5.99	6.05	6.11
14 3.03	3.70	4.11	4.41	4.64	4.83	4.99	5.13	5.25	5.36	5.46	5.55	5.64	5.71	5.79	5.85	5.91	5.97	6.03
15 3.01	3.67	4.08	4.37	4.59	4.78	4.94	5.08	5.20	5.31	5.40	5.49	5.57	5.65	5.72	5.78	5.85	5.90	5.96
16 3.00	3.65	4.05	4.33	4.56	4.74	4.90	5.03	5.15	5.26	5.35	5.44	5.52	5.59	5.66	5.73	5.79	5.84	5.90
17 2.98	3.63	4.02	4.30	4.52	4.70	4.86	4.99	5.11	5.21	5.31	5.39	5.47	5.54	5.61	5.67	5.73	5.79	5.84
18 2.97	3.61	4.00	4.28	4.49	4.67	4.82	4.96	5.07	5.17	5.27	5.35	5.43	5.50	5.57	5.63	5.69	5.74	5.79
19 2.96	3.59	3.98	4.25	4.47	4.64	4.79	4.92	5.04	5.14	5.23	5.31	5.39	5.46	5.53	5.59	5.65	5.70	5.75
20 2.95	3.58	3.96	4.23	4.44	4.62	4.77	4.90	5.01	5.11	5.20	5.28	5.36	5.43	5.49	5.55	5.61	5.66	5.71
24 2.92	3.53	3.90	4.17	4.37	4.54	4.68	4.81	4.91	5.01	5.10	5.18	5.25	5.32	5.38	5.44	5.49	5.54	5.59
30 2.89	3.49	3.84	4.10	4.30	4.46	4.60	4.72	4.82	4.92	5.00	5.08	5.15	5.21	5.27	5.33	5.38	5.43	5.47
40 2.86	3.44	3.79	4.04	4.23	4.39	4.52	4.63	4.73	4.82	4.90	4.98	5.04	5.11	5.16	5.22	5.27	5.31	5.36
60 2.83	3.40	3.74	3.98	4.16	4.31	4.44	4.55	4.65	4.73	4.81	4.88	4.94	5.00	5.06	5.11	5.15	5.20	5.24
120 2.80	3.36	3.68	3.92	4.10	4.24	4.36	4.47	4.56	4.64	4.71	4.78	4.84	4.90	4.95	5.00	5.04	5.09	5.13
∞ 2.77	3.31	3.63	3.86	4.03	4.17	4.29	4.39	4.47	4.55	4.62	4.68	4.74	4.80	4.84	4.89	4.93	4.97	5.01

α = 0.01

κ = 2	3	4	5	6	7	8	9	10	11	12	13	14	15	16	17	18	19	20
ν = 1 90.03	135.0	164.3	185.6	202.2	215.8	227.2	237.0	245.6	253.2	260.0	266.2	271.8	277.0	281.8	286.3	290.4	294.3	298.0
2 14.04	19.02	22.29	24.72	26.63	28.20	29.53	30.68	31.69	32.59	33.40	34.13	34.81	35.43	36.00	36.53	37.03	37.50	37.95
3 8.26	10.62	12.17	13.33	14.24	15.00	15.64	16.20	16.69	17.13	17.53	17.89	18.22	18.52	18.81	19.07	19.32	19.55	19.77
4 6.51	8.12	9.17	9.96	10.58	11.10	11.55	11.93	12.27	12.57	12.84	13.09	13.32	13.53	13.73	13.91	14.08	14.24	14.40
5 5.70	6.98	7.80	8.42	8.91	9.32	9.67	9.97	10.24	10.48	10.70	10.89	11.08	11.24	11.40	11.55	11.68	11.81	11.93
6 5.24	6.33	7.03	7.56	7.97	8.32	8.61	8.87	9.10	9.30	9.48	9.65	9.81	9.95	10.08	10.21	10.32	10.43	10.54
7 4.95	5.92	6.54	7.00	7.37	7.68	7.94	8.17	8.37	8.55	8.71	8.86	9.00	9.12	9.24	9.35	9.46	9.55	9.65
8 4.75	5.63	6.20	6.62	6.96	7.24	7.47	7.68	7.86	8.03	8.18	8.31	8.44	8.55	8.66	8.76	8.85	8.94	9.03
9 4.60	5.43	5.96	6.35	6.66	6.91	7.13	7.32	7.49	7.65	7.78	7.91	8.02	8.13	8.23	8.32	8.41	8.49	8.57
10 4.48	5.27	5.77	6.14	6.43	6.67	6.87	7.05	7.21	7.36	7.48	7.60	7.71	7.81	7.91	7.99	8.08	8.15	8.23
11 4.39	5.15	5.62	5.97	6.25	6.48	6.67	6.84	6.99	7.13	7.25	7.36	7.46	7.56	7.65	7.73	7.81	7.88	7.95
12 4.32	5.05	5.50	5.84	6.10	6.32	6.51	6.67	6.81	6.94	7.06	7.17	7.26	7.35	7.44	7.52	7.59	7.67	7.73
13 4.26	4.96	5.40	5.73	5.98	6.19	6.37	6.53	6.67	6.79	6.90	7.01	7.10	7.19	7.27	7.34	7.42	7.48	7.55
14 4.21	4.89	5.32	5.63	5.88	6.08	6.26	6.41	6.54	6.66	6.77	6.87	6.96	7.05	7.13	7.20	7.27	7.33	7.39
15 4.17	4.84	5.25	5.56	5.80	5.99	6.16	6.31	6.44	6.55	6.66	6.76	6.84	6.93	7.00	7.07	7.14	7.20	7.26
16 4.13	4.79	5.19	5.49	5.72	5.91	6.08	6.22	6.35	6.46	6.56	6.66	6.74	6.82	6.90	6.97	7.03	7.09	7.15
17 4.10	4.74	5.14	5.43	5.66	5.85	6.01	6.15	6.27	6.38	6.48	6.57	6.65	6.73	6.81	6.87	6.94	7.00	7.05
18 4.07	4.70	5.09	5.38	5.60	5.79	5.94	6.08	6.20	6.31	6.41	6.50	6.58	6.65	6.72	6.79	6.85	6.91	6.97
19 4.05	4.67	5.05	5.33	5.55	5.73	5.89	6.02	6.14	6.25	6.34	6.43	6.51	6.58	6.65	6.72	6.78	6.84	6.89
20 4.02	4.64	5.02	5.29	5.51	5.69	5.84	5.97	6.09	6.19	6.28	6.37	6.45	6.52	6.59	6.65	6.71	6.77	6.82
24 3.96	4.55	4.91	5.17	5.37	5.54	5.68	8.81	5.92	6.02	6.11	6.19	6.26	6.33	6.39	6.45	6.51	6.56	6.61
30 3.89	4.45	4.80	5.05	5.24	5.40	5.54	5.65	5.76	5.85	5.93	6.01	6.08	6.14	6.20	6.26	6.31	6.36	6.41
40 3.82	4.37	4.70	4.93	5.11	5.26	5.39	5.50	5.60	5.69	5.76	5.83	5.90	5.96	6.02	6.07	6.12	6.16	6.21
60 3.76	4.28	4.59	4.82	4.99	5.13	5.25	5.36	5.45	5.53	5.60	5.67	5.73	5.78	5.84	5.89	5.93	5.97	6.01
120 3.70	4.20	4.50	4.71	4.87	5.00	5.12	5.21	5.30	5.37	5.44	5.51	5.56	5.61	5.66	5.71	5.75	5.79	5.83
∞ 3.64	4.12	4.40	4.60	4.76	4.88	4.99	5.08	5.16	5.23	5.29	5.35	5.40	5.45	5.49	5.53	5.57	5.61	5.64

Index